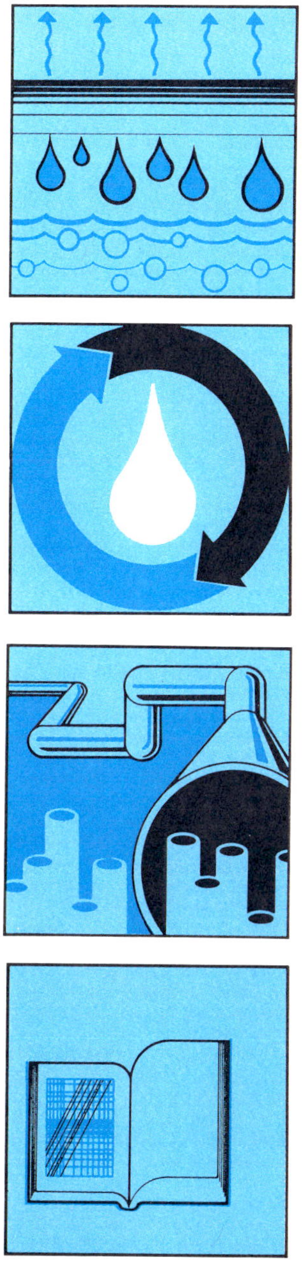

Hydraulic Institute Engineering Data Book

Second Edition

List of Members—1990

Afton Pumps, Inc. Houston, Texas
Ahlstrom Pumps Peacedale, Rhode Island
Ampco Metal Division Milwaukee, Wisconsin
AMW Industries Conway, Arizona
Aurora Pump
Unit of General Signal North Aurora, Illinois
Barrett, Haentjens & Company Hazelton, Pennsylvania
BGA International Salt Lake City, Utah
Blackmer Pump Grand Rapids, Michigan
Buffalo Pumps Inc. North Tonawanda, New York
Burks Pumps, Inc. Decatur, Illinois
Byron Jackson Products Division
BW/IP International Inc. Los Angeles, California
United Centrifugal Pumps San Jose, California
Carver Pump Company Muscatine, Iowa
Cascade Pump Company Sante Fe Springs, California
Chempump
Division of Crane Co. Warrington, Pennsylvania
Crane Deming Pumps Salem, Ohio
Cryodynamics Division
Ebara Int'l. Corporation Sacramento, California
Dresser Pump Division
Dresser Industries, Inc. Mountainside, New Jersey
The Durion Company, Inc. Dayton, Ohio
ESSCO Division Los Angeles, California
Fairbanks Morse Pump Corporation Kansas City, Kansas
Flygt Corporation Norwalk, Connecticut
Fluidpak, Inc. Cincinnati, Ohio
FMC Corp Jonesboro, Arkansas
Fybroc Division
MetPro Corporation, Hatfield, Pennsylvania
APV Gaulin, Inc. Everett, Massachusetts
Goulds Pumps, Inc. Seneca Falls, New York
Ingersoll-Rand Company Phillipsburg, New Jersey
IMO Delaval Inc. Lawrenceville, New Jersey
ITT A-C Pump (List Below) Cincinnati, Ohio
ITT Fluid Transfer Division Skokie, Illinois
Johnston Pump Company Brookshire, Texas
Krogh Pump Company San Francisco, California
Lawrence Pump and Engine Co.
Division of Stork Pump Inc. Lawrence, Massachusetts
Lawrence Pumps, Inc. Lawrence, Massachusetts
Layne & Bowler, Inc. Memphis, Tennessee
Chas. S. Lewis & Company, Inc. St. Louis, Missouri
M & W Pump Corporation Deerfield Beach, Florida
MetPro Corporation
Dean Pump Division Indianapolis, Indiana
Micro Pump Corp. Concord, California
Milton Roy Company
Flow Control Division Ivyland, Pennsylvania
National Pump Company, Inc. Forrest Park, Georgia
Oberdorfer Pumps, Inc. Syracuse, New York
PACO Pumps, Inc. Oakland, California
Patterson Pump Company
Subsidiary of Gorman Rupp Co. Toccoa, Georgia
Peabody Barnes, Inc. Mansfield, Ohio
Peabody Floway, Inc. Fresno, California
Peerless Pump Montebello, California
Price Pump Company Sonoma, California
Robbins & Myers, Inc.
Fluids Handling Group Springfield, Ohio
Roper Pump Company Commerce, Georgia
Scot Pump Division
Ardox Corporation Cedarburg, Wisconsin
Simmons Pump Corporation Willcox, Arizona
Sta-Rite Industries, Inc. Delevan, Wisconsin
Sunstrand Fluid Handling Arvada, Colorado
Taco Inc. Cranston, Rhode Island
Tuthill Corporation Chicago, Illinois
Union Pump Company Battle Creek, Michigan
Viking Pump, Inc. Cedar Falls, Iowa
Warren Pumps Inc. Warren, Massachusetts
Waukesha Pumps
AMCA Corp. Waukesha, Wisconsin
Wemco Sacramento, California
A. R. Wilfley & Sons, Inc. Denver, Colorado
Wilson-Snyder Pumps
A Unit of National-Oilwell Garland, Texas
Yeomens Chicago Corporation Melrose Park, Illinois
Zenith Pumps Division Waltham, Massachusetts

Hydraulic Institute
30200 Detroit Road
Cleveland, Ohio 44145

30200 Detroit Road
Cleveland, Ohio 44145

Printed in U.S.A.

SECTION IV—CHARACTERISTICS OF PIPING MATERIALS

SECTION V—USEFUL INFORMATION

SECTION VI—BIBLIOGRAPHY

HYDRAULIC INSTITUTE DATA BOOK - 2ND EDITION

COMMENTS

Suggestions, Correction

Reader participation is important to us. Your comments will be appreciated. We need your help so that this book serves our readership to the utmost.

Comments:

1.

2.

3.

4.

5.

6.

Tear out this sheet. Fold on the lines on the reverse side, tape shut and mail. Thanks.

Signature ______________________

Not Mandatory

Address ______________________

CUT ALONG DOTTED LINE

Fold

From:

PLACE
STAMP
HERE

CUT ALONG DOTTED LINE

To: Hydraulic Institute
Data Book Subcommittee
30200 Detroit Road
Cleveland, Ohio 44145-1967

Fold

Tape Shut

Section I

Preface

The Second Edition of the Engineering Data Book provides the user with updated technical information for those working with fluids.

As a companion to the Hydraulic Institute Standards, together with HI Test Standards, the Engineering Data Book complete the data for solving problems related to the pumping and transfer of fluids.

Origin

The Hydraulic Institute came into being on April 18–19, 1917, when sixteen manufacturers of industrial pumps met and formed The Hydraulic Society. In 1933 the Society was re-organized and the name changed to the Hydraulic Institute.

In 1917, when the Society was formed, the pump industry was faced with an urgent need to solve a number of engineering problems brought about by the production needs of World War I. Today, while the activity has grown manifold, engineering is still the basic interest. The most important manifestations of this interest appear in the development of standards, in engineering conferences, in sponsorship of cooperative research projects, and in cooperation with other engineering organizations.

Purpose and Aims

The purpose and aims of the Institute are best expressed in the following quotation taken from Article II of the By-Laws of the Institute:

"The Objects of the Institute are: To promote and further the interests of manufacturers of pumps, as well as the interests of the public in such matters as are involved in manufacturing, engineering, safety, transportation, and other problems of the industry, and to this end, among other things:

(a) To develop and publish standards for pumps.
(b) To collect and disseminate information of value to its members and to the public.
(c) To appear for its members before governmental departments and agencies and other bodies in regard to matters affecting the industry.
(d) To promote a spirit of cooperation among its members for the improved production, proper use, and increased distribution of pumps.
(e) To increase the amount and to improve the quality of pump service to the public.
(f) To support educational and research activities.
(g) To promote the business interests of its members but not to engage in business of the kind ordinarily carried on for profit or to perform particular services for its members or individual persons as distinguished from activities to improve the business conditions and lawful interests of all of its members."

Of parallel interest is the following quotation from Article III, Membership, of the By-Laws of the Institute:

"Corporations, firms and individuals engaged within the United States in the manufacture of pumps for sale in the open market, shall be eligible to apply for membership in the Institute."

The Hydraulic Institute appreciates its responsibility in this connection and will continue to welcome all suggestions of a constructive nature. Such recommendations will be taken into account in the development of future editions of the Handbook.

Section II

Characteristics of Fluids

IIA-1 PROPERTIES OF WATER AT VARIOUS TEMPERATURES FROM 32° TO 705.4° F

Temp. F	Temp. C	Specific Volume Cu Ft/Lb	SPECIFIC GRAVITY			Wt in Lb/Cu Ft	Vapor Pressure Psi Abs
			39.2 F Reference	60 F Reference	68 F Reference		
32	0	.01602	1.000	1.001	1.002	62.42	0.088
35	1.7	.01602	1.000	1.001	1.002	62.42	0.100
40	4.4	.01602	1.000	1.001	1.002	62.42	0.1217
50	10.0	.01603	.999	1.001	1.002	62.38	0.1781
60	15.6	.01604	.999	1.000	1.001	62.34	0.2563
70	21.1	.01606	.998	.999	1.000	62.27	0.3631
80	26.7	.01608	.996	.998	.999	62.19	0.5069
90	32.2	.01610	.995	.996	.997	62.11	0.6982
100	37.8	.01613	.993	.994	.995	62.00	0.9492
120	48.9	.01620	.989	.990	.991	61.73	1.692
140	60.0	.01629	.983	.985	.986	61.39	2.889
160	71.1	.01639	.977	.979	.979	61.01	4.741
180	82.2	.01651	.970	.972	.973	60.57	7.510
200	93.3	.01663	.963	.964	.966	60.13	11.526
212	100.0	.01672	.958	.959	.960	59.81	14.696
220	104.4	.01677	.955	.956	.957	59.63	17.186
240	115.6	.01692	.947	.948	.949	59.10	24.97
260	126.7	.01709	.938	.939	.940	58.51	35.43
280	137.8	.01726	.928	.929	.930	58.00	49.20
300	148.9	.01745	.918	.919	.920	57.31	67.01
320	160.0	.01765	.908	.909	.910	56.66	89.66
340	171.1	.01787	.896	.898	.899	55.96	118.01
360	182.2	.01811	.885	.886	.887	55.22	153.04
380	193.3	.01836	.873	.874	.875	54.47	195.77
400	204.4	.01864	.859	.860	.862	53.65	247.31
420	215.6	.01894	.846	.847	.848	52.80	308.83
440	226.7	.01926	.832	.833	.834	51.92	381.59
460	237.8	.0196	.817	.818	.819	51.02	466.9
480	248.9	.0200	.801	.802	.803	50.00	566.1
500	260.0	.0204	.785	.786	.787	49.02	680.8
520	271.1	.0209	.765	.766	.767	47.85	812.4
540	282.2	.0215	.746	.747	.748	46.51	962.5
560	293.3	.0221	.726	.727	.728	45.3	1133.1
580	304.4	.0228	.703	.704	.704	43.9	1325.8
600	315.6	.0236	.678	.679	.680	42.3	1542.9
620	326.7	.0247	.649	.650	.650	40.5	1786.6
640	337.8	.0260	.617	.618	.618	38.5	2059.7
660	348.9	.0278	.577	.577	.578	36.0	2365.4
680	360.0	.0305	.525	.526	.527	32.8	2708.1
700	371.1	.0369	.434	.435	.435	27.1	3093.7
705.4	374.1	.0503	.319	.319	.320	19.9	3206.2

Computed from Keenan & Keyes' Steam Table.

IIA-2 TEMPERATURE VS SPECIFIC GRAVITY FOR WATER

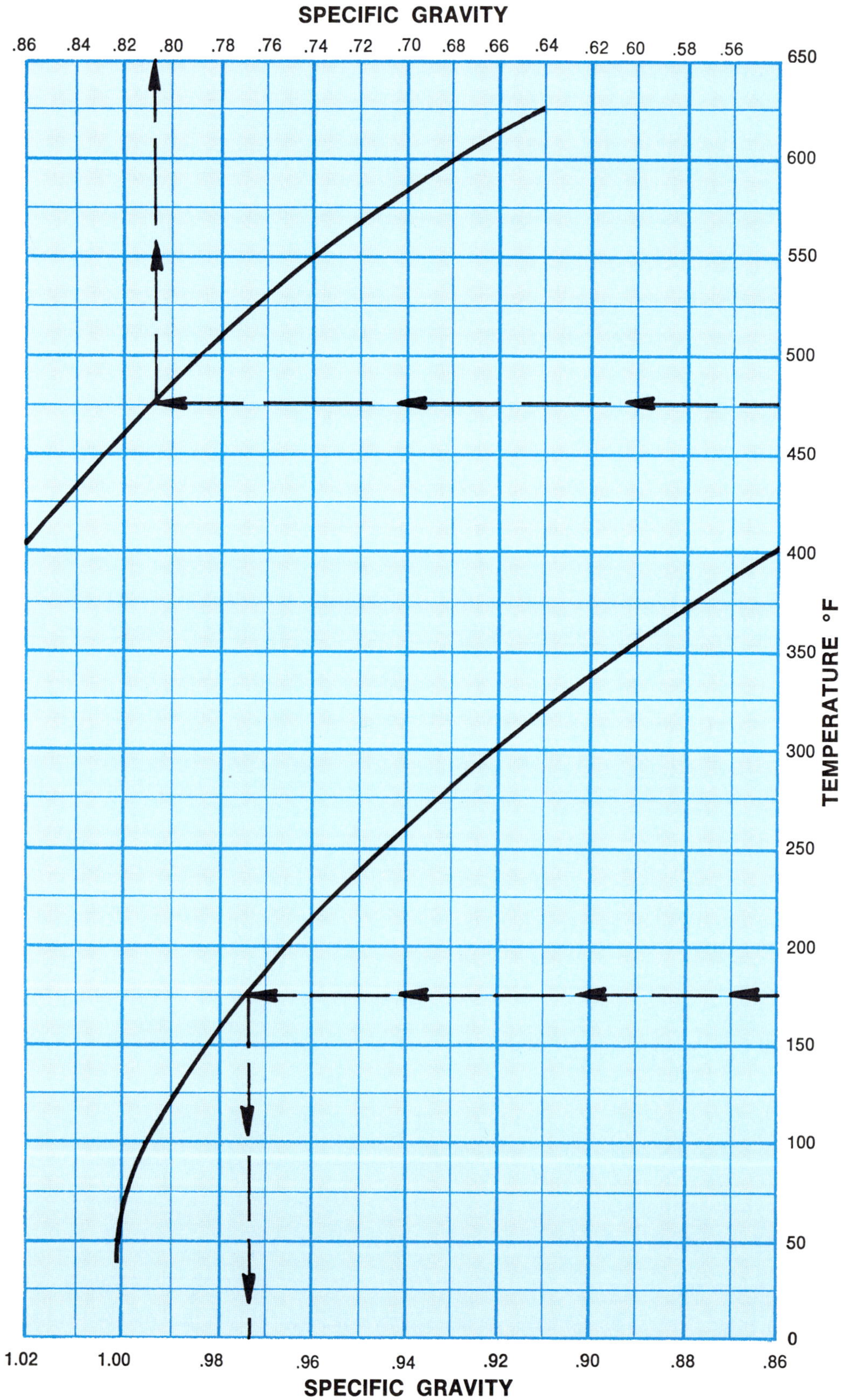

Reprinted with permission from the Allis-Chalmers Corporation.

IIB-1 DEGREES A.P.I. VS SPECIFIC GRAVITY FOR OIL

The relation of Degrees A.P.I. to Specific Gravity (G) is expressed by the following formula:

$$\text{Degrees A.P.I.} = \frac{141.5}{G} - 131.5,$$

$$G = \frac{141.5}{131.5 + \text{Degrees A.P.I.}}$$

The following tables are based on the weight of 1 gallon (U.S.) of oil with a volume of 231 cubic inches at 60 degrees Fahrenheit in air at 760 m.m. pressure and 50% humidity. Assumed weight of 1 gallon of water at 60° Fahrenheit in air is 8.32828 pounds.

To determine the resulting gravity by mixing oils of different gravities:

$$D = \frac{md_1 + nd_2}{m + n}$$

D = Density or Specific Gravity of mixture
m = Proportion of oil d_1 density
n = Proportion of oil of d_2 density
d_1 = Specific Gravity of m oil
d_2 = Specific Gravity of n oil

Degrees A.P.I.	Specific gravity at 60°-60°F.	Degrees A.P.I.	Specific gravity at 60°-60°F.	Degrees A.P.I.	Specific gravity at 60°-60°F.	Degrees A.P.I.	Specific gravity at 60°-60°F.	Degrees A.P.I.	Specific gravity at 60°-60°F.	Degrees A.P.I.	Specific gravity at 60°-60°F.	Degrees A.P.I.	Specific gravity at 60°-60°F.	Degrees A.P.I.	Specific gravity at 60°-60°F.
10.0	1.0000	15.0	.9659	20.0	.9340	25.0	.9042	30.0	.8762	35.0	.8498	40.0	.8251	45.0	.8017
10.1	.9993	15.1	.9652	20.1	.9334	25.1	.9036	30.1	.8756	35.1	.8493	40.1	.8246	45.1	.8012
10.2	.9986	15.2	.9646	20.2	.9328	25.2	.9030	30.2	.8751	35.2	.8488	40.2	.8241	45.2	.8008
10.3	.9979	15.3	.9639	20.3	.9321	25.3	.9024	30.3	.8745	35.3	.8483	40.3	.8236	45.3	.8003
10.4	.9972	15.4	.9632	20.4	.9315	25.4	.9018	30.4	.8740	35.4	.8478	40.4	.8232	45.4	.7999
10.5	.9965	15.5	.9626	20.5	.9309	25.5	.9013	30.5	.8735	35.5	.8473	40.5	.8227	45.5	.7994
10.6	.9958	15.6	.9619	20.6	.9303	25.6	.9007	30.6	.8729	35.6	.8468	40.6	.8222	45.6	.7990
10.7	.9951	15.7	.9613	20.7	.9297	25.7	.9001	30.7	.8724	35.7	.8463	40.7	.8217	45.7	.7985
10.8	.9944	15.8	.9606	20.8	.9291	25.8	.8996	30.8	.8718	35.8	.8458	40.8	.8212	45.8	.7981
10.9	.9937	15.9	.9600	20.9	.9285	25.9	.8990	30.9	.8713	35.9	.8453	40.9	.8208	45.9	.7976
11.0	.9930	16.0	.9593	21.0	.9279	26.0	.8984	31.0	.8708	36.0	.8448	41.0	.8203	46.0	.7972
11.1	.9923	16.1	.9587	21.1	.9273	26.1	.8978	31.1	.8702	36.1	.8443	41.1	.8198	46.1	.7967
11.2	.9916	16.2	.9580	21.2	.9267	26.2	.8973	31.2	.8697	36.2	.8438	41.2	.8193	46.2	.7963
11.3	.9909	16.3	.9574	21.3	.9260	26.3	.8967	31.3	.8692	36.3	.8433	41.3	.8189	46.3	.7958
11.4	.9902	16.4	.9567	21.4	.9254	26.4	.8961	31.4	.8686	36.4	.8428	41.4	.8184	46.4	.7954
11.5	.9895	16.5	.9561	21.5	.9248	26.5	.8956	31.5	.8681	36.5	.8423	41.5	.8179	46.5	.7949
11.6	.9888	16.6	.9554	21.6	.9242	26.6	.8950	31.6	.8676	36.6	.8418	41.6	.8174	46.6	.7945
11.7	.9881	16.7	.9548	21.7	.9236	26.7	.8944	31.7	.8670	36.7	.8413	41.7	.8170	46.7	.7941
11.8	.9874	16.8	.9541	21.8	.9230	26.8	.8939	31.8	.8665	36.8	.8408	41.8	.8165	46.8	.7936
11.9	.9868	16.9	.9535	21.9	.9224	26.9	.8933	31.9	.8660	36.9	.8403	41.9	.8160	46.9	.7932
12.0	.9861	17.0	.9529	22.0	.9218	27.0	.8927	32.0	.8654	37.0	.8398	42.0	.8155	47.0	.7927
12.1	.9854	17.1	.9522	22.1	.9212	27.1	.8922	32.1	.8649	37.1	.8393	42.1	.8151	47.1	.7923
12.2	.9847	17.2	.9516	22.2	.9206	27.2	.8916	32.2	.8644	37.2	.8388	42.2	.8146	47.2	.7918
12.3	.9840	17.3	.9509	22.3	.9200	27.3	.8911	32.3	.8639	37.3	.8383	42.3	.8142	47.3	.7914
12.4	.9833	17.4	.9503	22.4	.9194	27.4	.8905	32.4	.8633	37.4	.8378	42.4	.8137	47.4	.7909
12.5	.9826	17.5	.9497	22.5	.9188	27.5	.8899	32.5	.8628	37.5	.8373	42.5	.8132	47.5	.7905
12.6	.9820	17.6	.9490	22.6	.9182	27.6	.8894	32.6	.8623	37.6	.8368	42.6	.8128	47.6	.7901
12.7	.9813	17.7	.9484	22.7	.9176	27.7	.8888	32.7	.8618	37.7	.8363	42.7	.8123	47.7	.7896
12.8	.9806	17.8	.9478	22.8	.9170	27.8	.8883	32.8	.8612	37.8	.8358	42.8	.8118	47.8	.7892
12.9	.9799	17.9	.9471	22.9	.9165	27.9	.8877	32.9	.8607	37.9	.8353	42.9	.8114	47.9	.7887
13.0	.9792	18.0	.9465	23.0	.9159	28.0	.8871	33.0	.8602	38.0	.8348	43.0	.8109	48.0	.7883
13.1	.9786	18.1	.9459	23.1	.9153	28.1	.8866	33.1	.8597	38.1	.8343	43.1	.8104	48.1	.7879
13.2	.9779	18.2	.9452	23.2	.9147	28.2	.8860	33.2	.8591	38.2	.8338	43.2	.8100	48.2	.7874
13.3	.9772	18.3	.9446	23.3	.9141	28.3	.8855	33.3	.8586	38.3	.8333	43.3	.8095	48.3	.7870
13.4	.9765	18.4	.9440	23.4	.9135	28.4	.8849	33.4	.8581	38.4	.8328	43.4	.8090	48.4	.7865
13.5	.9759	18.5	.9433	23.5	.9129	28.5	.8844	33.5	.8576	38.5	.8324	43.5	.8086	48.5	.7861
13.6	.9752	18.6	.9427	23.6	.9123	28.6	.8838	33.6	.8571	38.6	.8319	43.6	.8081	48.6	.7857
13.7	.9745	18.7	.9421	23.7	.9117	28.7	.8833	33.7	.8565	38.7	.8314	43.7	.8076	48.7	.7852
13.8	.9738	18.8	.9415	23.8	.9111	28.8	.8827	33.8	.8560	38.8	.8309	43.8	.8072	48.8	.7848
13.9	.9732	18.9	.9408	23.9	.9106	28.9	.8822	33.9	.8555	38.9	.8304	43.9	.8067	48.9	.7844
14.0	.9725	19.0	.9402	24.0	.9100	29.0	.8816	34.0	.8550	39.0	.8299	44.0	.8063	49.0	.7839
14.1	.9718	19.1	.9396	24.1	.9094	29.1	.8811	34.1	.8545	39.1	.8294	44.1	.8058	49.1	.7835
14.2	.9712	19.2	.9390	24.2	.9088	29.2	.8805	34.2	.8540	39.2	.8289	44.2	.8054	49.2	.7831
14.3	.9705	19.3	.9383	24.3	.9082	29.3	.8800	34.3	.8534	39.3	.8285	44.3	.8049	49.3	.7826
14.4	.9698	19.4	.9377	24.4	.9076	29.4	.8794	34.4	.8529	39.4	.8280	44.4	.8044	49.4	.7822
14.5	.9692	19.5	.9371	24.5	.9071	29.5	.8789	34.5	.8524	39.5	.8275	44.5	.8040	49.5	.7818
14.6	.9685	19.6	.9365	24.6	.9065	29.6	.8783	34.6	.8519	39.6	.8270	44.6	.8035	49.6	.7813
14.7	.9679	19.7	.9358	24.7	.9059	29.7	.8778	34.7	.8514	39.7	.8265	44.7	.8031	49.7	.7809
14.8	.9672	19.8	.9352	24.8	.9053	29.8	.8772	34.8	.8509	39.8	.8260	44.8	.8026	49.8	.7805
14.9	.9665	19.9	.9346	24.9	.9047	29.9	.8767	34.9	.8504	39.9	.8256	44.9	.8022	49.9	.7800

(cont'd)

IIB-1 DEGREES A.P.I. VS SPECIFIC GRAVITY FOR OIL (Cont'd)

Degrees A.P.I.	Specific gravity at 60°-60°F.	Degrees A.P.I.	Specific gravity at 60°-60°F.	Degrees A.P.I.	Specific gravity at 60°-60°F.	Degrees A.P.I.	Specific gravity at 60°-60°F.	Degrees A.P.I.	Specific gravity at 60°-60°F.	Degrees A.P.I.	Specific gravity at 60°-60°F.	Degrees A.P.I.	Specific gravity at 60°-60°F.	Degrees A.P.I.	Specific gravity at 60°-60°F.
50.0	.7796	56.5	.7527	63.0	.7275	69.5	.7040	76.0	.6819	82.5	.6612	89.0	.6417	95.5	.6233
50.1	.7792	56.6	.7523	63.1	.7271	69.6	.7036	76.1	.6816	82.6	.6609	89.1	.6414	95.6	.6231
50.2	.7788	56.7	.7519	63.2	.7268	69.7	.7033	76.2	.6813	82.7	.6606	89.2	.6411	95.7	.6228
50.3	.7783	56.8	.7515	63.3	.7264	69.8	.7029	76.3	.6809	82.8	.6603	89.3	.6409	95.8	.6225
50.4	.7779	56.9	.7511	63.4	.7260	69.9	.7026	76.4	.6806	82.9	.6600	89.4	.6406	95.9	.6223
50.5	.7775	57.0	.7507	63.5	.7256	70.0	.7022	76.5	.6803	83.0	.6597	89.5	.6403	96.0	.6220
50.6	.7770	57.1	.7503	63.6	.7253	70.1	.7019	76.6	.6800	83.1	.6594	89.6	.6400	96.1	.6217
50.7	.7766	57.2	.7499	63.7	.7249	70.2	.7015	76.7	.6796	83.2	.6591	89.7	.6397	96.2	.6214
50.8	.7762	57.3	.7495	63.8	.7245	70.3	.7012	76.8	.6793	83.3	.6588	89.8	.6394	96.3	.6212
50.9	.7758	57.4	.7491	63.9	.7242	70.4	.7008	76.9	.6790	83.4	.6584	89.9	.6391	96.4	.6209
51.0	.7753	57.5	.7487	64.0	.7238	70.5	.7005	77.0	.6787	83.5	.6581	90.0	.6388	96.5	.6206
51.1	.7749	57.6	.7483	64.1	.7234	70.6	.7001	77.1	.6783	83.6	.6578	90.1	.6385	96.6	.6203
51.2	.7745	57.7	.7479	64.2	.7230	70.7	.6998	77.2	.6780	83.7	.6575	90.2	.6382	96.7	.6201
51.3	.7741	57.8	.7475	64.3	.7227	70.8	.6995	77.3	.6777	83.8	.6572	90.3	.6380	96.8	.6198
51.4	.7736	57.9	.7471	64.4	.7223	70.9	.6991	77.4	.6774	83.9	.6569	90.4	.6377	96.9	.6195
51.5	.7732	58.0	.7467	64.5	.7219	71.0	.6988	77.5	.6770	84.0	.6566	90.5	.6374	97.0	.6193
51.6	.7728	58.1	.7463	64.6	.7216	71.1	.6984	77.6	.6767	84.1	.6563	90.6	.6371	97.1	6190
51.7	.7724	58.2	.7459	64.7	.7212	71.2	.6981	77.7	.6764	84.2	.6560	90.7	.6368	97.2	.6187
51.8	.7720	58.3	.7455	64.8	.7208	71.3	.6977	77.8	.6761	84.3	.6557	90.8	.6365	97.3	.6184
51.9	.7715	58.4	.7451	64.9	.7205	71.4	.6974	77.9	.6757	84.4	.6554	90.9	.6362	97.4	.6182
52.0	.7711	58.5	.7447	65.0	.7201	71.5	.6970	78.0	.6754	84.5	.6551	91.0	.6360	97.5	.6179
52.1	.7707	58.6	.7443	65.1	.7197	71.6	.6967	78.1	.6751	84.6	.6548	91.1	.6357	97.6	.6176
52.2	.7703	58.7	.7440	65.2	.7194	71.7	.6964	78.2	.6748	84.7	.6545	91.2	.6354	97.7	.6174
52.3	.7699	58.8	.7436	65.3	.7190	71.8	.6960	78.3	.6745	84.8	.6542	91.3	.6351	97.8	.6171
52.4	.7694	58.9	.7432	65.4	.7186	71.9	.6957	78.4	.6741	84.9	.6539	91.4	.6348	97.9	.6168
52.5	.7690	59.0	.7428	65.5	.7183	72.0	.6953	78.5	.6738	85.0	.6536	91.5	.6345	98.0	.6166
52.6	.7686	59.1	.7424	65.6	.7179	72.1	.6950	78.6	.6735	85.1	.6533	91.6	.6342	98.1	.6163
52.7	.7682	59.2	.7420	65.7	.7175	72.2	.6946	78.7	.6732	85.2	.6530	91.7	.634	98.2	.6160
52.8	.7678	59.3	.7416	65.8	.7172	72.3	.6943	78.8	.6728	85.3	.6527	91.8	.6337	98.3	.6158
52.9	.7674	59.4	.7412	65.9	.7168	72.4	.6940	78.9	.6725	85.4	.6524	91.9	.6334	98.4	.6155
53.0	.7669	59.5	.7408	66.0	.7165	72.5	.6936	79.0	.6722	85.5	.6521	92.0	.6331	98.5	.6152
53.1	.7665	59.6	.7405	66.1	.7161	72.6	.6933	79.1	.6719	85.6	.6518	92.1	.6328	98.6	.6150
53.2	.7661	59.7	.7401	66.2	.7157	72.7	.6929	79.2	.6716	85.7	.6515	92.2	.6325	98.7	.6147
53.3	.7657	59.8	.7397	66.3	.7154	72.8	.6926	79.3	.6713	85.8	.6512	92.3	.6323	98.8	.6144
53.4	.7653	59.9	.7393	66.4	.7150	72.9	.6923	79.4	.6709	85.9	.6509	92.4	.6320	98.9	.6141
53.5	.7649	60.0	.7389	66.5	.7146	73.0	.6919	79.5	.6706	86.0	.6506	92.5	.6317	99.0	.6139
53.6	.7645	60.1	.7385	66.6	.7143	73.1	.6916	79.6	.6703	86.1	.6503	92.6	.6314	99.1	.6136
53.7	.7640	60.2	.7381	66.7	.7139	73.2	.6913	79.7	.6700	86.2	.6500	92.7	.6311	99.2	.6134
53.8	.7636	60.3	.7377	66.8	.7136	73.3	.6908	79.8	.6697	86.3	.6497	92.8	.6309	99.3	.6131
53.9	.7632	60.4	.7374	66.9	.7132	73.4	.6906	79.9	.6693	86.4	.6494	92.9	.6306	99.4	.6128
54.0	.7628	60.5	.7370	67.0	.7128	73.5	.6902	80.0	.6690	86.5	.6491	93.0	.6303	99.5	.6126
54.1	.7624	60.6	.7366	67.1	.7125	73.6	.6899	80.1	.6687	86.6	.6488	93.1	.6300	99.6	.6123
54.2	.7620	60.7	.7362	67.2	.7121	73.7	.6896	80.2	.6684	86.7	.6485	93.2	.6297	99.7	.6120
54.3	.7616	60.8	.7358	67.3	.7118	73.8	.6892	80.3	.6681	86.8	.6482	93.3	.6294	99.8	.6118
54.4	.7612	60.9	.7354	67.4	.7114	73.9	.6889	80.4	.6678	86.9	.6479	93.4	.6292	99.9	.6115
54.5	.7608	61.0	.7351	67.5	.7111	74.0	.6886	80.5	.6675	87.0	.6476	93.5	.6289	100.0	.6112
54.6	.7603	61.1	.7347	67.6	.7107	74.1	.6882	80.6	.6671	87.1	.6473	93.6	.6286		
54.7	.7599	61.2	.7343	67.7	.7103	74.2	.6879	80.7	.6668	87.2	.6470	93.7	.6283		
54.8	.7595	61.3	.7339	67.8	.7100	74.3	.6876	80.8	.6665	87.3	.6467	93.8	.6281		
54.9	.7591	61.4	.7335	67.9	.7096	74.4	.6872	80.9	.6662	87.4	.6464	93.9	.6278		
55.0	.7587	61.5	.7332	68.0	.7093	74.5	.6869	81.0	.6659	87.5	.6461	94.0	.6275		
55.1	.7583	61.6	.7328	68.1	.7089	74.6	.6866	81.1	.6656	87.6	.6458	94.1	.6272		
55.2	.7579	61.7	.7324	63.2	.7086	74.7	.6862	81.2	.6653	87.7	.6455	94.2	.6269		
55.3	.7575	61.8	.7320	68.3	.7082	74.8	.6859	81.3	.6650	87.8	.6452	94.3	.6267		
55.4	.7571	61.9	.7316	68.4	.7079	74.9	.6856	81.4	.6646	87.9	.6449	94.4	.6264		
55.5	.7567	62.0	.7313	68.5	.7075	75.0	.6852	81.5	.6643	88.0	.6446	94.5	.6261		
55.6	.7563	62.1	.7309	68.6	.7071	75.1	.6849	81.6	.6640	88.1	.6444	94.6	.6258		
55.7	.7559	62.2	.7305	68.7	.7068	75.2	.6846	81.7	.6637	88.2	.6441	94.7	.6256		
55.8	.7555	62.3	.7301	68.8	.7064	75.3	.6842	81.8	.6634	88.3	.6438	94.8	.6253		
55.9	.7551	62.4	.7298	68.9	.7061	75.4	.6839	81.9	.6631	88.4	.6435	94.9	.6250		
56.0	.7547	62.5	.7294	69.0	.7057	75.5	.6836	82.0	.6628	88.5	.6432	95.0	.6247		
56.1	.7543	62.6	.7290	69.1	.7054	75.6	.6832	82.1	.6625	88.6	.6429	95.1	.6244		
56.2	.7539	62.7	.7286	69.2	.7050	75.7	.6829	82.2	.6621	88.7	.6426	95.2	.6242		
56.3	.7535	62.8	.7283	69.3	.7047	75.8	.6826	82.3	.6618	88.8	.6423	95.3	.6239		
56.4	.7531	62.9	.7279	69.4	.7043	75.9	.6823	82.4	.6615	88.9	.6420	95.4	.6236		

IIB-2 SPECIFIC GRAVITIES VS DEGREES BAUMÉ

Specific Gravities at $\frac{60°}{60°}$ F. Corresponding to Degrees Baumé for Liquids Lighter than Water

Calculated from the formula, specific gravity $\frac{60°}{60°}$ *F.* $= \frac{140}{160 - \text{Deg. Bé}}$

Degrees Baumé	Specific gravity	Degrees Baumé	Specific gravity	Degrees Baumé	Specific gravity	Degrees Baumé	Specific gravity	Degrees Baumé	Specific gravity	Degrees Baumé	Specific gravity
10	1.0000	25	0.9032	40	0.8235	55	0.7568	70	0.7000	85	0.6512
11	0.9929	26	0.8974	41	0.8187	56	0.7527	71	0.6965	86	0.6482
12	0.9859	27	0.8917	42	0.8140	57	0.7487	72	0.6931	87	0.6452
13	0.9790	28	0.8861	43	0.8092	58	0.7447	73	0.6897	88	0.6422
14	0.9722	29	0.8805	44	0.8046	59	0.7407	74	0.6863	89	0.6393
15	0.9655	30	0.8750	45	0.8000	60	0.7368	75	0.6829	90	0.6364
16	0.9589	31	0.8696	46	0.7955	61	0.7330	76	0.6796	91	0.6335
17	0.9524	32	0.8642	47	0.7910	62	0.7292	77	0.6763	92	0.6306
18	0.9459	33	0.8589	48	0.7865	63	0.7254	78	0.6731	93	0.6278
19	0.9396	34	0.8537	49	0.7821	64	0.7216	79	0.6699	94	0.6250
20	0.9333	35	0.8485	50	0.7778	65	0.7179	80	0.6667	95	0.6222
21	0.9272	36	0.8434	51	0.7735	66	0.7143	81	0.6635	96	0.6195
22	0.9211	37	0.8383	52	0.7692	67	0.7107	82	0.6604	97	0.6167
23	0.9150	38	0.8333	53	0.7650	68	0.7071	83	0.6573	98	0.6140
24	0.9091	39	0.8284	54	0.7609	69	0.7035	84	0.6542	99	0.6114
										100	0.6087

Specific Gravities at $\frac{60°}{60°}$ F. Corresponding to Degrees Baumé for Liquids Heavier than Water

Calculated from the formula, specific gravity $\frac{60°}{60°}$ *F.* $= \frac{145}{145 - \text{Deg. Bé}}$

Degrees Baumé	Specific gravity	Degrees Baumé	Specific gravity	Degrees Baumé	Specific gravity	Degrees Baumé	Specific gravity	Degrees Baumé	Specific gravity	Degrees Baumé	Specific gravity
0	1.0000	12	1.0902	24	1.1983	36	1.3303	48	1.4948	60	1.7059
1	1.0069	13	1.0985	25	1.2083	37	1.3426	49	1.5104	61	1.7262
2	1.0140	14	1.1069	26	1.2185	38	1.3551	50	1.5263	62	1.7470
3	1.0211	15	1.1154	27	1.2288	39	1.3679	51	1.5426	63	1.7683
4	1.0284	16	1.1240	28	1.2393	40	1.3810	52	1.5591	64	1.7901
5	1.0357	17	1.1328	29	1.2500	41	1.3942	53	1.5761	65	1.8125
6	1.0432	18	1.1417	30	1.2609	42	1.4078	54	1.5934	66	1.8354
7	1.0507	19	1.1508	31	1.2719	43	1.4216	55	1.6111	67	1.8590
8	1.0584	20	1.1600	32	1.2832	44	1.4356	56	1.6292	68	1.8831
9	1.0662	21	1.1694	33	1.2946	45	1.4500	57	1.6477	69	1.9079
10	1.0741	22	1.1789	34	1.3063	46	1.4646	58	1.6667	70	1.9333
11	1.0821	23	1.1885	35	1.3182	47	1.4796	59	1.6860	—	——

IIB-3 — SOLIDS AND SLURRIES

Useful Formulas

a. The formula for specific gravity of a solids-liquids mixture or slurry, S_m is:

$$S_m = \frac{S_s \times S_1}{S_s + C_w (S_1 - S_m)}$$

where,

S_m = specific gravity of mixture or slurry
S_1 = specific gravity of liquid phase
S_s = specific gravity of solids phase
C_w = concentration of solids by weight
C_v = concentration of solids by volume

Example: If the liquid has a specific gravity of 1.2 and the concentration of solids by weight is 35% with the solids having a specific gravity of 2.2 then:

$$S_m = \frac{2.2 \times 1.2}{2.2 + .35\,(1.2 - 2.2)} = 1.43$$

b. Basic relationships among concentration and specific gravities of solid liquid mixtures are shown below.

In Terms Of	S_s, S_m, S_1	C_v	C_w
C_v	$\frac{S_m - S_1}{S_s - S_1}$		$C_w \frac{S_m}{S_s}$
C_w	$\frac{(S_m - S_1)}{(S_s - S_1)} \times \frac{S_s}{S_m}$	$C_v \frac{S_s}{S_m}$	

Where pumps are to be applied to mixtues which are both corrosive and abrasive, the predominant factor causing wear should be identified and the materials ofconstruction selected accordingly. This often results in a compromise and in many cases can only be decided as a result of test or operational experience.

For any slurry pump application a complete description of the mixture components is required in order to select the correct type of pump and materials of construction.

$$C_w = \frac{\text{weight of dry solids}}{\text{weight of dry solids} + \text{weight of liquid phase}}$$

$$C_v = \frac{\text{volume of dry solids}}{\text{volume of dry solids} + \text{volume of liquid phase}}$$

A nomograph for the relationship of concentration to specific gravity of dry solids in water is shown on page 12.

c. Slurry flow requirements can be determined from the expression:

$$Q_m = \frac{4 \text{ x dry solids (tons per hour)}}{C_w = S_m}$$

where,
Q_m = slurry flow (U.S. gallons per minute)
1 ton = 2000 lbs.

Example: 2,400 tons of dry solids is processed in 24 hours in water with a specific gravity of 1.0 and the concentration of solids by weight is 30% with the solids having a specific gravity of 2.7 then:

$$S_m = \frac{2.7 \times 1.0}{2.7 + .3\,(1 - 2.7)} = 1.23$$

$$Q_m = \frac{4 \times 100}{.3 \times 1.23} = 1{,}084 \text{ U.S. GPM}$$

d. Abrasive wear:

Wear increases rapidly when the particle hardness exceeds that of the metal surfaces being abraded. Always select metals with a higher relative hardness to that of the particle hardness. There is little to be gained by increasing the hardness of the metal unless it can be made to exceed that of the particles. The effective abrasion resistance of any metal will depend on its position on the mohs or knoop hardness scale. The relationships of various common ore minerals and metals is shown on page 11.

Wear increases rapidly when the particle size increases. The life of the pump parts can be extended by choosing the correct materials of construction.

Sharp angular particles cause about twice the wear of rounded particles.

Austinetic maganese steel is used when pumping large dense solids where the impact is high.

Hard irons are used to resist erosion and to a lesser extent impact wear.

Elastromeric materials are used when pumping concentrations of fine material but total head is usually restricted to about 100 ft. per stage.

Castable ceramic materials have excellent resistance to cutting erosion but impeller tip velocities are usually restricted to 100 ft./sec.

Classification of pumps according to partical size is shown on page 13.

IIB-4 APPROXIMATE COMPARISON OF HARDNESS VALUES OF COMMON ORES AND MINERALS

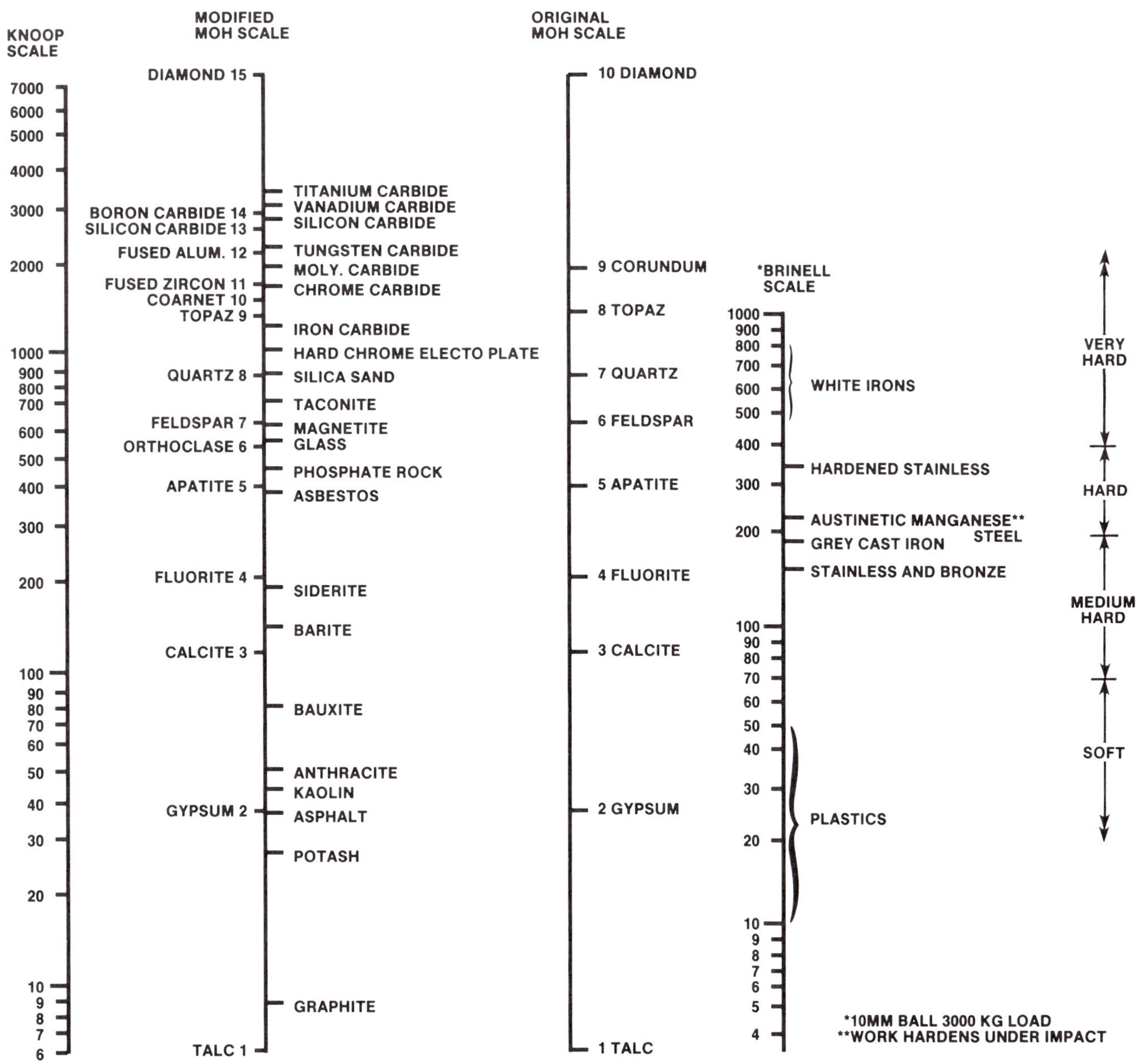

Reprinted with Permission of Goulds Pumps, Inc.

IIB-5 NOMOGRAPH OF THE RELATIONSHIP OF CONCENTRATION TO SPECIFIC GRAVITY IN AQUEOUS SLURRIES

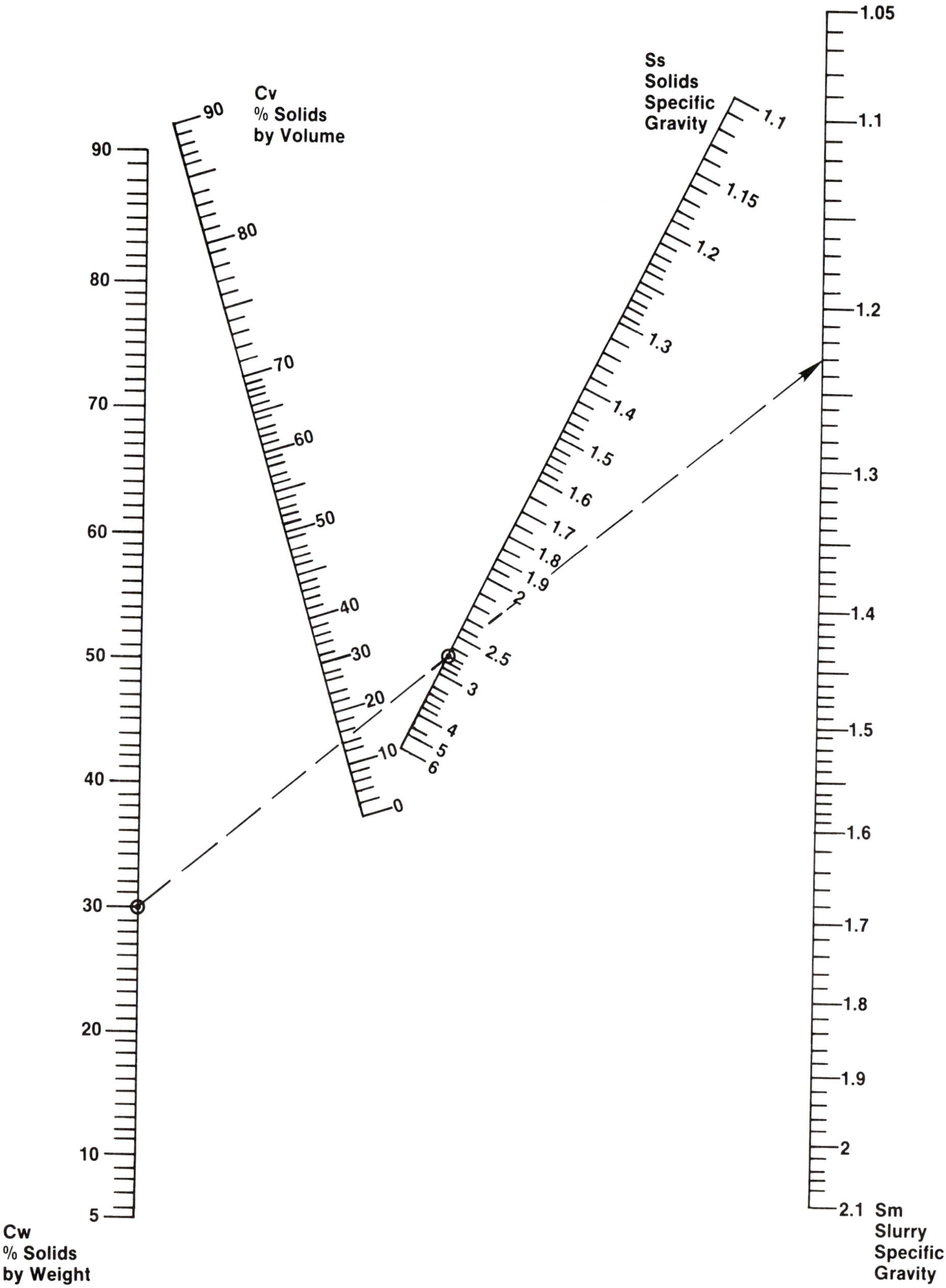

Characteristics of Fluids
Other Fluids

IIB6 CLASSIFICATION OF PUMPS ACCORDING TO SOLID SIZE

Tyler Standard Sleve Series Aperature Inch	MM	Mesh	Grade	Material
160	4060	—	Very large bolders	AUSTINETIC MANGANESE STEEL
80	2030	—	Large bolders	
40	1016	—	Medium bolders	
20	508	—	Small bolders	
10	254	—	Large cobbles	
3	76.2	—	Small cobbles	
2	50.8	—		HARD IRON
1.5	38.1	—	Very coarse gravel	
1.050	26.67	—		
.883	22.43	—	Coarse Gravel	
.742	18.85	—		
.624	15.85	—		
.525	13.33	-	Medium Gravel	
.441	11.20	—		
.371	9.423	—		
.321	7.925	2.5		
.263	6.680	3	Fine Gravel	
.221	5.613	3.5		
.185	4.699	4		
.156	3.962	5		
.131	3.327	6	Very Fine Gravel	RUBBER LINED (SMOOTH ROUND SOLIDS)
.110	2.794	7		
.093	2.362	8		
.078	1.981	9	Very Coarse Sand	
.065	1.651	10		RUBBER LINED (ANGULAR SOLIDS)
.055	1.397	12		
.046	1.168	14		
.039	.991	16	Coarse Sand	
.0328	.833	20		
.0276	.701	24		
.0232	.589	28		
.0195	.495	32	Medium Sand	
.0164	.417	35		
.0138	.351	42		
.0116	.295	48		
.0097	.248	60		
.0082	.204	65	Fine Sand	
.0069	.175	80		
.0058	.147	100		
.0049	.124	115		
.0041	.104	150		
.0035	.089	170	Silt Slmes	
.0029	.074	200		
.0024	.061	250		
.0021	.053	270		
.0017	.043	325		CERAMIC LINED
.0015	.038	400		
—	.025	*500	Pulvertized	
—	.020	*625		
—	.010	*1250		
—	.005	*2500		
—	.001	*12500	Mud Clay	
—	.0005			
—	.0024			

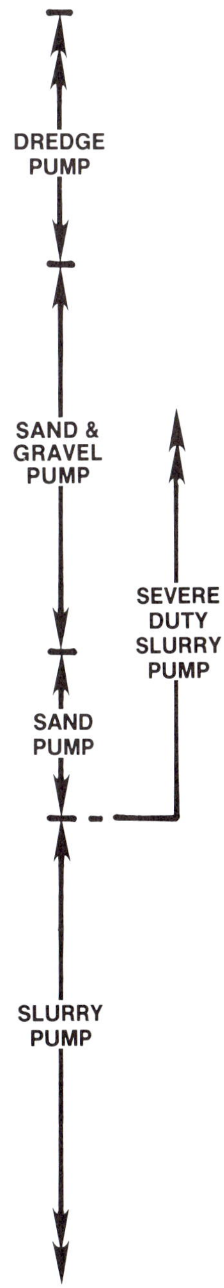

Note: This tabulation is for general guidance only since the selection of pump type and materials of construction also depends on the total head to be generated and the abrasivity of the slurry i.e. concentration, solids specific gravity, etc.

*Theoretical values Micron =.001 mm

Reprinted with Permission of Goulds Pumps, Inc.

IIB-7 VAPOR PRESSURE—LIQUID H_2

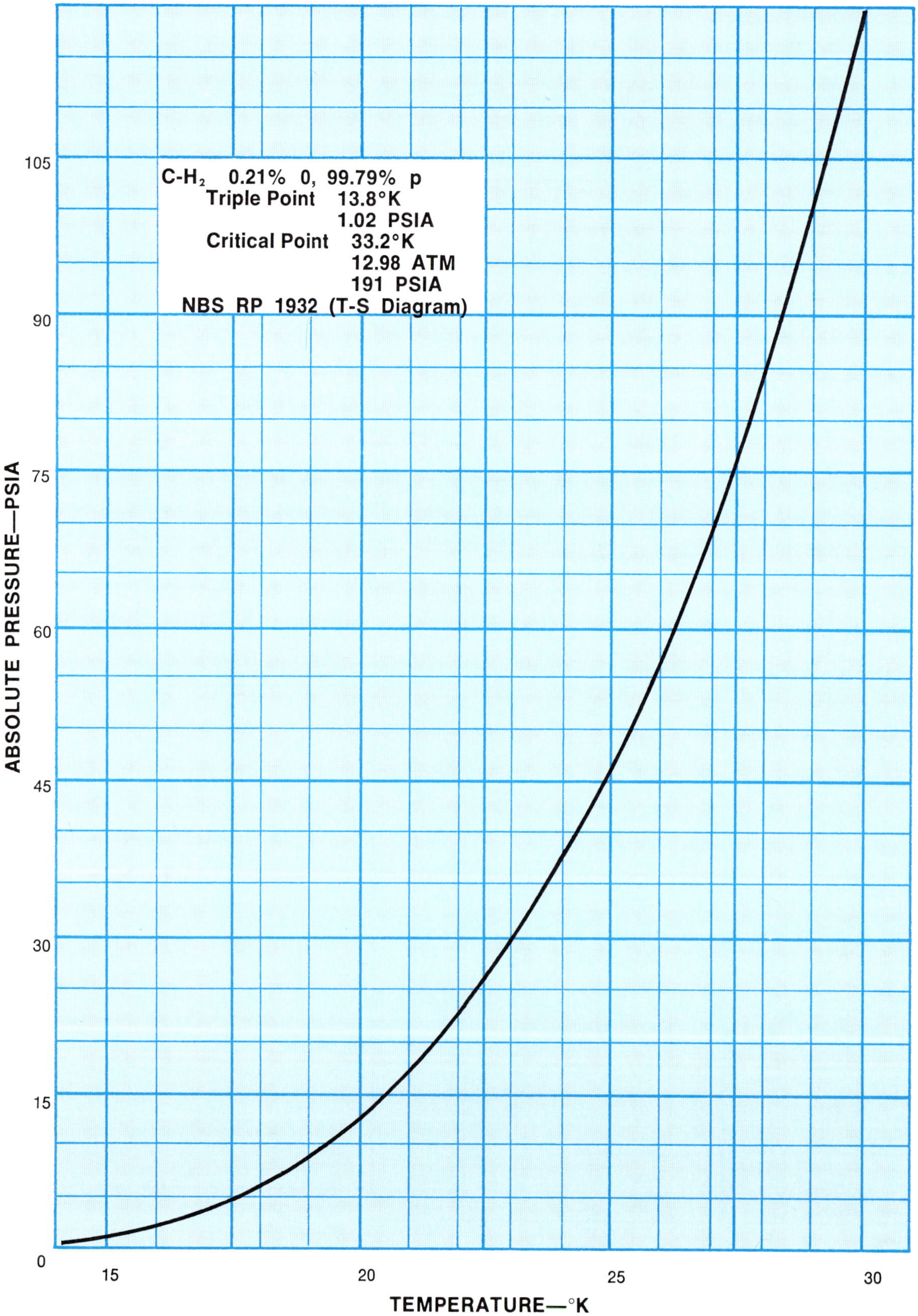

IIB-8 VAPOR PRESSURE OF HELIUM

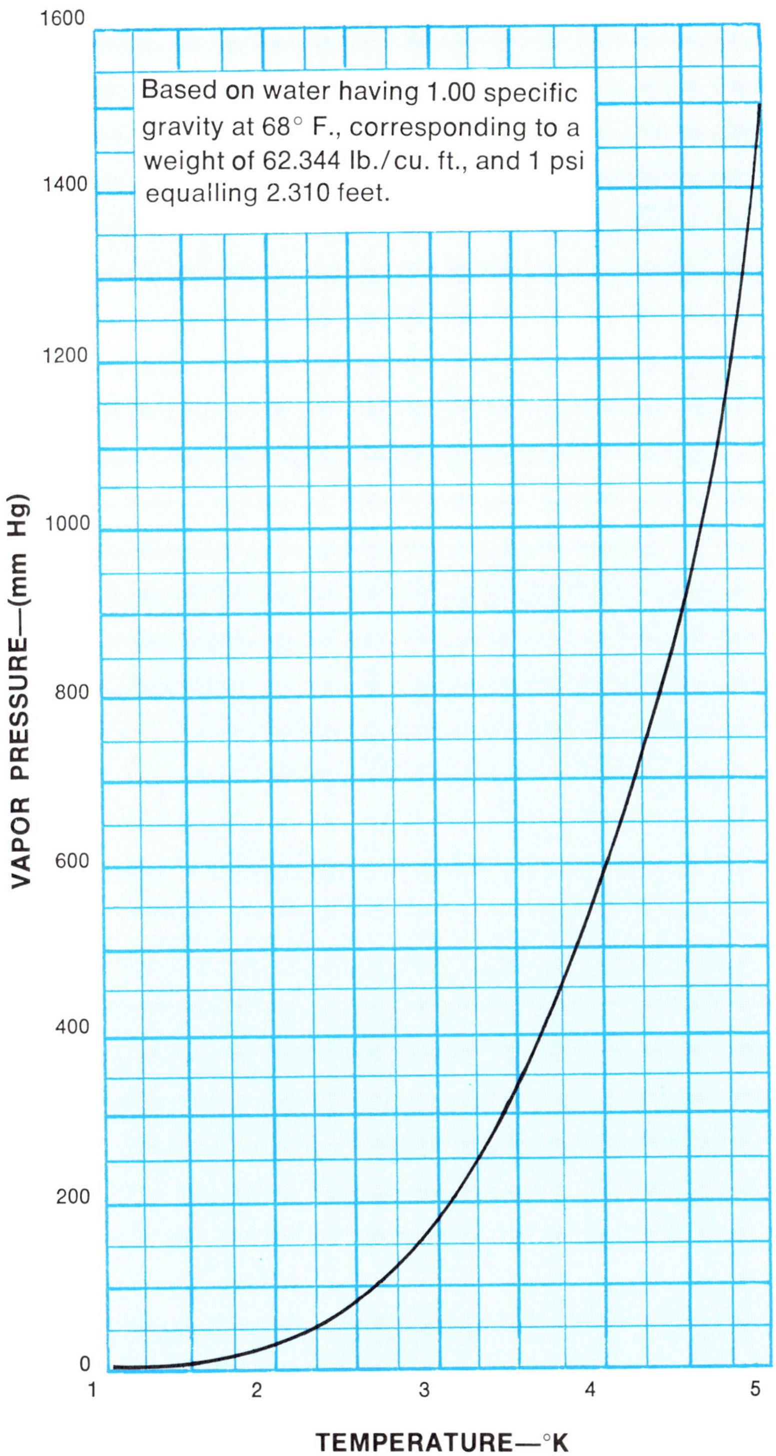

Reprinted with permission from the Byron Jackson Pump Division, Borg-Warner Corporation

IIB-9 VAPOR PRESSURE—VARIOUS LIQUIDS CHART 1—60° TO 240° F.

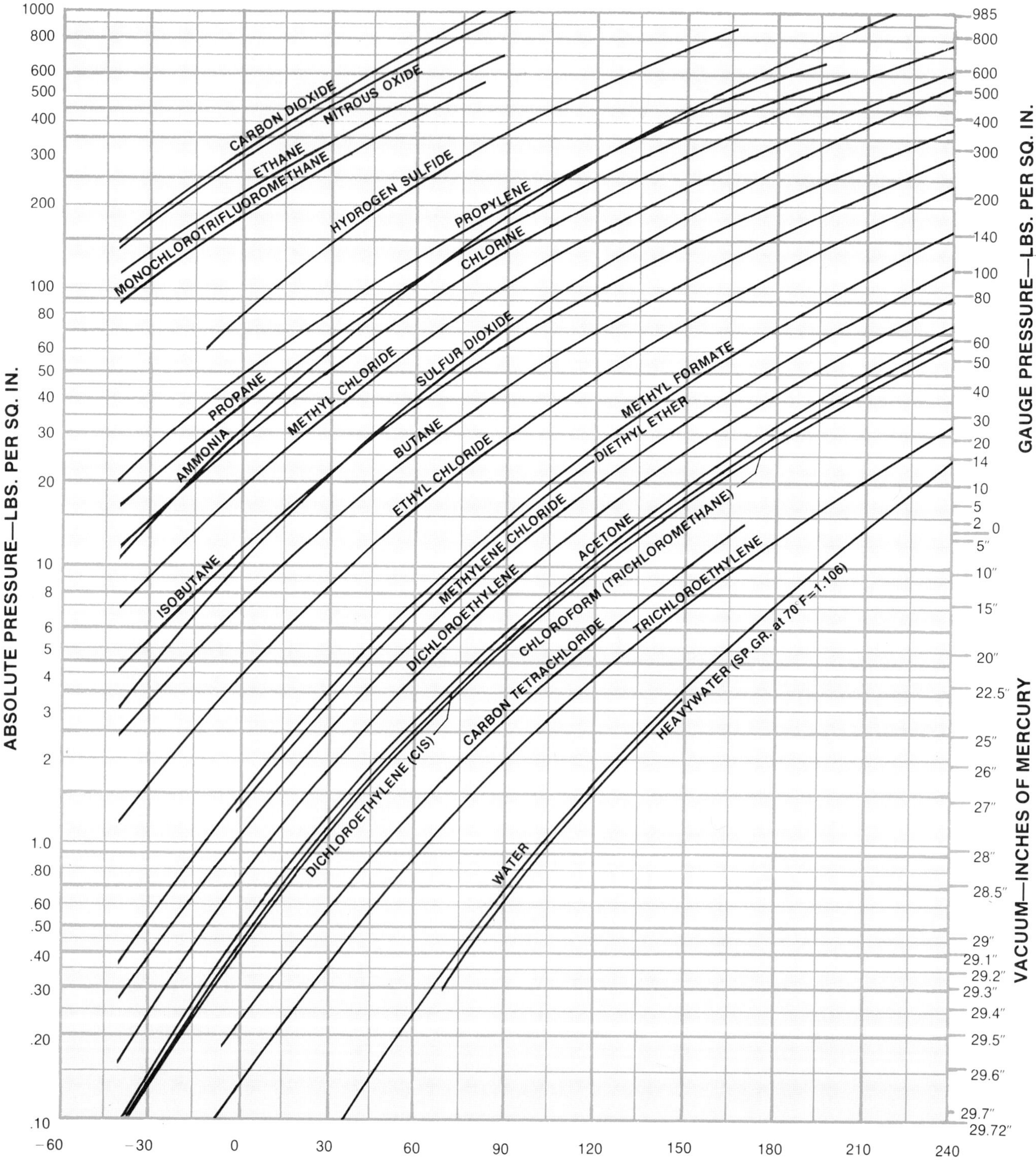

Reprinted with permission from
J. F. Pritchard & Company,
Kansas City, Missouri

IIB10 VAPOR PRESSURE—VARIOUS LIQUIDS CHART 2—180° TO 60° F.

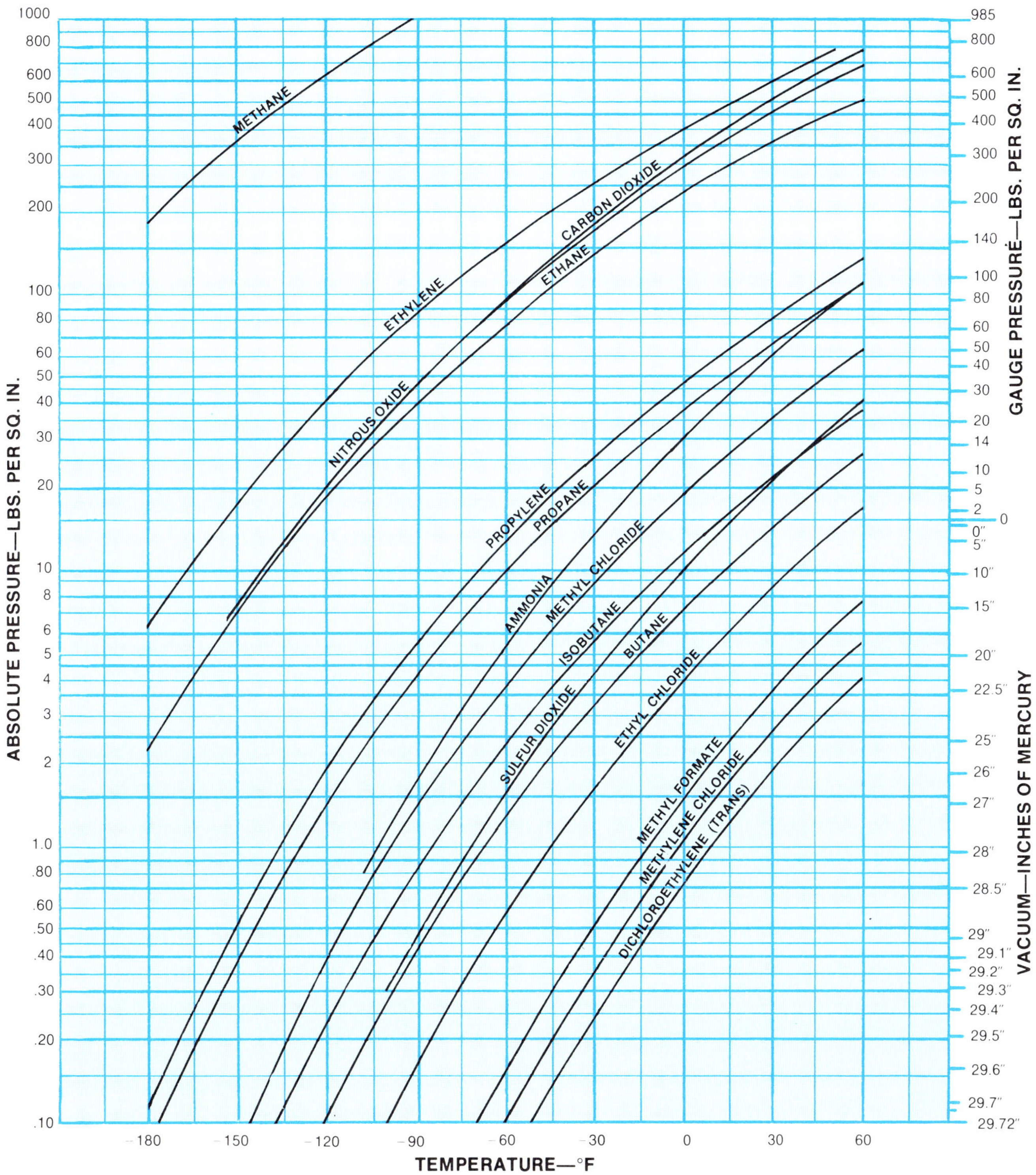

Reprinted with permission from
the Byron Jackson Pump Division,
Borg-Warner Corporation

Characteristics of Fluids
Viscosity

IIC-1 DEFINITIONS AND METHODS OF MEASUREMENT

The viscosity of a fluid (liquid or gas) is that property which tends to resist a shearing force. Since motion or flow of a fluid is produced by shearing forces, viscosity is associated with fluid motion. There is no relation between the viscosity and the specific gravity of most liquids. For instance, molasses having the same specific gravity (1.48) and the same Brix rating (90) may vary in viscosity from 128,000 to 303,000 Seconds Saybolt Universal (SSU).

There are two basic viscosity parameters; i.e., DYNAMIC (or ABSOLUTE) VISCOSITY and KINEMATIC VISCOSITY. The dynamic viscosity may be defined with the aid of Fig. 1 which shows two parallel plane surfaces of area (A) separated a distance (d) and the space between completely filled with fluid. A force (F) is applied to and in the plane of the upper surface, causing it to move with a velocity (V) parallel to the lower fixed surface. The velocity distribution will be linear over the distance (d) and experiments show that the slope of the velocity line $\left(\frac{V}{d}\right)$ will be directly proportional to the unit shearing force, $\tau = \frac{F}{A}$, for all "true" or "Newtonian" fluids. The proportionality factor (μ) is the dynamic viscosity. The foregoing may be expressed by the equations

$$\tau = \frac{\mathbf{F}}{\mathbf{A}} = \mu \frac{\mathbf{V}}{\mathbf{d}}$$

and

$$\mu = \frac{\tau}{\mathbf{V/d}}$$

Therefore, the dimensions of the dynamic viscosity are

$$\textbf{force}\ \frac{\textbf{time}}{\textbf{length}^2}$$

The unit of dynamic viscosity in English measure is the pound-second per square foot which is numerically identical with the slug per foot-second. The unit of dynamic viscosity in Metric measure is the dyne-second per square centimeter, called the POISE, which is numerically identical with the gram per centimeter-second. It usually is more convenient to express numerical values in CENTIPOISES such that 100 centipoises equal one POISE.

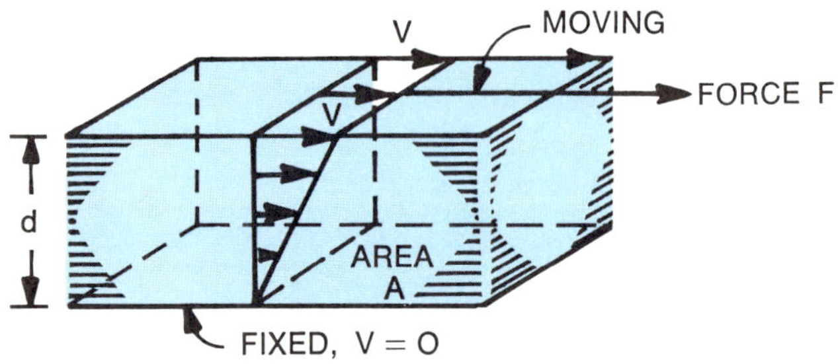

Figure 1

The kinematic viscosity (υ) may be obtained by dividing the dynamic viscosity (μ) by the mass density (ρ). The mass density is the specific weight (w) divided by the acceleration of gravity (g). These relationships may be expressed by the equation

$$\upsilon = \frac{\mu}{\mathbf{w/g}} = \frac{\mu}{\rho} \qquad (15)$$

Therefore, the dimensions of kinematic viscosity are

$$\frac{\text{length}^2}{\text{time}}$$

The unit of kinematic viscosity in English measure is the square foot per second. The unit of kinematic viscosity in Metric measure is the square centimeter per second called the STOKE. It usually is more convenient to express numerical values in CENTISTOKES such that 100 centistokes equal one stoke.

The distinction between the dynamic and the kinematic viscosity should be carefully noted so that the correct parameter will be used as required in computations. Some useful relationships are as follows:

1 pound = 444,823 dynes
1 foot = 30.4800 centimeters
1 square foot = 929.034 square centimeters
1 dyne-second per sq cm = 1 poise = 100 centipoises
1 sq cm/sec = 1 stoke = 100 centistokes
1 lb-sec/sq ft = 478.801 poises
= 47,880.1 centipoises

$$\mu\ \text{lb-sec/sq ft} = \frac{\mu}{47{,}880.1}\ \text{centipoises}$$

$$= 0.0000208855\ \text{centipoises}$$

$$\upsilon \text{ sq ft/sec} = \frac{\upsilon \text{ sq cm/sec}}{929.034}$$

$$= 0.00107639 \text{ stokes}$$

$$\upsilon = \frac{\mu}{\rho} = \frac{\mu}{w/g}$$

$$\upsilon \text{ sq ft/sec} = 0.000671970 \frac{\mu}{w}$$

where

υ = centipoises

and

w = lb/cu ft

g = 980.665 cm/sec/sec

= 32.1740 ft/sec/sec at sea level and approximately 45 degrees latitude

The viscosities of most fluids vary appreciably with changes in temperature. The influence of change in pressure usually is negligible.

The viscosities of fluids, such as mineral oil and water, are unaffected by the magnitude and kind of motion to which they may be subjected as long as the temperature remains constant. Thus the ratio of shear stress to shear rate is a constant for all shear rates, is independent of time, and zero shear rate exists only at zero shear stress; such a fluid is said to be **Newtonian.**

When the ratio of shear stress to shear rate increases as the shear rate increases, reversibly and independent of time, a fluid is said to be **dilatent.**

When the shear stress to shear rate ratio is constant for shear rates above zero, is independent of time, but when shear occurs only for shear stress above a fixed minimum greater than zero, a fluid is said to be **plastic.**

When the ratio of shear stress to shear rate decreases as shear rate increases, reversibly and independent of time, and zero shear rate occurs only at zero shear stress, a fluid is said to be **pseudo-plastic.**

When the ratio of shear stress to shear rate decreases as shear rate increases and is time dependent in that this ratio increases back to its "rest" value gradually with lapse of time at zero shear rate and stress, and decreases to a limit value gradually with lapse of time at constant shear rate, a fluid is said to be **thixotropic.**

When the shear stress to shear ratio rate is constant for all shear rates at any given instant of time, but increases with time, a fluid is said to be **rheopectic.**

Viscosity is measured by an instrument called a viscosimeter. A definite volume of fluid is allowed to flow through a capillary tube or orifice of specified proportions and the time of efflux noted. Instruments of the capillary type, such as the Ostwald, Bingham, and Ubbelohde viscosimeters are used primarily for fluids of low viscosity, such as water. Instruments of the orifice type are used commercially for more viscous fluids such as petroleum products, and the time of efflux of the sample is taken as a measure of the viscosity. The Saybolt viscosimeter is commonly used in the United States, the Saybolt Universal for fluids of medium viscosity and the Saybolt Furol for those of high viscosity. The viscosity is expressed in Seconds Saybolt Universal (SSU) or Seconds Saybolt Furol (SSF). The relationship between Saybolt Universal viscosities and kinematic viscosities in centistokes is given in "ASTM Conversion Tables for Kinematic and Saybolt Universal Viscosities"* or by the ASTM Standard, Designation: D446-85a.* Similar information for Saybolt Furol viscosities may be obtained from the ASTM Standard, Designation: D2161-87. The respective British counterparts of the Saybolt Universal and Saybolt Furol viscosimeters are the Redwood and Redwood Admiralty viscosimeıers. The Engler viscosimeter is used extensively on the continent of Europe. Viscosimeters such as the Brookfield are particularly useful with non-Newtonian fluids. There are many other viscosimeters for special purposes, discussion of which is beyond the scope of this Manual. Viscosity conversion tables for use with the above described viscosimeters are shown in Tables IIC-3 and 4. A viscosity blending chart for use with oils is shown in IIC-2. Let oil (A) have the higher viscosity and oil (B) the lower viscosity. Mark the viscosity of (A) and (B) on the right and left hand scales, respectively, and draw a straight line connecting the marks as shown. The viscosity of any blend of (A) and (B) will be shown by the intersection of the vertical line representing the percentage composition and the line described above. See examples 1 and 2.

*American Society for Testing Materials, 1916 Race St., Philadelphia, Pa. 19103.

IIC-2 VISCOSITY BLENDING CHART

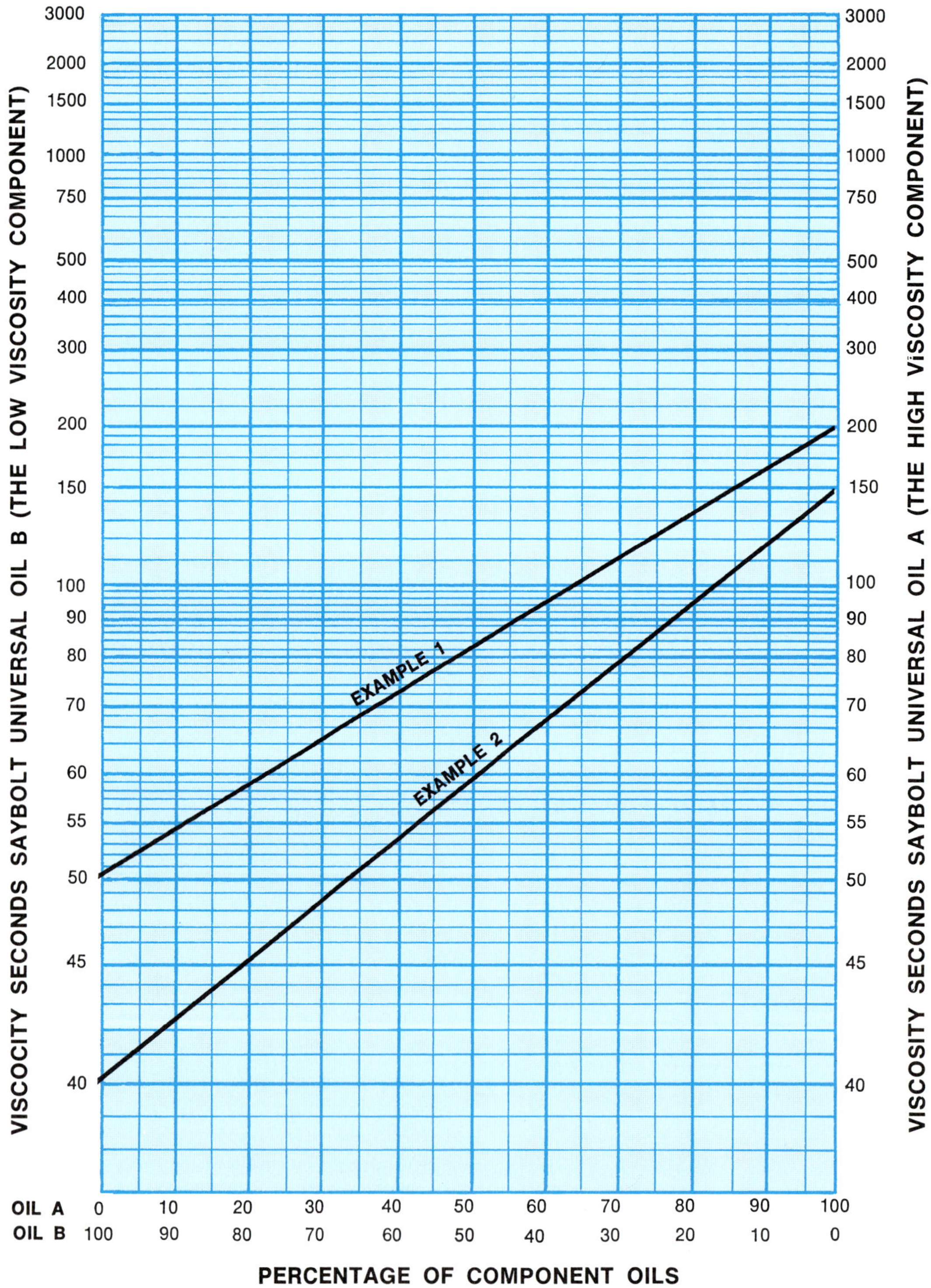

Characteristics of Fluids
Viscosity

IIC-3 —VISCOSITY CONVERSION TABLE

The following table will give an approximate comparison of various viscosity ratings so that if the viscosity is given in terms other than Saybolt Universal, it can be translated quickly by following horizontally to the Saybolt Universal column.

Seconds Saybolt Universal ssu	Kinematic Viscosity Centistokes *	Seconds Saybolt Furol ssf	Seconds Redwood 1 (Standard)	Seconds Redwood 2 (Admiralty)	Degrees Engler	Degrees Barbey	Seconds Parlin Cup #7	Seconds Parlin Cup #10	Seconds Parlin Cup #15	Seconds Parlin Cup #20	Seconds Ford Cup #3	Seconds Ford Cup #4
31	1.00	—	29	—	1.00	6200	—	—	—	—	—	—
35	2.56	—	32.1	—	1.16	2420	—	—	—	—	—	—
40	4.30	—	36.2	5.10	1.31	1440	—	—	—	—	—	—
50	7.40	—	44.3	5.83	1.58	838	—	—	—	—	—	—
60	10.3	—	52.3	6.77	1.88	618	—	—	—	—	—	—
70	13.1	12.95	60.9	7.60	2.17	483	—	—	—	—	—	—
80	15.7	13.70	69.2	8.44	2.45	404	—	—	—	—	—	—
90	18.2	14.44	77.6	9.30	2.73	348	—	—	—	—	—	—
100	20.6	15.24	85.6	10.12	3.02	307	—	—	—	—	—	—
150	32.1	19.30	128	14.48	4.48	195	—	—	—	—	—	—
200	43.2	23.5	170	18.90	5.92	144	40	—	—	—	—	—
250	54.0	28.0	212	23.45	7.35	114	46	—	—	—	—	—
300	65.0	32.5	254	28.0	8.79	95	52.5	15	6.0	3.0	30	20
400	87.60	41.9	338	37.1	11.70	70.8	66	21	7.2	3.2	42	28
500	110.0	51.6	423	46.2	14.60	56.4	79	25	7.8	3.4	50	34
600	132	61.4	508	55.4	17.50	47.0	92	30	8.5	3.6	58	40
700	154	71.1	592	64.6	20.45	40.3	106	35	9.0	3.9	67	45
800	176	81.0	677	73.8	23.35	35.2	120	39	9.8	4.1	74	50
900	198	91.0	762	83.0	26.30	31.3	135	41	10.7	4.3	82	57
1000	220	100.7	896	92.1	29.20	28.2	149	43	11.5	4.5	90	62
1500	330	150	1270	138.2	43.80	18.7	—	65	15.2	6.3	132	90
2000	440	200	1690	184.2	58.40	14.1	—	86	19.5	7.5	172	118
2500	550	250	2120	230	73.0	11.3	—	108	24	9	218	147
3000	660	300	2540	276	87.60	9.4	—	129	28.5	11	258	172
4000	880	400	3380	368	117.0	7.05	—	172	37	14	337	230
5000	1100	500	4230	461	146	5.64	—	215	47	18	425	290
6000	1320	600	5080	553	175	4.70	—	258	57	22	520	350
7000	1540	700	5920	645	204.5	4.03	—	300	67	25	600	410
8000	1760	800	6770	737	233.5	3.52	—	344	76	29	680	465
9000	1980	900	7620	829	263	3.13	—	387	86	32	780	520
10000	2200	1000	8460	921	292	2.82	—	430	96	35	850	575
15000	3300	1500	13700	—	438	2.50	—	650	147	53	1280	860
20000	4400	2000	18400	—	584	1.40	—	860	203	70	1715	1150

***Kinematic Viscosity (in centistokes)**

$$= \frac{\textbf{Absolute Viscosity (in centipoises)}}{\textbf{Density}}$$

When the Metric System terms centistokes and centipoises are used, the density is numerically equal to the specific gravity. Therefore, the following expression can be used which will be sufficiently accurate for most calculations:

Kinematic Viscosity (in centistokes)

$$= \frac{\textbf{Absolute Viscosity (in centipoises)}}{\textbf{Specific Gravity}}$$

When the English System units are used, the density must be used rather than the specific gravity.

For values of 70 centistokes and above, use the following conversion:

SSU = centistokes × 4.635

Above the range of this table and within the range of the viscosimeter, multiply the particular value by the following approximate factors to convert to SSU:

Viscosimeter	Factor
Saybolt Furol	10.
Redwood Standard	1.095
Redwood Admiralty	10.87
Engler-Degrees	34.5

Viscosimeter	Factor
Parlin cup #15	98.2
Parlin cup #20	187.0
Ford cup #4	17.4

IIC-4 –VISCOSITY CONVERSION TABLE

The following table will give an approximate comparison of various viscosity ratings so that if the viscosity is given in terms other than Saybolt Universal, it can be translated quickly by following horizontally to the Saybolt Universal column.

Seconds Saybolt Universal ssu	Kinematic Viscosity Centistokes*	Approx. Seconds Mac Michael	Approx. Gardner Holt Bubble	Seconds Zahn Cup #1	Seconds Zahn Cup #2	Seconds Zahn Cup #3	Seconds Zahn Cup #4	Seconds Zahn Cup #5	Seconds Demmler Cup #1	Seconds Demmler Cup #10	Approx. Seconds Stormer 100 gm Load	Seconds Pratt and Lambert "F"
31	1.00	—	—	—	—	—	—	—	—	—	—	—
35	2.56	—	—	—	—	—	—	—	—	—	—	—
40	4.30	—	—	—	—	—	—	—	1.3	—	—	—
50	7.40	—	—	—	—	—	—	—	2.3	—	2.6	—
60	10.3	—	—	—	—	—	—	—	3.2	—	3.6	—
70	13.1	—	—	—	—	—	—	—	4.1	—	4.6	—
80	15.7	—	—	—	—	—	—	—	4.9	—	5.5	—
90	18.2	—	—	—	—	—	—	—	5.7	—	6.4	—
100	20.6	125	—	38	18	—	—	—	6.5	—	7.3	—
150	32.1	145	—	47	20	—	—	—	10.0	1.0	11.3	—
200	43.2	165	A	54	23	—	—	—	13.5	1.4	15.2	—
250	54.0	198	A	62	26	—	—	—	16.9	1.7	19	—
300	65.0	225	B	73	29	—	—	—	20.4	2.0	23	—
400	87.0	270	C	90	37	—	—	—	27.4	2.7	31	7
500	110.0	320	D	—	46	—	—	—	34.5	3.5	39	8
600	132	370	F	—	55	—	—	—	41	4.1	46	9
700	154	420	G	—	63	22.5	—	—	48	4.8	54	9.5
800	176	470	—	—	72	24.5	—	—	55	5.5	62	10.8
900	198	515	H	—	80	27	18	—	62	6.2	70	11.9
1000	220	570	I	—	88	29	20	13	69	6.9	77	12.4
1500	330	805	M	—	—	40	28	18	103	10.3	116	16.8
2000	440	1070	Q	—	—	51	34	24	137	13.7	154	22
2500	550	1325	T	—	—	63	41	29	172	17.2	193	27.6
3000	660	1690	U	—	—	75	48	33	206	20.6	232	33.7
4000	880	2110	V	—	—	—	63	43	275	27.5	308	45
5000	1100	2635	W	—	—	—	77	50	344	34.4	385	55.8
6000	1320	3145	X	—	—	—	—	65	413	41.3	462	65.5
7000	1540	3670	—	—	—	—	—	75	481	48	540	77
8000	1760	4170	Y	—	—	—	—	86	550	55	618	89
9000	1980	4700	—	—	—	—	—	96	620	62	695	102
10000	2200	5220	Z	—	—	—	—	—	690	69	770	113
15000	3300	7720	Z2	—	—	—	—	—	1030	103	1160	172
20000	4400	10500	Z3	—	—	—	—	—	1370	137	1540	234

***Kinematic Viscosity (in centistokes)**

$$\text{Kinematic Viscosity (in centistokes)} = \frac{\text{Absolute Viscosity (in centipoises)}}{\text{Density}}$$

When the Metric System terms centistokes and centipoises are used, the density is numerically equal to the specific gravity. Therefore, the following expression can be used which will be sufficiently accurate for most calculations:

$$\text{Kinematic Viscosity (in centistokes)} = \frac{\text{Absolute Viscosity (in centipoises)}}{\text{Specific Gravity}}$$

When the English System units are used, the density must be used rather than the specific gravity.

For values of 70 centistokes and above, use the following conversion:

$$\text{SSU} = \text{centistokes} \times 4.635$$

Above the range of this table and within the range of the viscosimeter, multiply the particular value by the following approximate factors to convert to SSU:

Viscosimeter	Factor
Mac Michael	1.92 (approx.)
Demmler #1	14.6
Demmler #10	146.
Stormer	13. (approx.)

Section III

Fluid Flow

IIIA-1 —FLUID FLOW—GENERAL

SECTION 1—PIPE FRICTION

The resistance to the incompressible flow of any fluid in any pipe may be computed from the equation

$$h_f = f \frac{L}{D} \frac{V^2}{2g} \tag{1}$$

wherein

h_f = Frictional resistance in feet of fluid
L = Length of pipe in feet
D = Average internal diameter of pipe in feet
V = Average velocity in pipe in feet/second
g = Acceleration due to gravity in feet/second/second. Hereafter, the value 32.17 feet/second/second for sea level and 45 degrees latitude will be used.
f = Friction factor

SECTION 2—CONSTRUCTION OF CHARTS AND TABLES

The Colebrook Equation[9] 5

$$\frac{1}{\sqrt{f}} = -2 \log_{10} \left(\frac{\epsilon}{3.7D} + \frac{2.51}{R\sqrt{f}} \right) \tag{2}$$

offers a reliable means for computing the friction factor (*f*) to be used in Equation (1).

The Reynolds Number (R) is given by the equation

$$R = \frac{VD}{\upsilon} \tag{3}$$

wherein

V = Average velocity in feet/second
D = Average internal diameter in feet
υ = Kinematic viscosity of the fluid in square feet/second

References 1, 2, 3, 4, 5, 6 and 7 in Section VI were studied to obtain the best value of the roughness parameter (ϵ) and the probable variations in the friction factors for new pipes. The probable variations in (*f*) for some classes of new clean pipe are given in Table A.

TABLE A

Material	ϵ, feet	Probable maximum variation in *f* Per Cent
Steel or wrought iron	0.00015	−5 to +10
Asphalt-dipped cast iron	0.0004	−5 to + 5
Galvanized iron	0.0005	0 to +10

Equation (2) was combined with Equation (1) and solutions carried out for each kind and size of pipe. These were used to construct large-scale logarithmic plots from which the values of (h_f) shown in Section IIIB, Tables 1–31 incl., were obtained.

SECTION 3—OLD PIPES

A study of References 8, 9, 10, 11 in Section VI showed that the problem of estimating the friction factors for old pipes or allowing for the deterioration of new pipes is beyond the scope of this Handbook.

The deterioration of pipes with age depends on the particular chemical properties of the fluid and the metal with which it is in contact. It is recommended that prior experience be considered and local water supply officials be consulted where it is necessary to estimate the friction losses in old pipes or to allow for the aging of new pipes. References 8, 9, 10, 11 of Section VI may be consulted for general information on the subject. For commercial installations, it is recommended that 15 percent be added to the values shown in Tables 1–31.

SECTION 4—TABLES OF FRICTION LOSS FOR WATER, EXPLANATION

Frictional resistances for water flowing in new, clean steel pipe (Schedule 40)* or in asphalt-dipped cast iron pipe are given in Section IIIB, Tables 1–31 incl., herein.**

The tables show the discharge in cubic feet per second, the average velocity in feet per second, and the velocity head in feet for any fluid in a circular pipe of the same diameter as that specified in each table for rates of flow in gallons per minute. The values of the friction head (h_f) in feet of fluid per 100 feet of pipe apply to any fluid having a kinematic viscosity, $\nu = 0.00001216$ square feet per second (1.130 centistokes) which is the value for pure fresh water at 60° F.*** The friction heads are average values for pipes having the $\frac{\epsilon}{D}$ values given in the tables, where (ϵ) is a linear measure of the absolute roughness of the pipe walls and (D) is the internal diameter of the pipe. Further information on the roughness parameter is given in Section IV.

The tabulated values of (h_f) are in feet of pure fresh water (60° F) per 100 feet of new clean steel pipe (Schedule 40)* or of new clean asphalt-dipped cast iron pipe as specified.

No allowance has been made for age, differences in diameter resulting from manufacturing tolerances, or any abnormal conditions of interior surface. Any factor of safety must be estimated from the local conditions and the requirements of each particular installation. An example illustrating the use of the tables will be found in Section IIIB.

SECTION 5—VALVES AND FITTINGS

The resistance to flow caused by a valve or fitting may be computed from the equation

$$h = k \frac{V^2}{2g} \qquad (4)$$

wherein

h = Frictional resistance in feet of fluid
V = Average velocity in feet/second in a pipe of corresponding diameter
g = 32.17 feet/second/second
k = Resistance coefficient for valve or fitting

Values of (k) for valves and fittings may be found in Section IIIB. Reference to the literature will reveal wide differences in the published values of (k) for all types of valves and fittings. The available data are inconclusive. As indicated in Section IIIB, flanged valves and fittings usually exhibit lower resistance coefficients than screwed valves and fittings. The resistance coefficients decrease with the increasing size of most valves and fittings.

Cast iron flanged elbows and drainage-type elbows may be expected to approximate pipe bends. Values of the resistance coefficient (k) may be taken from Section IIIB. The solid line curves in Fig. IIIB-5A are given by Reference 12a of Section VI with the range of scatter of the test points as indicated. The broken line curves may be used as a guide to probable resistance coefficients for intermediate values of the relative roughness factor, $\frac{\epsilon}{D}$. A value of $\epsilon = 0.00085$ feet will be satisfactory for uncoated cast iron and cast steel elbows. Resistance coefficients for pipe bends with less than 90 degree deflection angles as reported by Wasielewski[9] 13 are shown in Fig. IIIB-5B. The curves shown are for smooth surfaces but may be used as a guide to approximating the resistance coefficients for surfaces of moderate roughness such as clean steel and cast iron. Figs. IIIB-5A and IIIB-5B in Section IIIB are not reliable below $\frac{R}{D} = 1$, where R is the radius of the elbow in feet. The approximate radius of a flanged elbow may be obtained by subtracting the flange thickness from the center-to-face dimension. The center-to-face dimension for a reducing elbow is usually identical with that of an elbow of the same straight size as the larger end.

The resistance coefficients for miter bends as reported by Shubert[9] 12b are shown in Section IIIB, Table 33. The rough pipe used in the Shubert investigation had a relative roughness factor, $\frac{\epsilon}{D}$ of about 0.0022. Reference 12b of Section VI should be consulted for information on the variation of the resistance coefficients with variation in the Reynolds Number.

*Note 1: ANSI B36.10M-1985 of the American National Standards Institute.

**The Tables should not be applied to problems involving the pumping of fluid-solid mixtures. It is suggested that prior experience be considered and that qualified sources be consulted where it is necessary to estimate the friction losses in pipes carrying solids in suspension.
For paper stock suspensions, see Section IIIB.

***For easy references, values of GPM and h_f are shown in shaded column.

The resistance to flow caused by a sudden enlargement may be computed from the equation

$$h = k \frac{(V_1 - V_2)^2}{2g}$$

$$= k \left(1 - \frac{A_1}{A_2}\right)^2 \frac{V_1^2}{2g}$$

$$= k \left[1 - \left(\frac{D_1}{D_2}\right)^2\right]^2 \frac{V_1^2}{2g}$$

$$= k \left[\left(\frac{D_2}{D_1}\right)^2 - 1\right]^2 \frac{V_2^2}{2g} \qquad (5)$$

wherein

- h = Frictional resistance in feet of fluid
- V_1 = Average velocity in feet/second in the smaller (upstream) pipe
- A_1 = Internal cross-sectional area of the smaller pipe in square feet
- D_1 = Internal diameter of the smaller pipe in feet
- V_2, A_2, D_2 = Corresponding values for the larger (downstream) pipe
- g = 32.17 feet/second/second
- k = Resistance coefficient, usually taken as unity since the variation is almost always less than ±3 per cent.

Equation (5) is useful for computing the resistance to flow caused by conical increasers and diffusers. Values of (k) for conical increasers based on data reported by Gibson[9] 14 are given in Section IIIB, Fig. IIIB-6 or may be computed by the equation

$$k = 3.50 \left(\tan \frac{\theta}{2}\right)^{1.22} \qquad (6)$$

wherein

θ = total conical angle of the increaser in degrees

Equation (6) applies only to values of θ between 7.5 and 35 degrees. Noteworthy is the fact that above 50 degrees a sudden enlargement will be as good or better than a conical increaser. Values of (k) for conical diffusers as reported by Reference 11 of Section VI are shown in Section IIIB, Fig. IIIB-6. The values shown include the entrance mouthpiece which accounts in part for the increase over Gibson's values for conical increasers. Resistance coefficients for reducers, as reported by Russell are given in Section IIIB, Fig. IIIB-7.

SECTION 6—FRICTION FACTOR DIAGRAMS

As previously stated, the resistance to the incompressible flow of any fluid in any pipe may be computed from the equation (1)

$$h_f = f \frac{L}{D} \frac{V^2}{2g}$$

Values of (f) may be obtained directly from Fig. IIIA-1 where the pipe is new clean asphalt-dipped cast iron, from Fig. IIIA-2 where the pipe is new clean steel of Schedule 40 wall thickness, or from Fig. IIIA-3 which applies to any size and type of surface. The probable variations in (f) for some classes of new clean pipe are given in Table A. It will be necessary to know the approximate value of the relative roughness factor, $\frac{\epsilon}{D}$ to enter Fig. IIIA-3 and this can be obtained, for several kinds of pipes, from Fig. IIIA-4.

If the fluid is fresh water at 60° F or atmospheric air at 60° F the scales at the top of Figs. IIIA-1–3 incl., may be used. For convenience in air and water computations only, the scale reading is the product of the average velocity in feet/second and the internal diameter in inches, (VD″).

For other fluids or temperatures the scales at the bottom of Figs. IIIA-1–3 incl., must be used. The scale reading is the Reynolds Number, R, given by equation (3).

The data for Figs. IIIA-1–3 incl., were obtained directly from solutions of Equation (2). Figs. IIIA-4 and IIIA-5 were taken from Reference 2 with minor changes.

Values of the kinematic viscosity (ν) at various temperatures are given in Fig. IIIA-5 for a number of different fluids. The Reynolds Number also may be obtained directly from Fig. IIIA-5 with the aid of the quantity (VD″) mentioned above. The tracer line shows that for kerosene at 175° F flowing with an average velocity of 12.5 ft/sec in a pipe of 4 inches internal diameter, the Reynolds Number would be 3.5×10^5. In cases where viscosities are obtained from sources other than Fig. IIIA-5, it is *absolutely essential* that they be expressed in *sq ft/sec* in order that they may be used with Fig. IIIA-5 or Equation (3). Kinematic viscosities measured in stokes or centistokes may be converted to (ν) in square feet/second by the formula.

$$\nu = \mathbf{0.00107639 \times stokes}$$
$$= \mathbf{0.0000107639 \times centistokes} \quad (7)$$

For further information on viscosity, see Section IIC.

If the Reynolds Number is less than 2000, the flow is laminar and the friction factor for any fluid in any pipe is given by the equation

$$f = \frac{64}{R} \quad (8)$$

If the Reynolds Number is above 4000, the flow will usually be turbulent and the curves in Figs. IIIA 1–3 incl., apply. The range R = 2000–4000 is called the critical zone in which the flow may be highly unstable and the friction factor indeterminate.

Example 1. Use of Friction Factor Diagrams. A pump delivers kerosene at 70° F through 512 feet of new 2-inch steel pipe (Schedule 40) and screwed fittings to a tank (See Fig. IIIA-6). The tank pressure is maintained constant at 50 psi at the level of the discharge pipe. It is required to estimate the suction head (h_s), the discharge head (h_d), and the reading of the discharge pressure gauge (P_d) when the rate of flow is 100 gpm.

SOLUTION: a. **Suction Line.** Table 9, Section IIIB, shows the average velocity (V) to be 9.56 ft/sec and the velocity head, $\frac{V^2}{2g}$ to be 1.42 feet of kerosene. Fig. IIIA-5 shows the specific gravity of kerosene to be 0.813 at 60/60 and the kinematic viscosity (ν) to be 0.000023 sq ft/sec at 70° F. Section IVA, shows the internal diameter of the pipe to be 0.1723 ft. The Reynolds Number, by Equation (3), will be

$$R = \frac{VD}{\nu}$$
$$= \frac{9.56 \times 0.1723}{0.000023}$$
$$= 71{,}600$$

As an alternate procedure, determine the quantity (VD″) and use Fig. IIIA-5 at 70° F. Proceed from the intersection of the curve labelled kerosene and 70° F to the (VD″) value of 19.8. The resulting Reynolds number is aproximately 7×10^4.

This is above the critical zone and the flow will be turbulent. The friction factor, from Fig. IIIA-2, will be $f = 0.0226$. The head loss due to pipe friction, by Equation (1), will be

$$h_f = f \frac{L}{D} \frac{V^2}{2g}$$
$$= 0.0226 \left(\frac{27}{0.1723}\right) 1.42$$
$$= 5.03 \text{ feet of kerosene}$$

The resistance coefficients for the fittings, from Section IIIB, Tables 32(a), 32(b) and 32(c), will be

Square-edged inlet,	k = 0.5
Wedge-disc gate valve,	k = 0.17

The head loss in the fittings, by Equation (4) will be

$$h = k \frac{V^2}{2g}$$
$$= (0.5 + 0.17)\ 1.42$$
$$= 0.95 \text{ feet of kerosene}$$

The suction head (h_s) will be

$$h_s = (67.00 - 62.52) - 5.03 - 0.95$$
$$= 4.48 - 5.98$$
$$= -1.50 \text{ feet of kerosene}$$

Therefore, the pump actually operates with a *suction lift* of 1.50 feet of kerosene.

b. **Discharge Line.** The head loss due to pipe friction, by Equation (1), will be

$$h_f = f \frac{L}{D} \frac{V^2}{2g}$$
$$= 0.0226 \left(\frac{485}{0.1723}\right) 1.42$$
$$= 90.3 \text{ feet of kerosene}$$

The resistance coefficients for the fittings, from Section IIIB, Tables 32(a), 32(b) and 32(c), will be

Swing check valve,	k = 2.2
Wedge-disc gate valve,	k = 0.17
Regular screwed elbow,	k = 1.0
Sudden enlargement,	k = 1.0

The head loss in the fittings, by Equation (4), will be

$$h = k \frac{V^2}{2g}$$
$$= [2.2 + 0.17 + 2\,(1.0) + 1.0]\ 1.42$$
$$= 7.63 \text{ feet of kerosene}$$

The specific gravity of kerosene at 70° F relative to water at 60° F may be estimated from Section IIC, to be 0.809. The specific weight of water at 60° F is 62.34 lb/cu. ft. from Section IIA. Therefore, the pressure head at discharge into the pressure tank will be

$$\text{Tank pressure head} = \frac{50 \times 144}{0.809 \times 62.34}$$
$$\cong 142.8 \text{ feet of kerosene}$$

The discharge head (h_d) will be

$$h_d = (62.00 - 62.52) + 90.3 + 7.63 + 142.8$$
$$\cong 240 \text{ feet of kerosene}$$

The pressure head at the centerline of the pipe where the gauge is connected to the discharge pipe will be the pressure head in the discharge tank plus the head loss in the pipe and fittings minus the velocity head at the point where the gauge is connected. No difference in elevations is involved because the pipe is horizontal. Therefore,

Required pressure head
$$= 142.8 + 90.3 + 7.63 - 1.42$$
$$\cong 239 \text{ feet of kerosene}$$

The pressure head at the center of the gauge will be 1 foot greater than this value because the gauge is 1 foot below the center of the pipe. Therefore, the reading of the discharge pressure gauge will be

$$P_d = \frac{(239 + 1)\ 62.34 \times 0.809}{144}$$
$$\cong 84 \text{ psi}$$

Example 2. Use of Friction Factor Diagrams. A simplified boiler feed pump installation is shown in Fig. IIIA-7. The rate of flow is 1450 gpm at the pump. It is required to compute the net positive suction head (h_{sv}), the total suction head (h_s), the total discharge head (h_d) and the total head (H).

SOLUTION: **a. Suction Line.** The following properties of 10-inch Schedule 40 steel pipe may be obtained from Section IVA.

Diameter, D = 10.020 inches = 0.8350 feet

Area, A = 0.5476 square feet

Relative roughness factor, $\frac{\epsilon}{D} = 0.000180$

The discharge, $Q = \frac{1450}{7.481 \times 60} = 3.230$ cu ft/sec*

The velocity, $V = \frac{3.230}{0.5476} = 5.90$ ft/sec*

*These also may be obtained from Section IIIB, Table 17, by interpolation.

The velocity head, $\frac{V^2}{2g} = \frac{(5.90)^2}{64.34} = 0.541$ feet

The kinematic viscosity, from Fig. IIIA-5, will be

$\nu = 0.0000026$ square feet/second

The Reynolds Number, by Equation (3), will be

$$R = \frac{VD}{\nu} = \frac{5.90 \times 0.8350}{0.0000026} = 1{,}890{,}000 = 1.89 \times 10^6$$

The friction factor may be taken from Fig. IIIA-2, $f = 0.0139$. The head loss due to pipe friction, by Equation (1), will be

$$h_f = f \frac{L}{D} \frac{V^2}{2g}$$
$$= 0.0139 \left(\frac{230}{0.8350}\right) 0.541$$
$$= 2.07 \text{ feet}$$

The resistance coefficients for the fittings may be obtained from Section IIIB, Tables 32(a), 32(b) and 32(c), as follows:

Entrance, k = 0.5 (assuming a square-edged inlet)

Wedge-disc gate valve, k = 0.06

From Section IIIB, Fig. IIIB-5a

For 90 degree pipe bend,

$$\frac{R}{D} = \frac{50}{10.020}$$
$$\cong 5$$

and

$$\frac{\epsilon}{D} = 0.000180$$

so that $k \cong 0.10$

For reducing elbow,

$$\text{Average } \frac{R}{D} = \frac{9.5}{9} = 1.06$$

The roughness parameter for cast fittings may be taken as,

$$\epsilon = 0.00085 \times 12 = 0.0102 \text{ inches}$$

Therefore,

$$\text{Average } \frac{\epsilon}{D} = \frac{0.0102}{9} = 0.0011$$

and

$$k \cong 0.38 \text{ for a straight size elbow}$$

The effect of the reduction of area in a reducing elbow is difficult to estimate. It usually reduces the loss below that to be expected in an equivalent straight size elbow. No further allowance will be made here because the elbow loss is already small.

The total resistance coefficient for the fittings in the suction line will be

Total k = 0.5 + 2 (0.06) + 4 (0.10) + 0.38 = 1.40

The head loss due to the fittings, by Equation (4), will be

$$\mathbf{h} = \mathbf{k}\,\frac{\mathbf{V^2}}{\mathbf{2g}}$$
$$= 1.40 \times 0.541$$
$$= 0.76 \text{ feet}$$

The total head loss in the suction line

$$= 2.07 + 0.76 = 2.83 \text{ feet.}$$

The net positive suction head, (h_{sv}), is the total suction head in feet of liquid absolute determined at the suction nozzle and referred to datum less the vapor pressure of the liquid in feet absolute. The datum is the centerline of the pump for horizontal pumps. The pressure in the direct contact heater is 30 psi absolute which is equal to the vapor pressure of water at 250° F. Therefore, the heater may be considered to be at atmospheric pressure for the purpose of computing the net positive suction head. The net positive suction head will be

$$h_{sv} = (132.00 - 67.00) - 2.83$$
$$= 65.00 - 2.83 = 62.17$$
$$\cong 62 \text{ feet of water at } 250^\circ \text{ F}$$

The total suction head is given by

$$h_s = h_{sv} + h_{vpa} - h_a$$

wherein

h_s = total suction head in feet (positive)
h_{sv} = net positive suction head in feet
h_{vpa} = vapor pressure of liquid in feet absolute
h_a = atmospheric pressure in feet absolute

Assuming the atmospheric pressure to be 14.7 psia, and remembering that the vapor pressure of the water in the direct contact heater is 30 psia,

$$h_a = 144\,\frac{p_a}{w}$$
$$= \frac{144 \times 14.7}{58.82}$$
$$= 36.0 \text{ feet}$$

and

$$h_{vpa} = 144\,p_{vpa}/w$$
$$= \frac{144 \times 30}{58.82}$$
$$= 73.5 \text{ feet}$$

Wherein w = 58.82 lb/cu ft at T = 250° F from Section IIA. Therefore, the total suction head is

$$h_s = 62.17 + 73.5 - 36.0$$
$$= 99.67$$
$$\cong 100 \text{ feet of water at } 250^\circ \text{ F.}$$

b. Discharge Line (250° F). The following properties of 8-inch Schedule 80 steel pipe may be obtained from Section IVA.

Diameter, D = 7.625 inches = 0.6354 feet

Area, A = 0.3171 square feet

$$\text{Relative roughness, } \frac{\epsilon}{D} = 0.000236$$

$$\text{The velocity, } V = \frac{Q}{A} = \frac{3.230}{0.3171} = 10.19 \text{ ft/sec}$$

$$\text{The velocity head, } \frac{V^2}{2g} = \frac{(10.19)^2}{64.34} = 1.61 \text{ feet}$$

The Reynolds Number,

$$\mathbf{R} = \frac{\mathbf{VD}}{\nu}$$
$$= \frac{10.19 \times 0.6354}{0.0000026}$$

$$= 2{,}490{,}000$$
$$= 2.49 \times 10^6$$

The friction factor, from Fig. IIIA-2 will be f = 0.0145.

The head loss due to pipe friction, by Equation (1), will be

$$h_f = f\frac{L}{D}\,\frac{V^2}{2g}$$
$$= 0.0145\left(\frac{108}{0.6354}\right)1.61$$
$$= 3.97 \text{ feet}$$

The head loss in the 6-inch by 8-inch discharge elbow may be estimated from Section IVA Tables for 6-inch Schedule 80 pipe, as follows:

$$D = 5.761 \text{ inches}$$
$$A = 0.1810 \text{ square feet}$$

Average diameter of elbow

$$= \frac{5.761 + 7.625}{2}$$
$$= 6.69 \text{ inches}$$

$$\text{Average}\frac{R}{D} = \frac{11.75}{6.69}$$
$$= 1.76$$

As before,

$$\frac{\epsilon}{D} = \frac{0.0102}{6.69} = 0.00152$$

and from Section IIIB, Fig. IIIB-5a:

$$k \cong 0.28$$

The loss caused by changing the direction of the flow will be

$$h = k\frac{V^2}{2g} = 0.28 \times 1.61 = 0.45 \text{ feet}$$

An additional loss will be caused by the use of the elbow as an increaser. This may be estimated as a sudden enlargement loss as it is unlikely that the elbow will act as favorably as a straight increaser.

The inlet velocity, $V_1 = \frac{Q}{A} = \frac{3.230}{0.1810} = 17.85$ ft/sec

As before,

$$V_2 = 10.19 \text{ ft/sec}$$

Therefore, by Equation (5)

$$h = \frac{(V_1 - V_2)^2}{2g} = \frac{(17.85 - 10.19)^2}{64.34} = 0.91 \text{ feet}$$

The total loss due to the elbow will be

$$h = 0.45 + 0.91 = 1.36 \text{ feet}$$

The head loss in the Venturi Meter will be partly in the meter nozzle and partly in the diffuser. The nozzle loss may be computed by the formula

$$\mathbf{h} = \left[\left(\frac{1}{C_v}\right)^2 - 1\right]\frac{V_2^2}{2g} \qquad (9)$$

wherein

- h = Head loss in feet of fluid
- C_v = Coefficient of velocity for nozzle
- V_2 = Velocity at throat (smallest cross-section) of nozzle in feet/second
- g = 32.17 feet/second/second

The coefficient of velocity has the same numerical value as the coefficient of discharge for most smooth nozzles such as are used in Venturi Meters. The throat velocity head may be obtained from the pipe velocity head,

$$\frac{V_1^2}{2g}$$

by

$$\frac{V_2^2}{2g} = \left(\frac{D_1}{D_2}\right)^4\frac{V_1^2}{2g}$$
$$= \left(\frac{7.625}{3.750}\right)^4 1.61$$
$$= 27.5 \text{ feet}$$

Therefore, the nozzle loss will be

$$h = \left[\left(\frac{1}{0.98}\right)^2 - 1\right]27.5$$
$$= (1.041 - 1)\,27.5$$
$$= 1.13 \text{ feet}$$

The loss in the 16-inch long diffuser may be estimated by Equation (6) or, from Section IIIB, Fig. IIIB-6 as follows:

$$\frac{D_2 - D_1}{2L} = \frac{7.625 - 3.750}{2 \times 16} = \frac{3.875}{32} = 0.121$$

for which

$$k = 0.266$$

and

$$h = k\left[\left(\frac{D_2}{D_1}\right)^2 - 1\right]^2 \frac{V_2^2}{2g}$$

$$= 0.266\left[\left(\frac{7.625}{3.750}\right)^2 - 1\right]^2 1.61$$

$$= 4.21 \text{ feet}$$

The head loss in the Venturi Meter will be

$$h = 1.13 + 4.21 = 5.34 \text{ feet}$$

Resistance coefficients for the remainder of the fittings may be obtained:

From Section IIIB, Tables 32(a), 32(b) and 32(c)

Swing check valve,	$k = 2.0$
Wedge-disc gate valve,	$k = 0.075$
Sudden enlargement at heater,	$k = 1.0$

From Section IIIB, Fig. IIIB-5a; 90 degree flanged elbow,

$$\frac{R}{D} = \frac{11.75}{7.625}$$

$$= 1.54$$

and

$$\frac{\epsilon}{D} = \frac{0.0102}{7.625} = 0.00134$$

for which

$$k \cong 0.29$$

The head loss in these fittings, by Equation (4) will be

$$h = [2.0 + 2(0.075) + 0.29 + 1.0]\ 1.61$$
$$= 3.44 \times 1.61 = 5.54 \text{ feet}$$

The total friction head loss in the 250° F discharge line will be

$$h = 3.97 + 1.36 + 5.34 + 5.54$$

$$= 16.21 \text{ feet of water at } 250° \text{ F}$$

c. Discharge Line (350° F). The volumetric rate of flow will be increased due to the expansion of water in the heater. The specific weight at each temperature, from Section IIA, will be:

at $T = 250°$ F, $w = 58.82$ lb/cu ft

at $T = 350°$ F, $w = 55.59$ lb/cu ft

The discharge, $Q = 3.230 \dfrac{58.82}{55.59} = 3.418$ cu ft/sec

The velocity, $V = \dfrac{Q}{A} = \dfrac{3.418}{0.3171} = 10.78$ ft/sec

The velocity head, $\dfrac{V^2}{2g} = \dfrac{(10.78)^2}{64.34} = 1.81$ feet

The kinematic viscosity, by Fig. IIIA-5, will be

$$\nu = 0.0000019 \text{ sq ft/sec}$$

The Reynolds Number,

$$R = \frac{VD}{\nu} = \frac{10.78 \times 0.6354}{0.0000019} = 3{,}600{,}000 = 3.6 \times 10^6$$

The friction factor, from Fig. IIIA-3, will be

$$f = 0.0143 \text{ for } \frac{\epsilon}{D} = 0.000236$$

The head loss due to pipe friction, by Equation (1), will be

$$h_f = f\frac{L}{D}\frac{V^2}{2g} = 0.0143\frac{165}{0.6354}1.81 = 6.72 \text{ feet}$$

The resistance coefficients for the fittings may be obtained:

From Section IIIB, Tables 32(a), 32(b) and 32(c).

Entrance from heater,

$k = 0.5$ (assuming a square edged inlet)

Wedge-disc gate valve,

$k = 0.075$

Sudden enlargement to economizer,

$k = 1.0$

From Section III(B), Fig. IIIB-5a: 90 degree pipe bend,

$$\frac{R}{D} = \frac{40}{7.625}$$

$$= 5.25$$

and

$$\frac{\epsilon}{D} = 0.000236$$

for which

$$k = 0.10$$

For 90 degree flanged elbow,

k = 0.29, the same as for the 250° F discharge line.

The expansion loop may be considered as three 90 degree pipe bends, making a total of six pipe bends in the line.

The head loss in the fittings, by Equation (4), will be

$$h = [0.5 + 4\,(0.075) + 6\,(0.10) + 2\,(0.29) + 1.0]\ 1.81$$
$$= 5.39 \text{ feet}$$

The total head loss due to friction in the 350° F discharge line will be

$$h = 6.72 + 5.39$$
$$= 12.11 \text{ feet of water at } 350° \text{ F}$$

exclusive of the loss in the pressure regulator.

All head losses should be expressed in feet of water at the pump temperature (250° F) before computing the pump discharge head. The necessary conversions, based on the specific weights, are as follows:

Discharge line, 350° F,

$$\text{Friction loss} = 12.11\ \frac{55.59}{58.82}$$
$$= 11.44 \text{ feet of water at } 250° \text{ F}$$

Heater and pressure regulator,

$$\text{Friction loss} = \frac{(40 + 50)\ 144}{58.82}$$
$$= 220 \text{ feet of water at } 250° \text{ F}$$

Static Head friction loss

$$= (127.00 - 77.00)\ \frac{55.59}{58.82} + (77.00 - 67.00)$$
$$= 47.25 + 10.00$$
$$= 57.25 \text{ feet of water at } 250° \text{ F}$$

Economizer pressure head friction loss

$$= \frac{900 \times 144}{58.82}$$
$$= 2203 \text{ feet of water at } 250° \text{ F}$$

The total discharge head will be

$$h_d = 57.25 + 2203 + 16.21 + 11.44 + 220$$
$$= 2507.90$$
$$\cong 2508 \text{ feet of water at } 250° \text{ F}$$

The total head for the pump will be

$$H = h_d - h_s$$
$$= 2508 - 100$$
$$\cong 2408 \text{ feet of water at } 250° \text{ F}$$

It should be noted that the pipe friction and fitting losses in the discharge line are a very small portion of the total discharge head. Therefore, the degree of refinement of the preceding computations is unwarranted for an installation of this type. The indicated procedure may be useful in the solution of other problems where the hydraulic losses are more important.

Fig. IIIA-2 FRICTION FACTORS FOR STEEL OR WROUGHT IRON PIPE

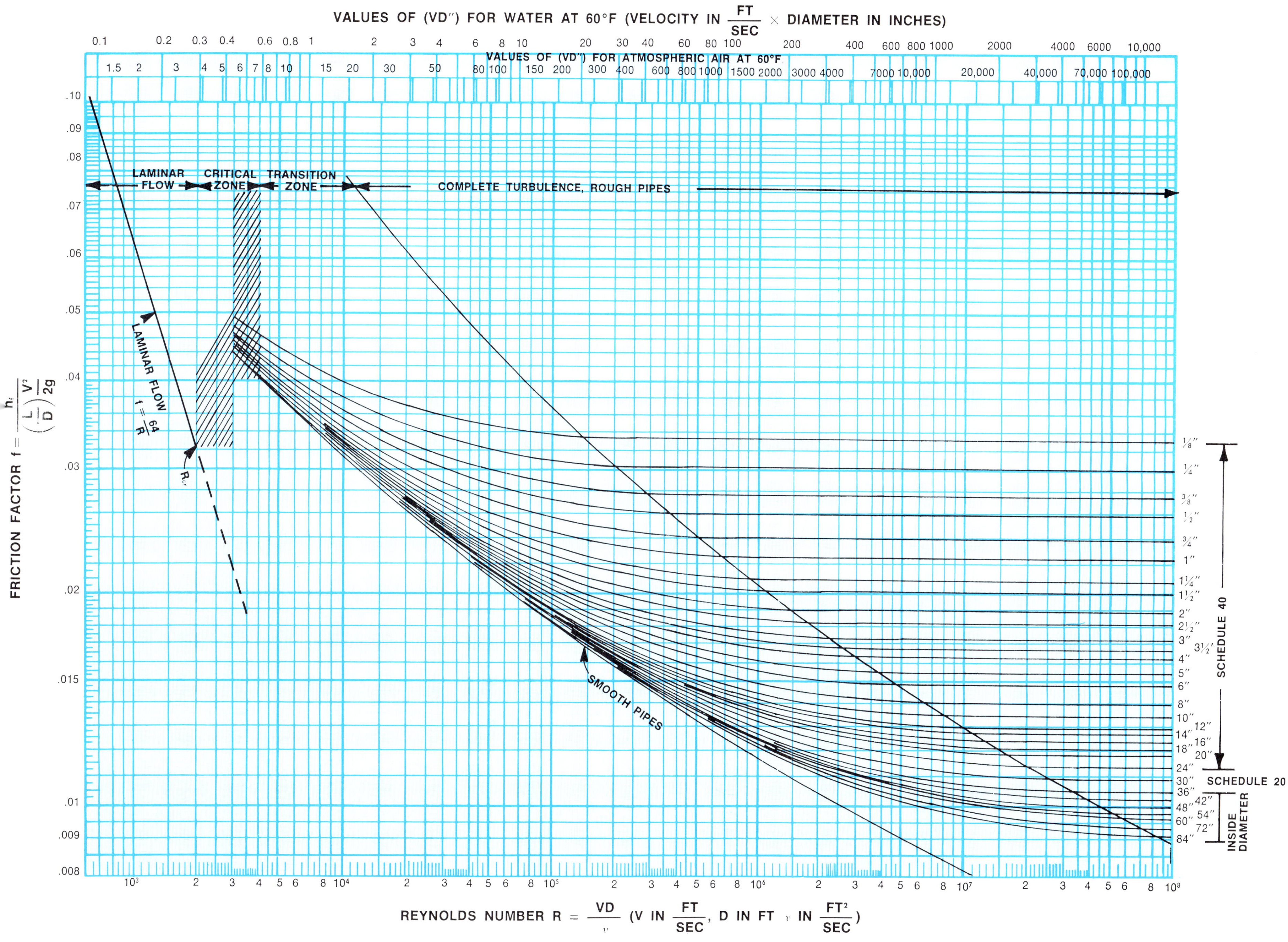

Fig. IIIA-1 FRICTION FACTORS FOR ASPHALT-DIPPED CAST IRON PIPE

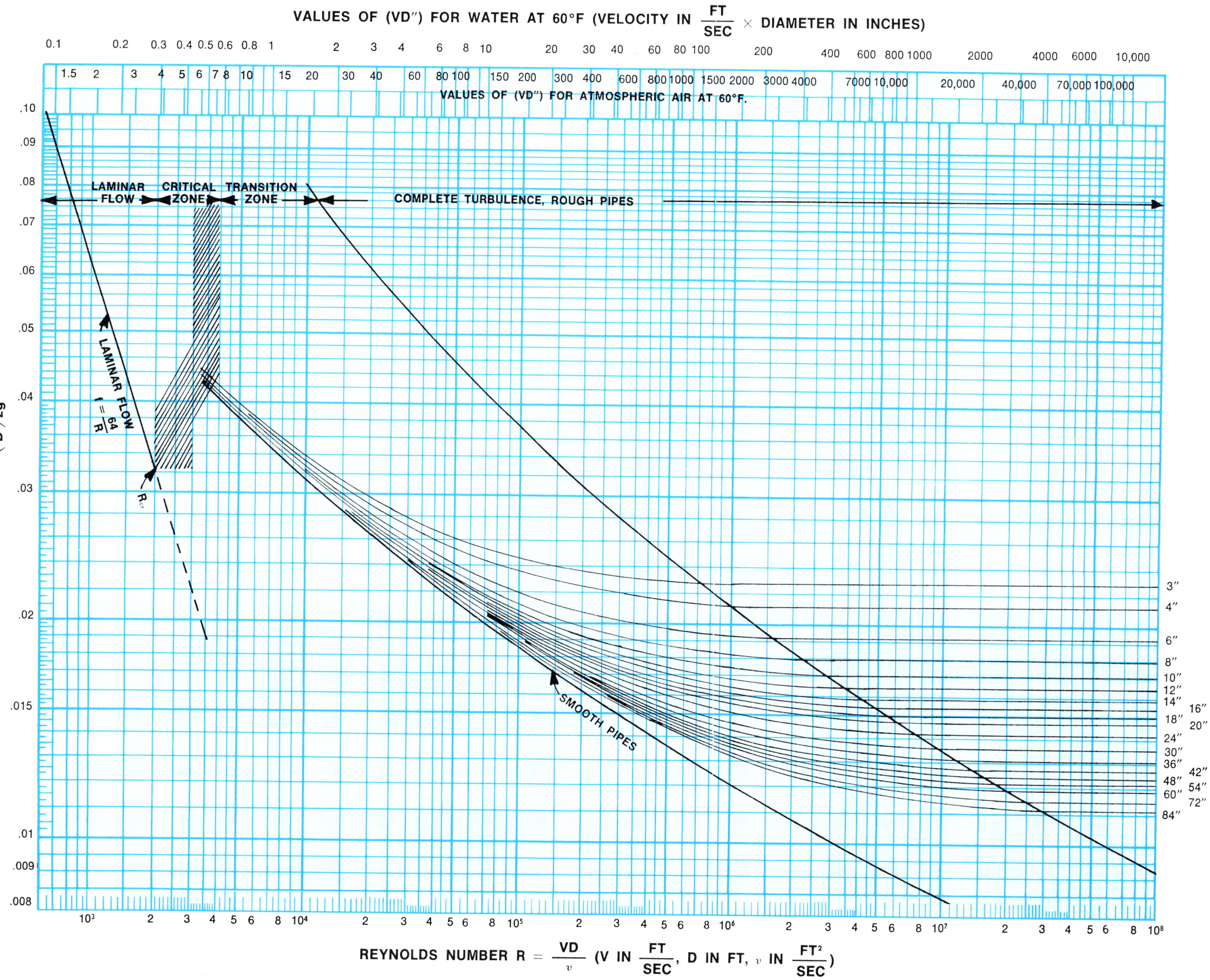

Fig. IIIA-3 FRICTION FACTORS FOR ANY KIND AND SIZE OF PIPE

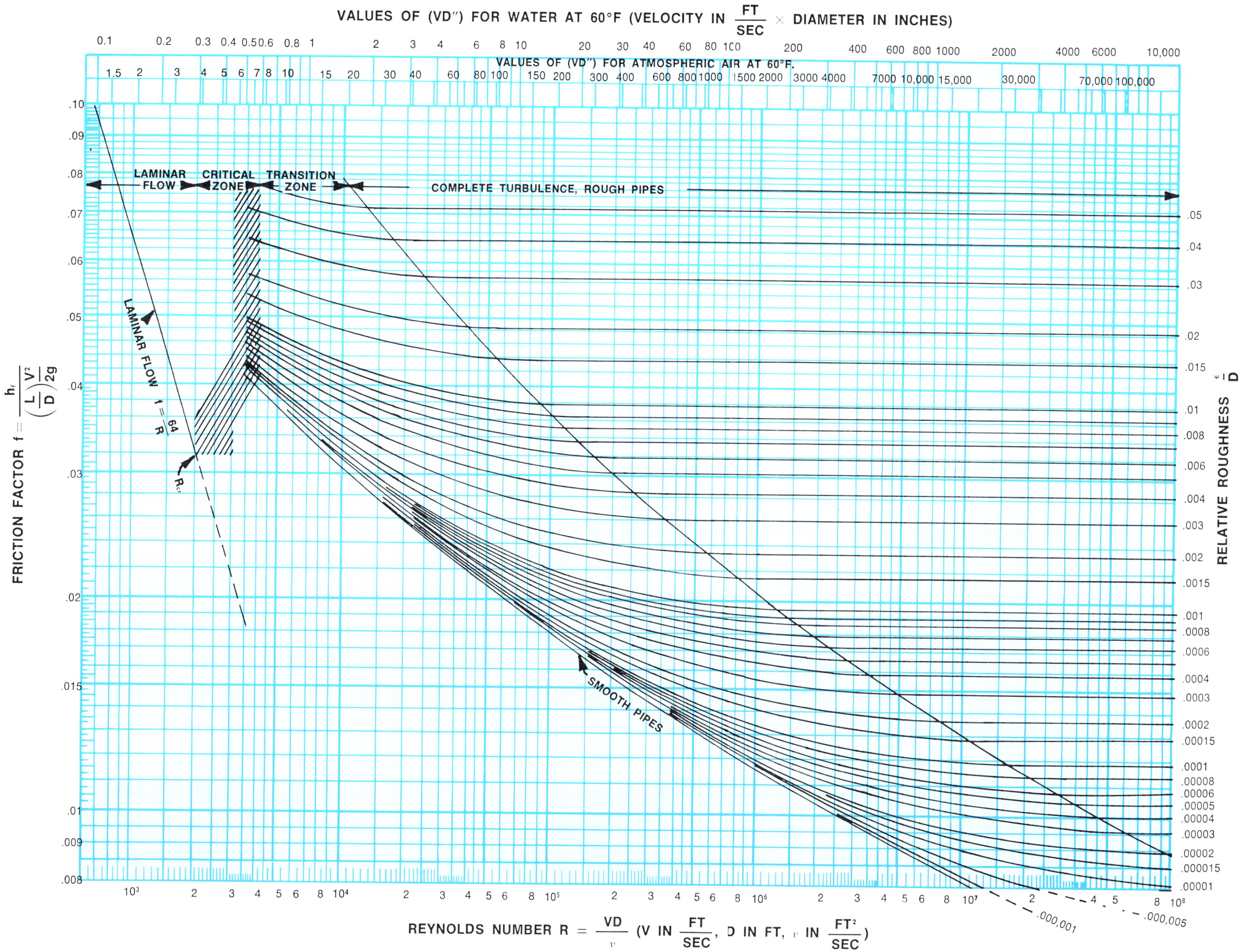

Fig. IIIA-4 RELATIVE ROUGHNESS FACTORS FOR NEW CLEAN PIPES

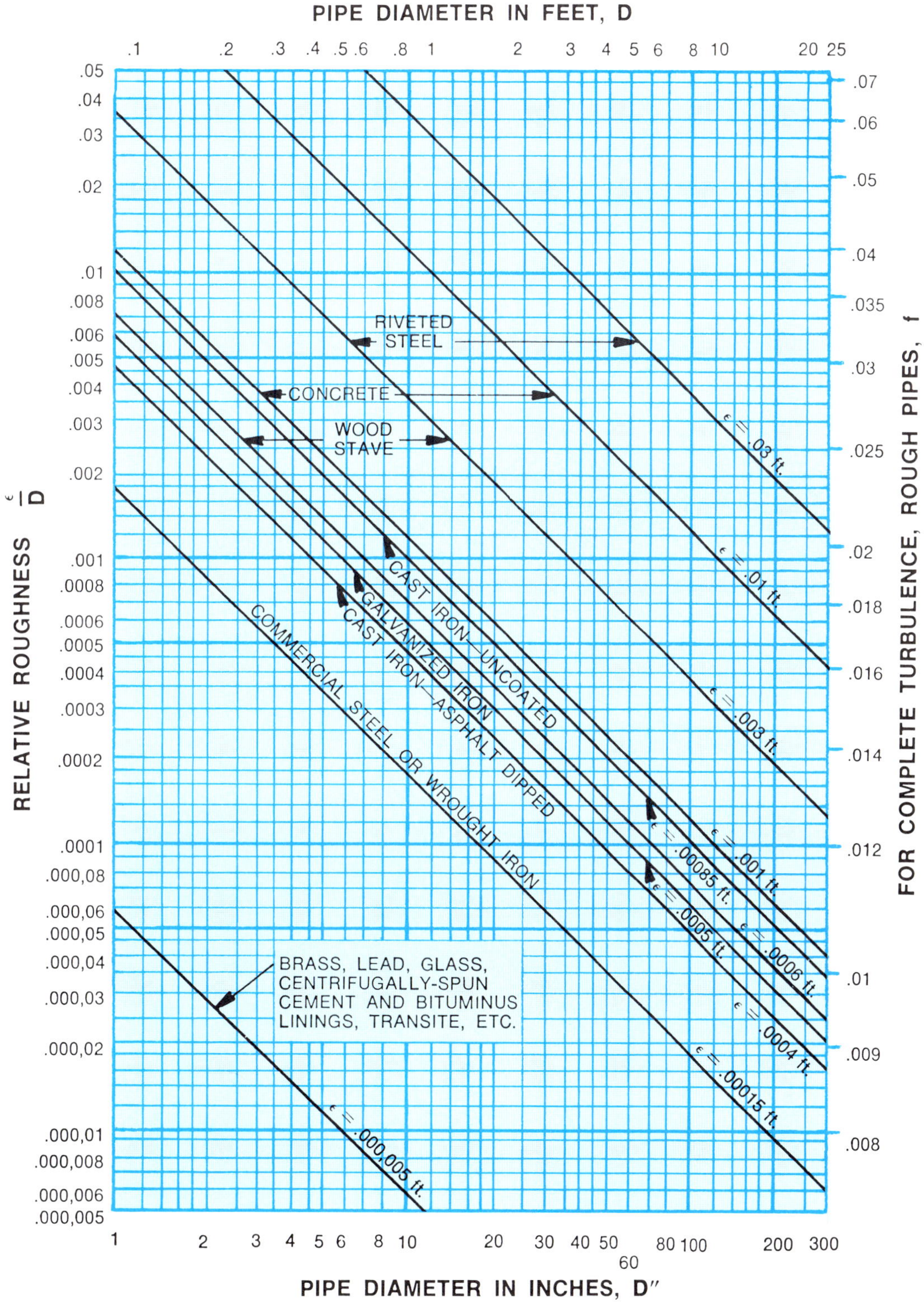

Fig. IIIA-5 KINEMATIC VISCOSITY AND REYNOLDS NUMBER CHART

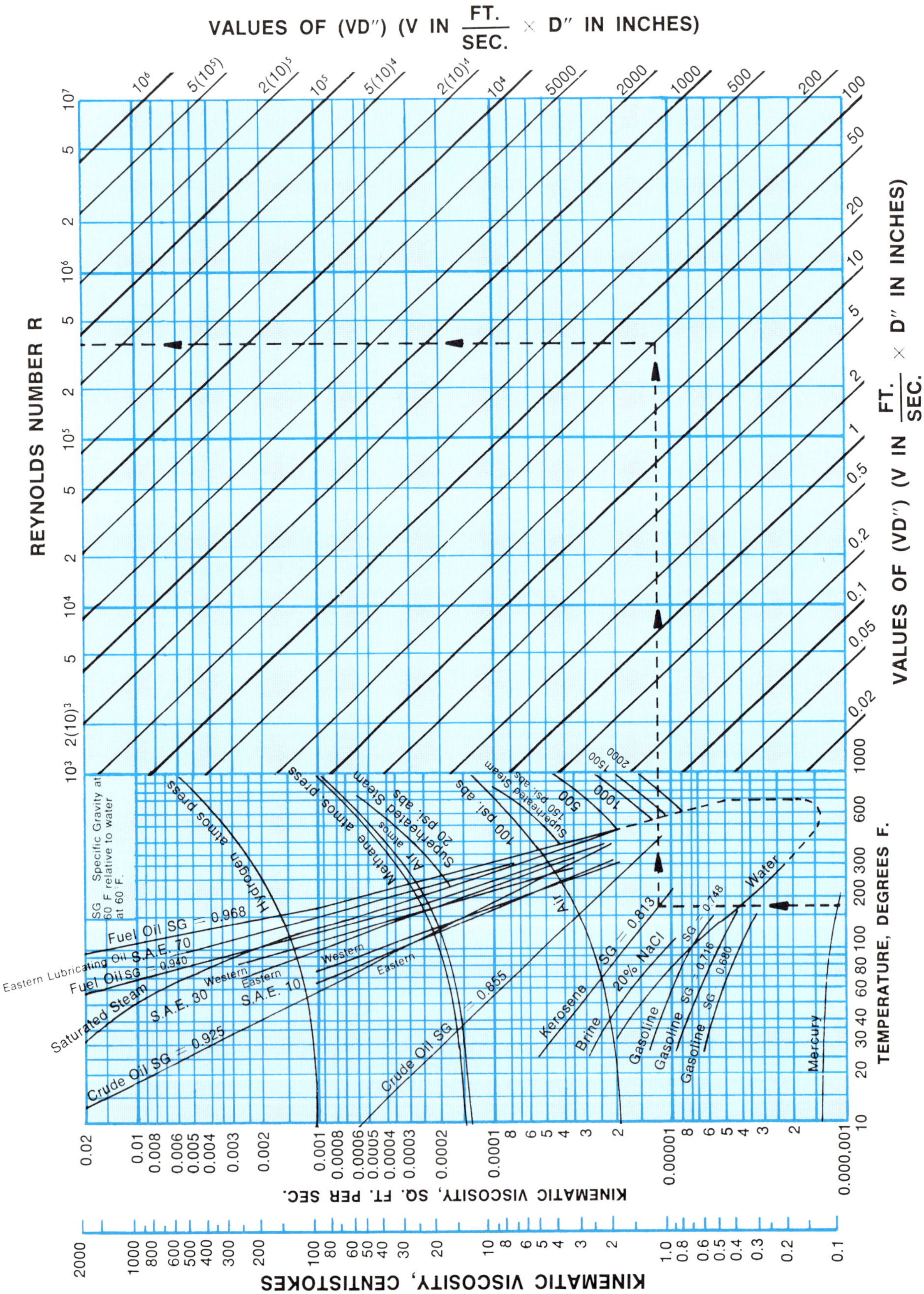

Fig. IIIA-6 INSTALLATION FOR EXAMPLE No. 1

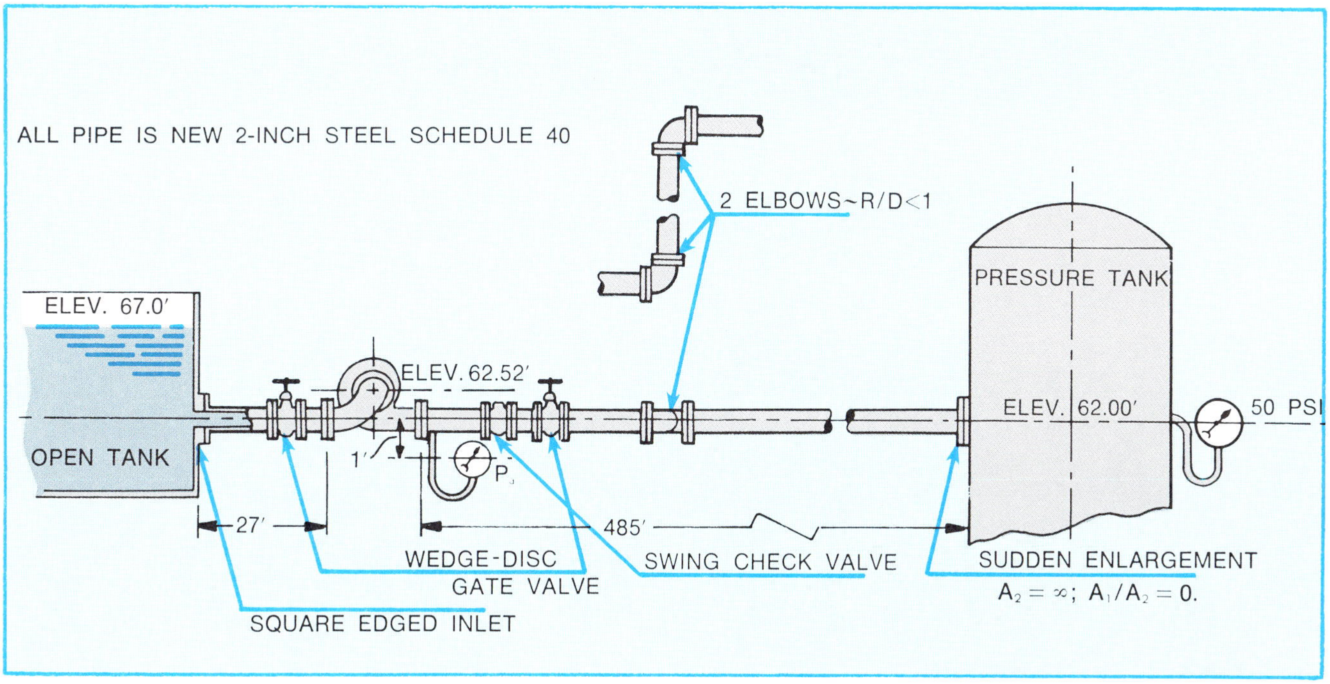

Fig. IIIA-7 INSTALLATION FOR EXAMPLE No. 2

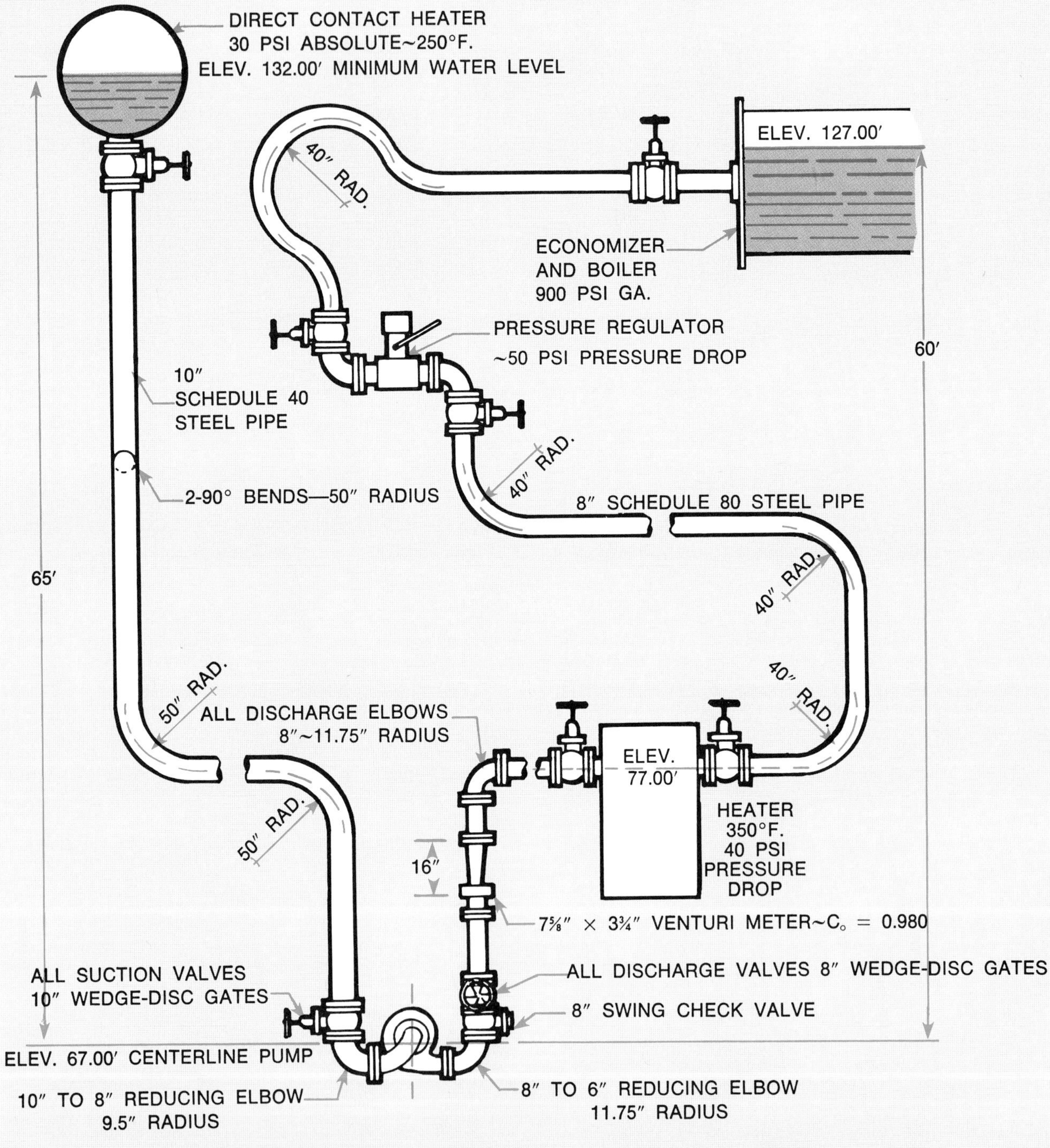

TOTAL LENGTH OF SUCTION PIPE 230 FT. INCLUDING FITTINGS
TOTAL LENGTH OF 250° DISCHARGE PIPE 108 FT. INCLUDING FITTINGS
TOTAL LENGTH OF 350° DISCHARGE PIPE 165 FT. INCLUDING FITTINGS

IIIB-1—FRICTION LOSS—WATER

Example 3. Use of Tables for Water. A pump draws water (60° F) from a sump and delivers it to an elevated tank through 1250 feet of new 4-inch diameter asphalt-dipped cast iron pipe (See Fig. IIIB-1). The suction pipe is vertical, 5 feet long and equipped with a foot valve. The discharge pipe contains two regular 90 degree flanged elbows with $\frac{R}{D} = 1.4$, a swing check valve, and an open wedge-disc gate valve. It is required to find the suction lift (h_s), the discharge head (h_d), and the reading of the discharge pressure gauge (P_d), when the rate of flow is 200 gallons per minute.

SOLUTION: **a. Suction Line.** The following information may be obtained from Table 13 for a flow of 200 gallons per minute:

$$\text{Velocity head, } \frac{V^2}{2g} = 0.405 \text{ feet}$$

Pipe friction loss, h_f = 2.74 feet per 100 feet of pipe.

The resistance coefficient for the foot valve may be obtained from Tables 32(b) and 32(c):

Foot valve, $k = 0.8$

The head loss due to pipe friction will be

$$h_f = 2.74 \frac{5}{100} = 0.14 \text{ feet}$$

and, by Equation (4), Section IIIA, the head loss in the foot valve will be

$$h = k \frac{V^2}{2g} = (0.8 \times 0.405) = 0.324 \text{ feet}$$

The pump must raise the water from elevation 24.00 to elevation 28.62. Therefore, the suction lift will be

$$h_s = (28.62 - 24.00) + 0.32 + 0.14 = 5.08 \text{ feet.}$$

b. Discharge Line. The head loss due to pipe friction in the discharge line will be

$$h_f = 2.74 \frac{1250}{100} = 34.25 \text{ feet}$$

The resistance coefficients for the fittings may be obtained from Tables 32(a), 32(b) and 32(c), as follows:

*Regular flanged elbow, $k = 0.31$
Swing check valve, $k = 2.0$
Wedge-disc gate valve, $k = 0.17$
Sudden enlargement, $k = 1.0$

The total resistance coefficient for the fittings will be

$$k = 2 \times 0.31 + 2 + 0.17 + 1.0 = 3.79$$

and, by Equation (4), Section IIIA, the head loss due to the fittings will be

$$h = k \frac{V^2}{2g} = 3.79 \times 0.405 = 1.53 \text{ feet}$$

The pump must raise the water from elevation 28.62 to elevation 289.00. Therefore, the discharge head will be

$$h_d = (289.00 - 28.62) + 1.53 + 34.25 \cong 296 \text{ feet}$$

The pressure head in the discharge pipe at the point where the pressure gauge is attached will be equal to the static head plus the friction losses in the discharge pipe minus the velocity head. The static head = 265 − 7 = 258 feet. The other quantities have been computed above. Therefore, the required pressure head = 258 + 1.53 + 34.25 − 0.405 ≅ 293 feet. The pressure head at the center of the gauge will be 2 feet less than this value because of the 2-foot vertical connection. The specific weight of water at 60° F is 62.34 pounds per cubic foot from Section IIA. Therefore, the reading of the discharge pressure gauge will be

$$P_d = \frac{(293 - 2) \times 62.34}{144} = 126 \text{ psi}$$

*(NOTE: It is preferable to use Fig. IIIA-5 for flanged elbows when the $\frac{R}{D}$ ratio is greater than unity. The use of Fig. IIIA-5 is illustrated in Section IIIA, Example 4.)

FIG. IIIB-1 INSTALLATION FOR EXAMPLE NO. 3

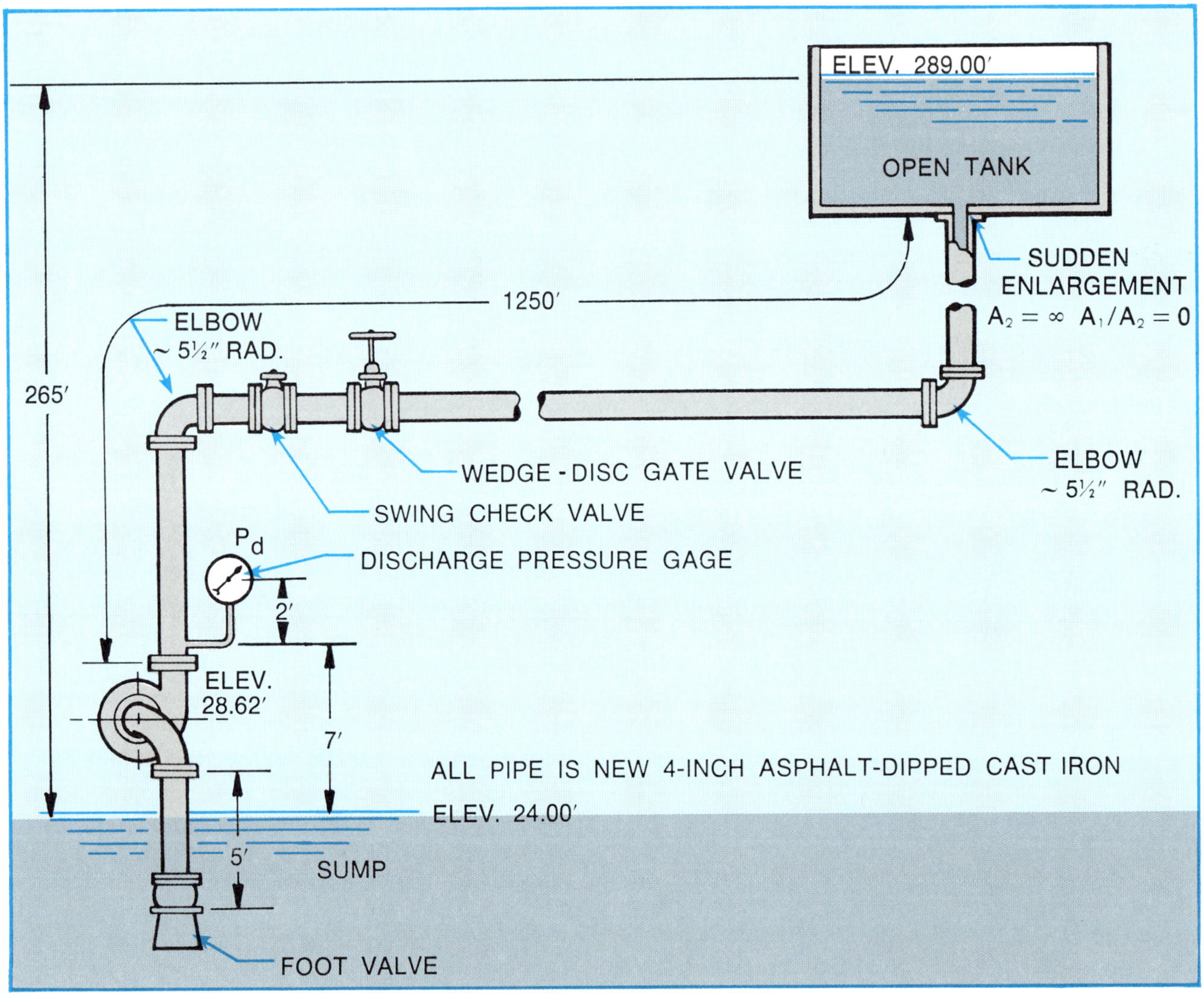

Friction Loss - Water

Fig. IIIB-2—FRICTION LOSS FOR WATER IN FEET PER FOOT OF STAINLESS STEEL TUBING AND IN FEET FOR SANITARY FITTINGS.

Capacity in U.S. G.P.M.	O.D. — 1″ I.D. — 0.902″			O.D. — 1½″ I.D. — 1.402″			O.D. — 2″ I.D. — 1.870″			O.D. — 2½″ I.D. — 2.370″			O.D. — 3″ I.D. — 2.870″			O.D. — 4″ I.D. — 3.834″		
	Tubing	Elbow	Tee	Tubing	Elbow	Tee	Tubing	Elbow	Tee	Tubing	Elbow	Tee	Tubing	Elbow	Tee	Tubing	Elbow	Tee
2	.01	.01	.1															
4	.025	.02	.2															
5	.035	.025	.25															
10	.12	.05	.4	.02	.01	.15	.005	.015	.1									
15	.25	.1	.8	.04	.02	.25	.013	.02	.15									
20	.43	.22	1.5	.06	.03	.3	.02	.025	.2	.005	.02	.1	.003	.02	.06			
25	.66	.4	2.3	.08	.04	.4	.025	.03	.25	.006	.03	.15	.004	.03	.08			
30	.93	.7	3.3	.105	.06	.55	.035	.05	.3	.008	.05	.2	.005	.04	.1			
35	1.22	1.25	5.2	.135	.09	.8	.04	.06	.4	.011	.06	.25	.006	.05	.13			
40				.17	.11	1.0	.05	.08	.5	.015	.07	.3	.007	.06	.15			
45				.21	.16	1.3	.063	.1	.6	.02	.09	.35	.008	.065	.18			
50				.25	.2	1.6	.073	.12	.7	.022	.1	.4	.01	.07	.2			
60				.34	.35	2.2	.1	.18	.9	.03	.12	.45	.015	.08	.25			
80				.57	.76	3.7	.16	.3	1.5	.05	.15	.55	.02	.1	.4			
100				.85	1.35	5.8	.23	.44	2.3	.075	.18	.6	.03	.11	.5	.008	.04	.1
120				1.18	2.05	9.1	.32	.64	3.3	.105	.21	1.0	.04	.13	.6	.01	.05	.15
140							.42	.85	4.5	.14	.23	1.25	.05	.16	.8	.013	.06	.2
160							.54	1.13	5.8	.17	.28	1.6	.07	.2	1.1	.015	.07	.25
180							.67	1.45	7.4	.205	.31	2.0	.08	.21	1.3	.02	.08	.3
200							.81	1.82	9.0	.245	.35	2.5	.1	.26	1.6	.025	.09	.4
220							.95	2.22	11.0	.29	.41	3.0	.12	.3	1.9	.028	.1	.5
240							1.10	2.63	13.5	.34	.48	3.7	.14	.33	2.2	.035	.11	.55
260										.39	.53	4.5	.165	.39	2.5	.04	.115	.6
280										.45	.61	5.3	.19	.42	2.8	.045	.12	.65
300										.515	.7	6.2	.22	.5	3.1	.05	.13	.7
350										.68	1.05	8.5	.28	.67	4.1	.07	.15	.9
400										.86	1.55	11.0	.36	.88	5.2	.085	.18	1.2
450										1.05	2.25	13.5	.44	1.1	6.6	.105	.2	1.5
500													.54	1.4	8.0	.13	.23	1.75
550													.64	1.7	9.5	.15	.27	2.1
600													.75	2.05	10.2	.175	.3	2.5
650													.87	2.41	13.0	.2	.34	2.8
700													1.0	2.8	15.0	.23	.4	3.4
750																.26	.43	3.8
800																.3	.5	4.4
850																.33	.56	5.0
900																.37	.62	5.7
950																.41	.7	6.3
1000																.45	.8	7.0
1100																.53	1.06	8.6

NOTES: 1. FOR ELBOWS — R/D = 1.5

2. FLOW THRU TEES — FLOW A TO B PORT C CAPPED OFF.

3. TEST MEDIUM — WATER AT 70° F

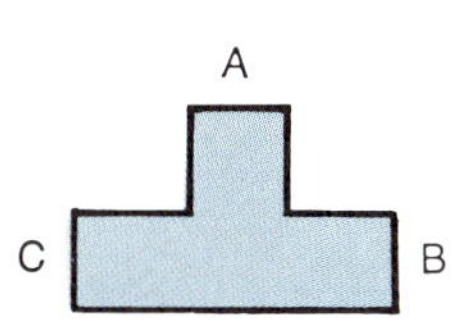

IIIB-3 FRICTION LOSSES IN DEEP WELL VERTICAL TURBINE PUMP COLUMNS AND DISCHARGE HEADS

Column Friction Loss

The column friction chart Fig. IIIB-3(a) should be used to determine the loss in head due to column friction. This chart has been compiled from data on head losses where the flow is between the inside diameter of the column pipe and the outside diameter of the shaft-enclosing tube.

For open line shafting, the losses shown on Fig. IIIB-3 should be used by assuming the losses equal to those indicated on the chart for a shaft-enclosing tube of a size that would normally enclose the open line shaft in question.

The diagonals on Fig. IIIB-3(a) are labeled to show nominal diameters (in inches of the outer pipe column and the inner shaft enclosing tube). For the outer pipe columns, the calculations used in constructing the chart were based on inside diameters, which are closest to the nominal size for pipe up to and including 12 inches, (for example, 10 inch = 10.2 inch inside diameter); in sizes 14 inches and larger the diameters shown are equivalent to the outside diameter of pipe with 3⁄8 inch wall thickness (for example, 16 inch = 15¼ inch inside diameter). For the inner columns (shaft-enclosing tubes), the calculations were based on the outside diameters of standard or extra-heavy pipe. Thus, "8 × 2" on the chart is actually 8.071 × 2⅜, and 16 × 3 is 15¼ × 3½.

Discharge Head Loss

The discharge head loss chart Fig. IIIB-3(b) should be used to determine the hydraulic losses in the discharge head. Losses in discharge heads vary with the size of the head, the design of the head, and the size of tubing or shaft, column and discharge pipe used. Fig. IIIB-3(b) represents estimated average losses. Where extreme accuracy is imperative, actual loss measurements in the discharge head to be used—with the correct tubing or shaft, column, and discharge pipe—should be specified on the bid request by the purchaser.

Fluid Flow

Friction Loss - Water

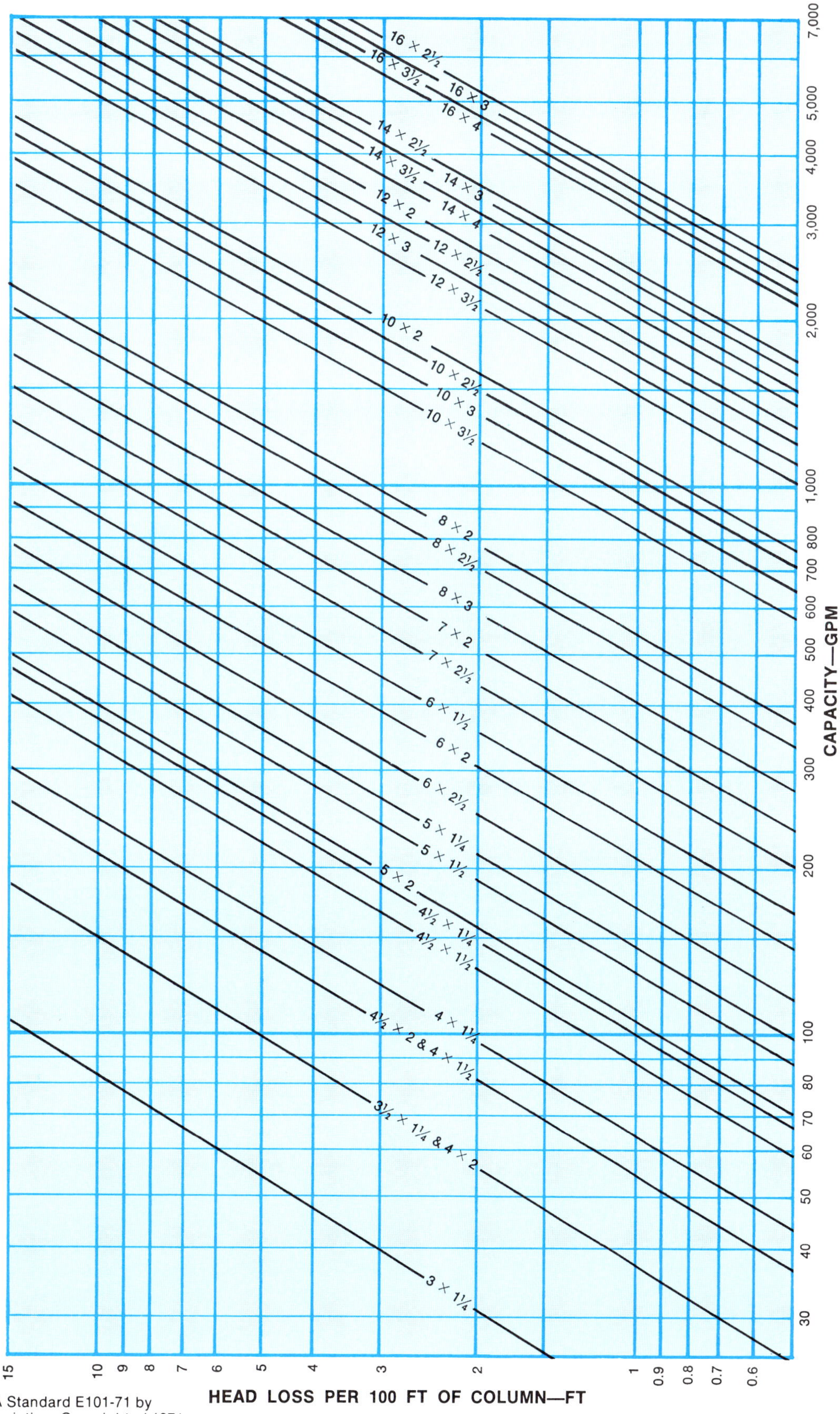

Fig. IIIB-3(a) FRICTION LOSS CHART FOR STANDARD PIPE COLUMN

Fig. IIIB-3(b) HEAD LOSS IN DISCHARGE HEADS

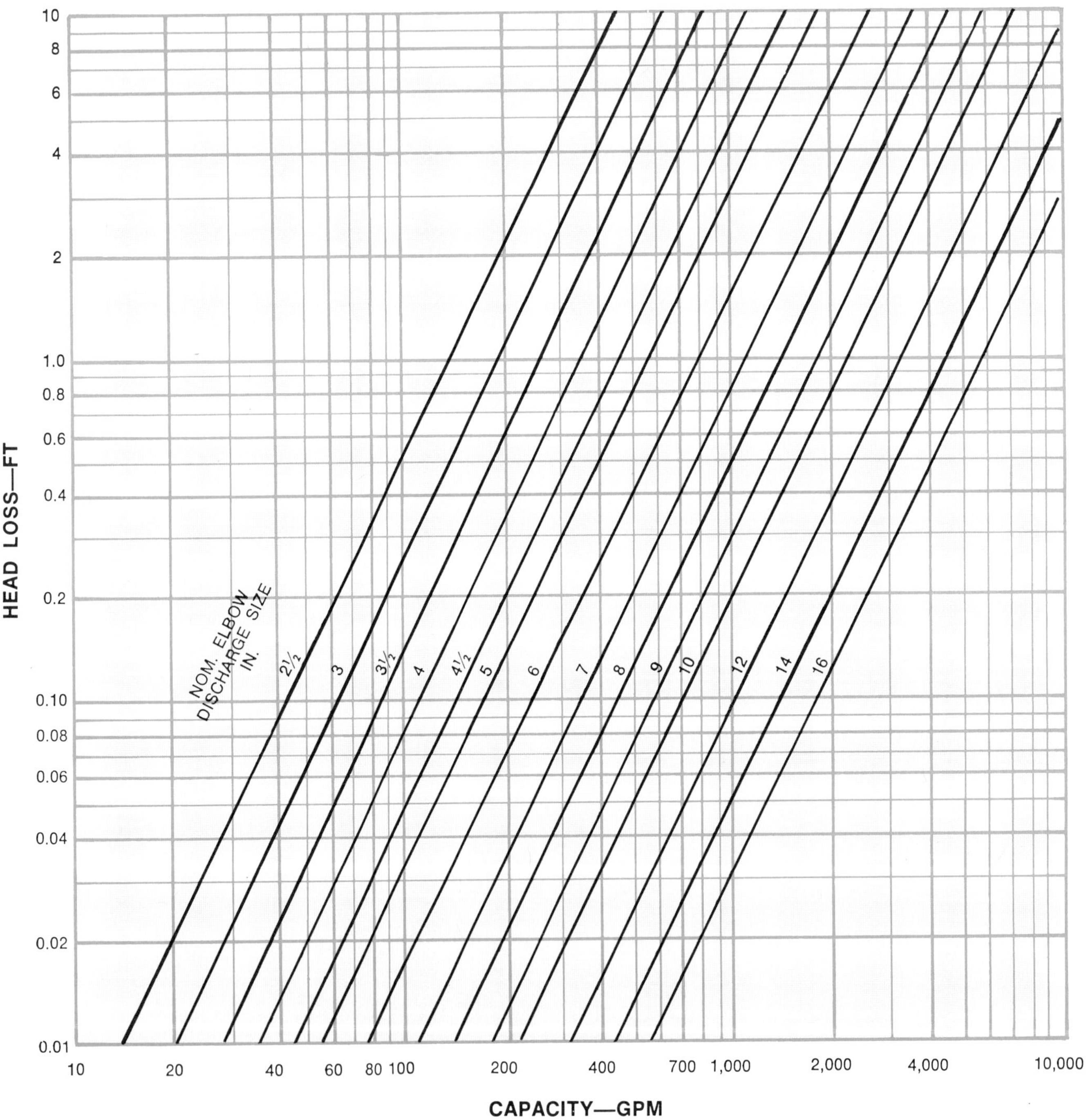

IIIB-4 FRICTION LOSS FOR WATER IN FEET PER 100 FEET OF PIPE

TABLE 1

⅛ INCH NOMINAL		STEEL SCHEDULE 40 ID = 0.269 INCHES ϵ/D = 0.00669		
DISCHARGE CFS	DISCHARGE GPM	V ft/sec	$V^2/2g$ feet	h_f feet per 100 feet of pipe
.0000446	0.02	0.113	0.000198	0.272
.0000891	0.04	0.226	0.000792	0.543
.000134	0.06	0.339	0.00178	0.815
.000178	0.08	0.452	0.00317	1.087
.000223	0.10	0.565	0.00495	1.359
.000267	0.12	0.677	0.00713	1.630
.000312	0.14	0.790	0.00971	1.902
.000356	0.16	0.903	0.0128	2.174
.000401	0.18	1.02	0.0160	2.445
.000446	0.20	1.13	0.0198	2.717
TRANSITION TO TURBULENT FLOW				
.000668	0.3	1.69	0.0446	9.70
.000891	0.4	2.26	0.0792	16.2
.00111	0.5	2.82	0.124	24.2
.00134	0.6	3.39	0.178	33.8
.00156	0.7	3.95	0.243	44.8
.00178	0.8	4.52	0.317	57.4
.00201	0.9	5.08	0.401	71.6
.00223	1.0	5.65	0.495	87.0
.00267	1.2	6.77	0.713	122
.00312	1.4	7.90	0.971	164
.00356	1.6	9.03	1.28	212
.00401	1.8	10.2	1.60	265
.00446	2.0	11.3	1.98	324

TABLE 2

¼ INCH NOMINAL		STEEL SCHEDULE 40 ID = 0.364 INCHES ϵ/D = 0.00495		
DISCHARGE CFS	DISCHARGE GPM	V ft/sec	$V^2/2g$ feet	h_f feet per 100 feet of pipe
0.000111	0.05	0.154	0.000369	0.203
0.000223	0.10	0.308	0.001477	0.405
0.000334	0.15	0.462	0.00332	0.608
0.000446	0.20	0.617	0.00591	0.810
0.000557	0.25	0.771	0.00923	1.013
TRANSITION TO TURBULENT FLOW				
0.000891	0.4	1.23	0.0236	3.7
0.00134	0.6	1.85	0.0532	7.6
0.00178	0.8	2.47	0.0946	12.7
0.00223	1.0	3.08	0.1477	19.1
0.00267	1.2	3.70	0.213	26.7
0.00312	1.4	4.32	0.290	35.3
0.00356	1.6	4.93	0.378	45.2
0.00401	1.8	5.55	0.479	56.4
0.00446	2.0	6.17	0.591	69.0
0.00557	2.5	7.71	0.923	105
0.00668	3.0	9.25	1.330	148
0.00780	3.5	10.79	1.810	200
0.00891	4.0	12.33	2.36	259
0.0100	4.5	13.87	2.99	326
0.0111	5.0	15.42	3.69	398

NOTE: No allowance has been made for age, differences in diameter, or any abnormal condition of interior surface. Any factor of safety must be estimated from the local conditions and the requirements of each particular installation.

IIIB-4 FRICTION LOSS FOR WATER IN FEET PER 100 FEET OF PIPE

TABLE 3

3/8 INCH NOMINAL		STEEL SCHEDULE 40 ID = 0.493 INCHES ϵ/D = 0.00365		
DISCHARGE		V	$V^2/2g$	h_f
CFS	GPM	ft/sec	feet	feet per 100 feet of pipe
0.00134	0.6	1.01	0.0158	1.74
0.00178	0.8	1.34	0.0281	2.89
0.00223	1.0	1.68	0.0440	4.30
0.00334	1.5	2.52	0.0988	8.93
0.00446	2.0	3.36	0.176	15.0
0.00557	2.5	4.20	0.274	22.6
0.00668	3.0	5.04	0.395	31.8
0.00780	3.5	5.88	0.538	42.6
0.00891	4.0	6.72	0.702	54.9
0.0100	4.5	7.56	0.889	68.4
0.0111	5.0	8.40	1.097	83.5
0.0123	5.5	9.24	1.33	100
0.0134	6.0	10.08	1.58	118
0.0149	6.5	10.9	1.85	137
0.0156	7.0	11.8	2.15	158
0.0167	7.5	12.6	2.47	181
0.0178	8.0	13.4	2.81	205
0.0189	8.5	14.3	3.17	231
0.0201	9.0	15.1	3.56	258
0.0212	9.5	16.0	3.96	286
0.0223	10.0	16.8	4.39	316

TABLE 4

1/2 INCH NOMINAL		STEEL SCHEDULE 40 ID = 0.622 INCHES ϵ/D = 0.00289		
DISCHARGE		V	$V^2/2g$	h_f
CFS	GPM	ft/sec	feet	feet per 100 feet of pipe
0.00156	0.7	0.739	0.00849	0.740
0.00223	1.0	1.056	0.0173	1.86
0.00334	1.5	1.58	0.0390	2.85
0.00446	2.0	2.11	0.0693	4.78
0.00557	2.5	2.64	0.1083	7.16
0.00668	3.0	3.17	0.156	10.0
0.00780	3.5	3.70	0.212	13.3
0.00891	4.0	4.22	0.277	17.1
0.0100	4.5	4.75	0.351	21.3
0.0111	5.0	5.28	0.433	25.8
0.0123	5.5	5.81	0.524	30.9
0.0134	6.0	6.34	0.624	36.5
0.0149	6.5	6.86	0.732	42.4
0.0156	7.0	7.39	0.849	48.7
0.0167	7.5	7.92	0.975	55.5
0.0178	8.0	8.45	1.109	62.7
0.0189	8.5	8.98	1.25	70.3
0.0201	9.0	9.50	1.40	78.3
0.0212	9.5	10.03	1.56	86.9
0.0223	10.0	10.56	1.73	95.9
0.0245	11.0	11.6	2.10	115
0.0267	12.0	12.7	2.49	136
0.0290	13.0	13.7	2.93	159
0.0312	14.0	14.8	3.40	183
0.0334	15.0	15.8	3.90	209

NOTE: No allowance has been made for age, differences in diameter, or any abnormal condition of interior surface. Any factor of safety must be estimated from the local conditions and the requirements of each particular installation.

IIIB-4 FRICTION LOSS FOR WATER IN FEET PER 100 FEET OF PIPE

TABLE 5

¾ INCH NOMINAL		STEEL SCHEDULE 40 ID = 0.824 INCHES ε/D = 0.00218		
DISCHARGE CFS	DISCHARGE GPM	V ft/sec	$V^2/2g$ feet	h_f feet per 100 feet of pipe
0.00223	1.0	0.602	0.00563	0.260
0.00334	1.5	0.903	0.0127	0.730
0.00446	2.0	1.20	0.0225	1.21
0.00557	2.5	1.50	0.0352	1.80
0.00668	3.0	1.81	0.0506	2.50
0.00780	3.5	2.11	0.0689	3.30
0.00891	4.0	2.41	0.0900	4.21
0.0100	4.5	2.71	0.114	5.21
0.0111	5.0	3.01	0.141	6.32
0.0134	6.0	3.61	0.203	8.87
0.0156	7.0	4.21	0.276	11.8
0.0178	8.0	4.81	0.360	15.0
0.0201	9.0	5.42	0.456	18.8
0.0223	10	6.02	0.563	23.0
0.0245	11	6.62	0.681	27.6
0.0267	12	7.22	0.810	32.6
0.0290	13	7.82	0.951	37.8
0.0312	14	8.42	1.103	43.5
0.0334	15	9.03	1.27	49.7
0.0356	16	9.63	1.44	56.3
0.0379	17	10.23	1.63	63.1
0.0401	18	10.8	1.82	70.3
0.0423	19	11.4	2.03	78.0
0.0446	20	12.0	2.25	86.1
0.0490	22	13.2	2.72	104
0.0535	24	14.4	3.24	122
0.0579	26	15.6	3.80	143
0.0624	28	16.8	4.41	164
0.0668	30	18.1	5.06	187

TABLE 6

1 INCH NOMINAL		STEEL SCHEDULE 40 ID = 1.049 INCHES ε/D = 0.00172		
DISCHARGE CFS	DISCHARGE GPM	V ft/sec	$V^2/2g$ feet	h_f feet per 100 feet of pipe
0.00223	1	0.371	0.00214	0.114
0.00446	2	0.742	0.00857	0.379
0.0068	3	1.114	0.01927	0.772
0.00891	4	1.48	0.0343	1.295
0.0111	5	1.86	0.0535	1.93
0.0134	6	2.23	0.0771	2.68
0.0156	7	2.60	0.1049	3.56
0.0178	8	2.97	0.137	4.54
0.0201	9	3.34	0.173	5.65
0.0223	10	3.71	0.214	6.86
0.0267	12	4.45	0.308	9.62
0.0312	14	5.20	0.420	12.8
0.0356	16	5.94	0.548	16.5
0.0401	18	6.68	0.694	20.6
0.0446	20	7.42	0.857	25.1
0.0490	22	8.17	1.036	30.2
0.0535	24	8.91	1.23	35.6
0.0579	26	9.65	1.45	41.6
0.0624	28	10.39	1.68	47.9
0.0668	30	11.1	1.93	54.6
0.0713	32	11.9	2.19	61.8
0.0758	34	12.6	2.48	69.4
0.0802	36	13.4	2.78	77.4
0.0847	38	14.1	3.09	86.0
0.0891	40	14.8	3.43	95.0
0.0936	42	15.6	3.78	104.5
0.0980	44	16.3	4.15	114
0.102	46	17.1	4.53	124
0.107	48	17.8	4.93	135
0.111	50	18.6	5.35	146
0.123	55	20.4	6.48	176
0.134	60	22.3	7.71	209
0.145	65	24.1	9.05	245
0.156	70	26.0	10.49	283
0.167	75	27.8	12.0	324
0.178	80	29.7	13.7	367
0.189	85	31.6	15.5	413
0.201	90	33.4	17.3	462
0.212	95	35.3	19.3	513
0.223	100	37.1	21.4	567

NOTE: No allowance has been made for age, differences in diameter, or any abnormal condition of interior surface. Any factor of safety must be estimated from the local conditions and the requirements of each particular installation.

Fluid Flow

Friction Loss - Water

IIIB-4 FRICTION LOSS FOR WATER IN FEET PER 100 FEET OF PIPE

TABLE 7

1¼ INCH NOMINAL		STEEL SCHEDULE 40 ID = 1.380 INCHES ϵ/D = 0.00130		
DISCHARGE CFS	DISCHARGE GPM	V ft/sec	$V^2/2g$ feet	h_f feet per 100 feet of pipe
0.00334	1.5	0.322	0.00161	0.0623
0.00446	2	0.429	0.00286	0.102
0.00668	3	0.644	0.00644	0.207
0.00891	4	0.858	0.01144	0.342
0.0111	5	1.073	0.0179	0.508
0.0134	6	1.29	0.0257	0.704
0.0156	7	1.50	0.0350	0.930
0.0178	8	1.72	0.0458	1.18
0.0201	9	1.93	0.0579	1.46
0.0223	10	2.15	0.0715	1.77
0.0267	12	2.57	0.103	2.48
0.0312	14	3.00	0.140	3.28
0.0356	16	3.43	0.183	4.20
0.0401	18	3.86	0.232	5.22
0.0446	20	4.29	0.286	6.34
0.0490	22	4.72	0.346	7.58
0.0535	24	5.15	0.412	8.92
0.0579	26	5.58	0.483	10.37
0.0624	28	6.01	0.561	11.9
0.0668	30	6.44	0.644	13.6
0.0713	32	6.86	0.732	15.3
0.0758	34	7.29	0.827	17.2
0.0802	36	7.72	0.927	19.2
0.0847	38	8.15	1.032	21.3
0.0891	40	8.58	1.14	23.5
0.0936	42	9.01	1.26	25.8
0.0980	44	9.44	1.38	28.2
0.102	46	9.87	1.51	30.7
0.107	48	10.30	1.65	33.3
0.111	50	10.7	1.79	36.0
0.123	55	11.8	2.16	43.2
0.134	60	12.9	2.57	51.0
0.145	65	13.9	3.02	59.6
0.156	70	15.0	3.50	68.8
0.167	75	16.1	4.02	78.7
0.178	80	17.2	4.58	89.2
0.189	85	18.2	5.17	100.2
0.201	90	19.3	5.79	112
0.212	95	20.4	6.45	124
0.223	100	21.5	7.15	138
0.245	110	23.6	8.65	166
0.267	120	25.7	10.3	197
0.290	130	27.9	12.1	230
0.312	140	30.0	14.0	267
0.334	150	32.2	16.1	306

TABLE 8

1½ INCH NOMINAL		STEEL SCHEDULE 40 ID = 1.610 INCHES ϵ/D = 0.00112		
DISCHARGE CFS	DISCHARGE GPM	V ft/sec	$V^2/2g$ feet	h_f feet per 100 feet of pipe
0.00334	1.5	0.236	0.000868	0.0300
0.00446	2	0.315	0.00154	0.0492
0.00668	3	0.473	0.00347	0.0988
0.00891	4	0.630	0.00618	0.164
0.0111	5	0.788	0.00965	0.242
0.0134	6	0.946	0.0139	0.333
0.0156	7	1.103	0.0189	0.439
0.0178	8	1.26	0.0247	0.558
0.0201	9	1.42	0.0313	0.689
0.0223	10	1.58	0.0386	0.829
0.0267	12	1.89	0.0556	1.16
0.0312	14	2.21	0.0756	1.53
0.0356	16	2.52	0.0988	1.96
0.0401	18	2.84	0.125	2.42
0.0446	20	3.15	0.154	2.94
0.0490	22	3.47	0.187	3.52
0.0535	24	3.78	0.222	4.14
0.0579	26	4.10	0.261	4.81
0.0624	28	4.41	0.303	5.51
0.0668	30	4.73	0.347	6.26
0.0713	32	5.04	0.395	7.07
0.0758	34	5.36	0.446	7.92
0.0802	36	5.67	0.500	8.82
0.0847	38	5.99	0.577	9.78
0.0891	40	6.30	0.618	10.79
0.0936	42	6.62	0.681	11.8
0.0980	44	6.93	0.747	12.9
0.102	46	7.25	0.817	14.0
0.107	48	7.56	0.889	15.2
0.111	50	7.88	0.965	16.4
0.123	55	8.67	1.17	19.7
0.134	60	9.46	1.39	23.2
0.145	65	10.24	1.63	27.1
0.156	70	11.03	1.89	31.3
0.167	75	11.8	2.17	35.8
0.178	80	12.6	2.47	40.5
0.189	85	13.4	2.79	45.6
0.201	90	14.2	3.13	51.0
0.212	95	15.0	3.48	56.5
0.223	100	15.8	3.86	62.2
0.245	110	17.3	4.67	74.5
0.267	120	18.9	5.56	88.3
0.290	130	20.5	6.52	103
0.312	140	22.1	7.56	119
0.334	150	23.6	8.68	137
0.356	160	25.2	9.88	156
0.379	170	26.8	11.15	175
0.401	180	28.4	12.50	196
0.423	190	29.9	13.9	218
0.446	200	31.5	15.4	241

NOTE: No allowance has been made for age, differences in diameter, or any abnormal condition of interior surface. Any factor of safety must be estimated from the local conditions and the requirements of each particular installation.

Fluid Flow
Friction Loss - Water

IIB-4 FRICTION LOSS FOR WATER IN FEET PER 100 FEET OF PIPE

TABLE 9

2 INCH NOMINAL		STEEL SCHEDULE 40 ID = 2.067 INCHES ϵ/D = 0.00087		
DISCHARGE CFS	DISCHARGE GPM	V ft/sec	$V^2/2g$ feet	h_f feet per 100 feet of pipe
0.00446	2	0.191	0.000568	0.0151
0.00668	3	0.287	0.00128	0.0302
0.00891	4	0.382	0.00227	0.0497
0.0111	5	0.478	0.00355	0.0731
0.0134	6	0.574	0.00511	0.1004
0.0156	7	0.669	0.00696	0.131
0.0178	8	0.765	0.00909	0.166
0.0201	9	0.860	0.0115	0.205
0.0223	10	0.956	0.0142	0.248
0.0267	12	1.15	0.0205	0.343
0.0312	14	1.34	0.0278	0.453
0.0356	16	1.53	0.0364	0.578
0.0401	18	1.72	0.0460	0.717
0.0446	20	1.91	0.0568	0.868
0.0490	22	2.10	0.0688	1.03
0.0535	24	2.29	0.0818	1.20
0.0579	26	2.49	0.0960	1.39
0.0624	28	2.68	0.111	1.60
0.0668	30	2.87	0.128	1.82
0.0780	35	3.35	0.174	2.42
0.0891	40	3.82	0.227	3.10
0.100	45	4.30	0.288	3.85
0.111	50	4.78	0.355	4.67
0.123	55	5.26	0.430	5.59
0.134	60	5.74	0.511	6.59
0.145	65	6.21	0.600	7.69
0.156	70	6.69	0.696	8.86
0.167	75	7.17	0.799	10.1
0.178	80	7.65	0.909	11.4
0.189	85	8.13	1.03	12.8
0.201	90	8.60	1.15	14.2
0.212	95	9.08	1.28	15.8
0.223	100	9.56	1.42	17.4
0.245	110	10.52	1.72	20.9
0.267	120	11.5	2.05	24.7
0.290	130	12.4	2.40	28.8
0.312	140	13.4	2.78	33.2
0.334	150	14.3	3.20	38.0
0.356	160	15.3	3.64	43.0
0.379	170	16.3	4.11	48.4
0.401	180	17.2	4.60	54.1
0.423	190	18.2	5.13	60.1
0.446	200	19.1	5.68	66.3
0.490	220	21.0	6.88	80.0
0.535	240	22.9	8.18	95.0
0.579	260	24.9	9.60	111
0.624	280	26.8	11.14	128
0.668	300	28.7	12.8	146
0.713	320	30.6	14.5	166
0.758	340	32.5	16.4	187
0.802	360	34.4	18.4	209
0.847	380	36.3	20.5	233
0.891	400	38.2	22.7	258

TABLE 10

2½ INCH NOMINAL		STEEL SCHEDULE 40 ID = 2.469 INCHES ϵ/D = 0.000729		
DISCHARGE CFS	DISCHARGE GPM	V ft/sec	$V^2/2g$ feet	h_f feet per 100 feet of pipe
0.00668	3	0.201	0.000628	0.0129
0.00891	4	0.268	0.00112	0.0213
0.0134	6	0.402	0.00251	0.0432
0.0178	8	0.536	0.00447	0.0712
0.0223	10	0.670	0.00698	0.105
0.0267	12	0.804	0.0100	0.145
0.0312	14	0.938	0.0137	0.191
0.0356	16	1.07	0.0179	0.243
0.0401	18	1.21	0.0226	0.300
0.0446	20	1.34	0.0279	0.362
0.0490	22	1.47	0.0338	0.430
0.0535	24	1.61	0.0402	0.502
0.0579	26	1.74	0.0472	0.580
0.0624	28	1.88	0.0547	0.663
0.0668	30	2.01	0.0628	0.753
0.0780	35	2.35	0.0855	1.00
0.0891	40	2.68	0.112	1.28
0.100	45	3.02	0.141	1.60
0.111	50	3.35	0.174	1.94
0.123	55	3.69	0.211	2.32
0.134	60	4.02	0.251	2.72
0.145	65	4.36	0.295	3.16
0.156	70	4.69	0.342	3.63
0.167	75	5.03	0.393	4.13
0.178	80	5.36	0.447	4.66
0.189	85	5.70	0.504	5.22
0.201	90	6.03	0.565	5.82
0.212	95	6.37	0.630	6.45
0.223	100	6.70	0.698	7.11
0.245	110	7.37	0.844	8.51
0.267	120	8.04	1.00	10.0
0.290	130	8.71	1.18	11.7
0.312	140	9.38	1.37	13.5
0.334	150	10.05	1.57	15.4
0.356	160	10.7	1.79	17.4
0.379	170	11.4	2.02	19.6
0.401	180	12.1	2.26	21.9
0.423	190	12.7	2.52	24.2
0.446	200	13.4	2.79	26.7
0.490	220	14.7	3.38	32.2
0.535	240	16.1	4.02	38.1
0.579	260	17.4	4.72	44.5
0.624	280	18.8	5.47	51.3
0.668	300	20.1	6.28	58.5
0.780	350	23.5	8.55	79.2
0.891	400	26.8	11.2	103
1.003	450	30.2	14.1	130
1.114	500	33.5	17.4	160
1.225	550	36.9	21.1	193
1.337	600	40.2	25.1	230

NOTE: No allowance has been made for age, differences in diameter, or any abnormal condition of interior surface. Any factor of safety must be estimated from the local conditions and the requirements of each particular installation.

IIIB-4 FRICTION LOSS FOR WATER IN FEET PER 100 FEET OF PIPE

TABLE 11

3 INCH NOMINAL		STEEL SCHEDULE 40 ID = 3.068 INCHES ϵ/D = 0.000587			ASPHALT-DIPPED CAST IRON ID = 3.00 INCHES ϵ/D = 0.00160		
DISCHARGE CFS	DISCHARGE GPM	V ft/sec	$V^2/2g$ feet	h_f feet per 100 feet of pipe	V ft/sec	$V^2/2g$ feet	h_f feet per 100 feet of pipe
0.0111	5	0.217	0.000732	0.0112	0.227	0.000800	0.0128
0.0223	10	0.434	0.00293	0.0372	0.454	0.00320	0.0435
0.0334	15	0.651	0.00659	0.0762	0.681	0.00720	0.0900
0.0446	20	0.868	0.0117	0.126	0.908	0.0128	0.1510
0.0557	25	1.085	0.0183	0.189	1.13	0.0200	0.2280
0.0668	30	1.30	0.0263	0.262	1.36	0.0288	0.320
0.0780	35	1.52	0.0359	0.347	1.59	0.0392	0.427
0.0891	40	1.74	0.0468	0.443	1.82	0.0512	0.549
0.100	45	1.95	0.0593	0.547	2.04	0.0648	0.683
0.111	50	2.17	0.0732	0.662	2.27	0.0800	0.830
0.123	55	2.39	0.0885	0.789	2.50	0.0968	0.993
0.134	60	2.60	0.105	0.924	2.72	0.115	1.170
0.145	65	2.82	0.124	1.07	2.95	0.135	1.36
0.156	70	3.04	0.143	1.22	3.18	0.157	1.56
0.167	75	3.25	0.165	1.39	3.40	0.180	1.78
0.178	80	3.47	0.187	1.57	3.63	0.205	2.02
0.189	85	3.69	0.211	1.76	3.86	0.231	2.28
0.201	90	3.91	0.237	1.96	4.08	0.259	2.55
0.212	95	4.12	0.264	2.17	4.31	0.289	2.82
0.223	100	4.34	0.2927	2.39	4.54	0.320	3.10
0.245	110	4.77	0.354	2.86	4.99	0.387	3.73
0.267	120	5.21	0.421	3.37	5.45	0.461	4.40
0.290	130	5.64	0.495	3.92	5.90	0.541	5.13
0.312	140	6.08	0.574	4.51	6.35	0.627	5.93
0.334	150	6.51	0.659	5.14	6.81	0.720	6.80
0.356	160	6.94	0.749	5.81	7.26	0.820	7.71
0.379	170	7.38	0.846	6.53	7.72	0.925	8.70
0.401	180	7.81	0.948	7.28	8.17	1.04	9.73
0.423	190	8.25	1.06	8.07	8.62	1.16	10.80
0.446	200	8.68	1.17	8.90	9.08	1.28	11.9
0.490	220	9.55	1.42	10.7	9.98	1.55	14.3
0.535	240	10.4	1.69	12.6	10.9	1.84	17.0
0.579	260	11.3	1.98	14.7	11.8	2.16	19.8
0.624	280	12.2	2.29	16.9	12.7	2.51	22.8
0.668	300	13.0	2.63	19.2	13.6	2.88	26.1
0.713	320	13.9	3.00	22.0	14.5	3.28	29.7
0.758	340	14.8	3.38	24.8	15.4	3.70	33.6
0.802	360	15.6	3.79	27.7	16.3	4.15	37.8
0.847	380	16.5	4.23	30.7	17.2	4.62	42.2
0.891	400	17.4	4.68	33.9	18.2	5.12	46.8
0.936	420	18.2	5.16	37.3	19.1	5.65	51.5
0.980	440	19.1	5.67	40.9	20.0	6.20	56.4
1.025	460	20.0	6.19	44.6	20.9	6.77	61.5
1.069	480	20.8	6.74	48.5	21.8	7.38	66.8
1.114	500	21.7	7.32	52.5	22.7	8.00	72.3
1.225	550	23.9	8.85	63.2	25.0	9.68	87
1.337	600	26.0	10.5	74.8	27.2	11.5	102
1.448	650	28.2	12.4	87.5	29.5	13.5	121
1.560	700	30.4	14.3	101	31.8	15.7	142
1.671	750	32.5	16.5	116	34.0	18.0	162
1.782	800	34.7	18.7	131	36.3	20.5	184
1.894	850	36.9	21.1	148	38.6	23.1	207
2.005	900	39.1	23.7	165	40.8	25.9	232
2.117	950	41.2	26.4	184	43.1	28.9	258
2.228	1 000	43.4	29.27	204	45.4	32.0	285

NOTE: No allowance has been made for age, differences in diameter, or any abnormal condition of interior surface. Any factor of safety must be estimated from the local conditions and the requirements of each particular installation.

IIIB-4 FRICTION LOSS FOR WATER IN FEET PER 100 FEET OF PIPE

TABLE 12

3½ INCH NOMINAL		STEEL SCHEDULE 40 ID = 3.548 INCHES ϵ/D = 0.000507		
DISCHARGE CFS	DISCHARGE GPM	V ft/sec	$V^2/2g$ feet	h_f feet per 100 feet of pipe
0.0111	5	0.162	0.000409	0.00562
0.0223	10	0.323	0.00164	0.0186
0.0334	15	0.487	0.00368	0.0377
0.0446	20	0.649	0.00655	0.0630
0.0557	25	0.811	0.0102	0.0938
0.0668	30	0.974	0.0147	0.130
0.0780	35	1.14	0.0200	0.172
0.0891	40	1.30	0.0262	0.219
0.100	45	1.46	0.0331	0.271
0.111	50	1.62	0.0409	0.328
0.134	60	1.95	0.0589	0.455
0.156	70	2.27	0.0802	0.604
0.178	80	2.60	0.105	0.773
0.201	90	2.92	0.133	0.959
0.223	100	3.25	0.164	1.17
0.245	110	3.57	0.198	1.39
0.267	120	3.89	0.236	1.64
0.290	130	4.22	0.277	1.90
0.312	140	4.54	0.321	2.18
0.334	150	4.87	0.368	2.48
0.356	160	5.19	0.419	2.80
0.379	170	5.52	0.473	3.15
0.401	180	5.84	0.530	3.50
0.423	190	6.17	0.591	3.87
0.446	200	6.49	0.655	4.27

3½ INCH NOMINAL		STEEL SCHEDULE 40 ID = 3.548 INCHES ϵ/D = 0.000507		
DISCHARGE CFS	DISCHARGE GPM	V ft/sec	$V^2/2g$ feet	h_f feet per 100 feet of pipe
0.490	220	7.14	0.792	5.12
0.535	240	7.79	0.943	6.04
0.579	260	8.44	1.11	7.04
0.624	280	9.09	1.28	8.11
0.668	300	9.74	1.47	9.26
0.713	320	10.4	1.68	10.48
0.758	340	11.0	1.89	11.8
0.802	360	11.7	2.12	13.2
0.847	380	12.3	2.36	14.6
0.891	400	13.0	2.62	16.2
0.936	420	13.6	2.89	17.8
0.980	440	14.3	3.17	19.4
1.025	460	14.9	3.46	21.2
1.069	480	15.6	3.77	23.0
1.114	500	16.2	4.09	25.0
1.225	550	17.8	4.95	30.1
1.337	600	19.5	5.89	35.6
1.448	650	21.1	6.91	41.6
1.560	700	22.7	8.02	48.0
1.671	750	24.3	9.20	54.9
1.782	800	26.0	10.5	62.3
1.894	850	27.6	11.8	70.1
2.005	900	29.2	13.3	78.4
2.117	950	30.8	14.8	87.2
2.228	1 000	32.5	16.4	96.4
2.451	1 100	35.7	19.8	116.1
2.674	1 200	38.9	23.6	138
2.896	1 300	42.2	27.7	162
3.119	1 400	45.4	32.1	187
3.342	1 500	48.7	36.8	214

NOTE: No allowance has been made for age, differences in diameter, or any abnormal condition of interior surface. Any factor of safety must be estimated from the local conditions and the requirements of each particular installation.

Fluid Flow

Friction Loss - Water

IIIB-4 FRICTION LOSS FOR WATER IN FEET PER 100 FEET OF PIPE

TABLE 13

4 INCH NOMINAL		STEEL SCHEDULE 40 ID = 4.026 INCHES ϵ/D = 0.000447			ASPHALT-DIPPED CAST IRON ID = 4.00 INCHES ϵ/D = 0.00120		
DISCHARGE CFS	DISCHARGE GPM	V ft/sec	$V^2/2g$ feet	h_f feet per 100 feet of pipe	V ft/sec	$V^2/2g$ feet	h_f feet per 100 feet of pipe
0.0111	5	0.126	0.000247	0.00310	0.128	0.000253	0.00325
0.0223	10	0.252	0.000987	0.01017	0.255	0.00101	0.01080
0.0446	20	0.504	0.00395	0.0344	0.511	0.00405	0.03700
0.0668	30	0.756	0.00888	0.0702	0.766	0.00912	0.0770
0.0981	40	1.01	0.0158	0.118	1.02	0.0162	0.131
0.111	50	1.26	0.0247	0.176	1.28	0.0253	0.199
0.134	60	1.51	0.0355	0.245	1.53	0.0365	0.278
0.156	70	1.76	0.0484	0.325	1.79	0.0496	0.370
0.178	80	2.02	0.0632	0.415	2.04	0.0648	0.476
0.201	90	2.27	0.0800	0.515	2.30	0.0820	0.594
0.223	100	2.52	0.0987	0.624	2.55	0.101	0.725
0.245	110	2.77	0.119	0.744	2.81	0.123	0.869
0.267	120	3.02	0.142	0.877	3.06	0.146	1.03
0.290	130	3.28	0.167	1.017	3.32	0.171	1.19
0.312	140	3.53	0.193	1.165	3.57	0.199	1.38
0.334	150	3.78	0.222	1.32	3.83	0.228	1.58
0.356	160	4.03	0.253	1.49	4.08	0.259	1.78
0.379	170	4.28	0.285	1.67	4.34	0.293	2.00
0.401	180	4.54	0.320	1.86	4.60	0.328	2.24
0.423	190	4.79	0.356	2.06	4.85	0.366	2.49
0.446	200	5.04	0.395	2.27	5.11	0.406	2.74
0.490	220	5.54	0.478	2.72	5.62	0.490	3.28
0.535	240	6.05	0.569	3.21	6.13	0.583	3.88
0.579	260	6.55	0.667	3.74	6.64	0.685	4.54
0.624	280	7.06	0.774	4.30	7.15	0.794	5.25
0.668	300	7.56	0.888	4.89	7.66	0.912	6.03
0.713	320	8.06	1.01	5.51	8.17	1.04	6.87
0.758	340	8.57	1.14	6.19	8.68	1.17	7.75
0.802	360	9.07	1.28	6.92	9.19	1.31	8.68
0.847	380	9.58	1.43	7.68	9.70	1.46	9.66
0.891	400	10.1	1.58	8.47	10.2	1.62	10.7
0.936	420	10.6	1.74	9.30	10.7	1.79	11.7
0.980	440	11.1	1.91	10.2	11.2	1.96	12.8
1.025	460	11.6	2.09	11.1	11.7	2.14	14.0
1.069	480	12.1	2.27	12.0	12.3	2.33	15.3
1.114	500	12.6	2.47	13.0	12.8	2.53	16.6
1.225	550	13.9	2.99	15.7	14.0	3.06	19.9
1.337	600	15.1	3.55	18.6	15.3	3.65	23.6
1.448	650	16.4	4.17	21.7	16.6	4.28	27.7
1.560	700	17.6	4.84	25.0	17.9	4.96	32.1
1.671	750	18.9	5.55	28.6	19.1	5.70	36.7
1.782	800	20.2	6.32	32.4	20.4	6.48	41.6
1.894	850	21.4	7.13	36.5	21.7	7.32	46.8
2.005	900	22.7	8.00	40.8	23.0	8.20	52.3
2.117	950	23.9	8.91	45.3	24.3	9.14	58.1
2.228	1 000	25.2	9.87	50.2	25.5	10.1	64.2
2.451	1 100	27.7	11.9	60.5	28.1	12.3	78.2
2.674	1 200	30.2	14.2	72.0	30.6	14.6	92.8
2.896	1 300	32.8	16.7	84.3	33.2	17.1	108.2
3.119	1 400	35.3	19.3	97.6	35.7	19.9	126
3.342	1 500	37.8	22.2	112	38.3	22.8	144
3.565	1 600	40.3	25.3	127	40.8	25.9	164
3.788	1 700	42.8	28.5	143	43.4	29.3	185
4.010	1 800	45.4	32.0	160	46.0	32.8	207
4.233	1 900	47.9	35.6	178	48.5	36.6	230
4.456	2 000	50.4	39.5	196	51.1	40.5	255

NOTE: No allowance has been made for age, differences in diameter, or any abnormal condition of interior surface. Any factor of safety must be estimated from the local conditions and the requirements of each particular installation.

IIIB-4 FRICTION LOSS FOR WATER IN FEET PER 100 FEET OF PIPE

TABLE 14

5 INCH NOMINAL — STEEL SCHEDULE 40, ID = 5.047 INCHES, ϵ/D = 0.000357

DISCHARGE CFS	DISCHARGE GPM	V ft/sec	$V^2/2g$ feet	h_f feet per 100 feet of pipe
0.0111	5	0.0802	0.0000999	0.00107
0.0223	10	0.160	0.000400	0.00348
0.0446	20	0.321	0.00160	0.0116
0.0668	30	0.481	0.00360	0.0237
0.0891	40	0.641	0.00639	0.0395
0.111	50	0.802	0.00999	0.0587
0.134	60	0.962	0.0144	0.0814
0.156	70	1.12	0.0196	0.1076
0.178	80	1.28	0.0256	0.137
0.201	90	1.44	0.0324	0.169
0.223	100	1.60	0.0400	0.204
0.267	120	1.92	0.0576	0.286
0.312	140	2.25	0.0783	0.380
0.356	160	2.57	0.102	0.487
0.401	180	2.89	0.129	0.606
0.446	200	3.21	0.160	0.736
0.490	220	3.53	0.193	0.879
0.535	240	3.85	0.230	1.035
0.579	260	4.17	0.270	1.20
0.624	280	4.49	0.313	1.38
0.668	300	4.81	0.360	1.58
0.713	320	5.13	0.409	1.78
0.758	340	5.45	0.462	2.00
0.802	360	5.77	0.518	2.22
0.847	380	6.09	0.577	2.46
0.891	400	6.41	0.639	2.72
0.936	420	6.74	0.705	2.98
0.980	440	7.06	0.774	3.26
1.02	460	7.38	0.846	3.55
1.07	480	7.70	0.921	3.85
1.11	500	8.02	0.999	4.16
1.23	550	8.82	1.21	4.98
1.34	600	9.62	1.44	5.88
1.45	650	10.4	1.69	6.87
1.56	700	11.2	1.96	7.93
1.67	750	12.0	2.25	9.05
1.78	800	12.8	2.56	10.22
1.89	850	13.6	2.89	11.5
2.01	900	14.4	3.24	12.9
2.12	950	15.2	3.61	14.3
2.23	1 000	16.0	4.00	15.8
2.45	1 100	17.6	4.84	19.0
2.67	1 200	19.2	5.76	22.5
2.90	1 300	20.8	6.75	26.3
3.12	1 400	22.5	7.83	30.4
3.34	1 500	24.1	8.99	34.8
3.56	1 600	25.7	10.2	39.5
3.79	1 700	27.3	11.6	44.5
4.01	1 800	28.8	12.9	49.7
4.23	1 900	30.5	14.4	55.2
4.46	2 000	32.1	16.0	61.0
4.68	2 100	33.7	17.6	67.1
4.90	2 200	35.3	19.3	73.5
5.12	2 300	36.9	21.1	80.1
5.35	2 400	38.5	23.0	87.0
5.57	2 500	40.1	25.0	94.2
5.79	2 600	41.7	27.0	102

NOTE: No allowance has been made for age, differences in diameter, or any abnormal condition of interior surface. Any factor of safety must be estimated from the local conditions and the requirements of each particular installation.

IIIB-4 FRICTION LOSS FOR WATER IN FEET PER 100 FEET OF PIPE

TABLE 15

6 INCH NOMINAL		STEEL SCHEDULE 40 ID = 6.065 INCHES ϵ/D = 0.000293			ASPHALT-DIPPED CAST IRON ID = 6.00 INCHES ϵ/D = 0.000800		
DISCHARGE							
CFS	GPM	V ft/sec	$V^2/2g$ feet	h_f feet per 100 feet of pipe	V ft/sec	$V^2/2g$ feet	h_f feet per 100 feet of pipe
0.0223	10	0.111	0.000192	0.00146	0.113	0.000200	0.00157
0.0446	20	0.222	0.000767	0.00487	0.227	0.000800	0.00523
0.0668	30	0.333	0.00172	0.00988	0.340	0.00180	0.01070
0.0891	40	0.444	0.00307	0.0164	0.454	0.00320	0.0179
0.111	50	0.555	0.00479	0.0244	0.567	0.00500	0.0268
0.134	60	0.666	0.00690	0.0337	0.681	0.00720	0.0374
0.156	70	0.777	0.00939	0.0445	0.794	0.00980	0.0496
0.178	80	0.888	0.0123	0.0564	0.908	0.0128	0.0635
0.201	90	0.999	0.0155	0.0698	1.02	0.0162	0.0789
0.223	100	1.11	0.0192	0.0843	1.13	0.0200	0.0958
0.267	120	1.33	0.0276	0.118	1.36	0.0288	0.130
0.312	140	1.55	0.0376	0.155	1.59	0.0392	0.178
0.356	160	1.78	0.0491	0.198	1.82	0.0512	0.229
0.401	180	2.00	0.0621	0.246	2.04	0.0648	0.282
0.446	200	2.22	0.0767	0.299	2.27	0.0800	0.346
0.490	220	2.44	0.0927	0.357	2.50	0.0968	0.415
0.535	240	2.66	0.110	0.419	2.72	0.115	0.490
0.579	260	2.89	0.130	0.487	2.95	0.135	0.570
0.624	280	3.11	0.150	0.560	3.18	0.157	0.655
0.668	300	3.33	0.172	0.637	3.40	0.180	0.745
0.713	320	3.55	0.196	0.719	3.63	0.205	0.846
0.758	340	3.78	0.222	0.806	3.86	0.231	0.952
0.802	360	4.00	0.240	0.898	4.08	0.259	1.06
0.847	380	4.22	0.277	0.993	4.31	0.289	1.18
0.891	400	4.44	0.307	1.09	4.54	0.320	1.30
0.936	420	4.66	0.338	1.20	4.76	0.353	1.43
0.980	440	4.89	0.371	1.31	4.99	0.387	1.57
1.025	460	5.11	0.405	1.42	5.22	0.423	1.71
1.07	480	5.33	0.442	1.54	5.45	0.461	1.86
1.11	500	5.55	0.479	1.66	5.67	0.500	2.02
1.23	550	6.11	0.580	1.99	6.24	0.605	2.42
1.34	600	6.66	0.690	2.34	6.81	0.720	2.84
1.45	650	7.22	0.810	2.73	7.37	0.845	3.33
1.56	700	7.77	0.939	3.13	7.94	0.980	3.87
1.67	750	8.33	1.08	3.57	8.51	1.12	4.45
1.78	800	8.88	1.23	4.03	9.08	1.28	5.06
1.89	850	9.44	1.38	4.53	9.64	1.44	5.69
2.01	900	9.99	1.55	5.05	10.2	1.62	6.34
2.12	950	10.5	1.73	5.60	10.8	1.80	7.02
2.23	1 000	11.1	1.92	6.17	11.3	2.00	7.73
2.45	1 100	12.2	2.32	7.41	12.5	2.42	9.80
2.67	1 200	13.3	2.76	8.76	13.6	2.88	11.2
2.90	1 300	14.4	3.24	10.2	14.7	3.38	13.0
3.12	1 400	15.5	3.76	11.8	15.9	3.92	15.1
3.34	1 500	16.7	4.31	13.5	17.0	4.50	17.4
3.56	1 600	17.8	4.91	15.4	18.2	5.12	19.8
3.79	1 700	18.9	5.54	17.3	19.3	5.78	22.3
4.01	1 800	20.0	6.21	19.4	20.4	6.48	24.8
4.23	1 900	21.1	6.92	21.6	21.6	7.22	27.6
4.46	2 000	22.2	7.67	23.8	22.7	8.00	30.5
4.68	2 100	23.3	8.45	26.2	23.8	8.82	33.6
4.90	2 200	24.4	9.27	28.8	25.0	9.68	36.8
5.12	2 300	25.5	10.1	31.4	26.1	10.6	40.1
5.35	2 400	26.6	11.0	34.2	27.2	11.5	43.5
5.57	2 500	27.8	12.0	37.0	28.4	12.5	47.1
5.79	2 600	28.9	13.0	39.9	29.5	13.5	51.0
6.02	2 700	30.0	14.0	42.9	30.6	14.6	55.2
6.24	2 800	31.1	15.0	46.1	31.8	15.7	59.6
6.46	2 900	32.2	16.1	49.4	32.9	16.8	64.1
6.68	3 000	33.3	17.2	52.8	34.0	18.0	68.8
7.13	3 200	35.5	19.6	59.9	36.3	20.5	78.0
7.58	3 400	37.8	22.2	67.4	38.6	23.1	88.0
8.02	3 600	40.0	24.8	75.5	40.8	25.9	98.7
8.47	3 800	42.2	27.7	84.1	43.1	28.9	110
8.91	4 000	44.4	30.7	93.1	45.4	32.0	122

Fluid Flow

Friction Loss - Water

IIIB-4 FRICTION LOSS FOR WATER IN FEET PER 100 FEET OF PIPE

TABLE 16

8 INCH NOMINAL		STEEL SCHEDULE 40 ID = 7.981 INCHES ϵ/D = 0.000226			ASPHALT-DIPPED CAST IRON ID = 8.00 INCHES ϵ/D = 0.00060		
DISCHARGE							
CFS	GPM	V ft/sec	$V^2/2g$ feet	h_f feet per 100 feet of pipe	V ft/sec	$V^2/2g$ feet	h_f feet per 100 feet of pipe
0.0223	10	0.0641	0.0000639	0.000401	0.0638	0.0000633	0.000399
0.0446	20	0.128	0.000256	0.001320	0.128	0.000253	0.001320
0.0668	30	0.192	0.000575	0.00266	0.191	0.000570	0.00269
0.0891	40	0.257	0.00102	0.00442	0.255	0.00101	0.00447
0.111	50	0.321	0.00160	0.00652	0.319	0.00158	0.00664
0.134	60	0.385	0.00230	0.00904	0.382	0.00228	0.00920
0.156	70	0.449	0.00313	0.01190	0.447	0.00310	0.01210
0.178	80	0.513	0.00409	0.0151	0.511	0.00405	0.0154
0.201	90	0.577	0.00518	0.0186	0.574	0.00513	0.0191
0.223	100	0.641	0.00639	0.0224	0.638	0.00633	0.0232
0.267	120	0.770	0.00920	0.0311	0.766	0.00911	0.0323
0.312	140	0.898	0.0125	0.0410	0.893	0.0124	0.0428
0.356	160	1.03	0.0164	0.0521	1.02	0.0162	0.0548
0.401	180	1.15	0.0207	0.0644	1.15	0.0205	0.0681
0.446	200	1.28	0.0256	0.0780	1.28	0.0253	0.0828
0.490	220	1.41	0.0309	0.0928	1.40	0.0306	0.0989
0.535	240	1.54	0.0368	0.1088	1.53	0.0365	0.1163
0.579	260	1.67	0.0432	0.1260	1.66	0.0428	0.135
0.624	280	1.80	0.0501	0.144	1.79	0.0496	0.155
0.668	300	1.92	0.0575	0.163	1.91	0.0570	0.176
0.713	320	2.05	0.0655	0.184	2.04	0.0648	0.198
0.758	340	2.18	0.0739	0.206	2.17	0.0732	0.222
0.802	360	2.31	0.0828	0.229	2.30	0.0820	0.248
0.847	380	2.44	0.0923	0.253	2.43	0.0914	0.275
0.891	400	2.57	0.102	0.279	2.55	0.101	0.304
1.003	450	2.89	0.129	0.348	2.87	0.128	0.380
1.11	500	3.21	0.160	0.424	3.19	0.158	0.464
1.23	550	3.53	0.193	0.507	3.51	0.191	0.557
1.34	600	3.85	0.230	0.597	3.83	0.228	0.658
1.45	650	4.17	0.271	0.694	4.15	0.267	0.767
1.56	700	4.49	0.313	0.797	4.47	0.310	0.884
1.67	750	4.81	0.360	0.907	4.79	0.356	1.01
1.78	800	5.13	0.409	1.02	5.11	0.405	1.14
1.89	850	5.45	0.462	1.147	5.42	0.457	1.29
2.01	900	5.77	0.518	1.27	5.74	0.513	1.44
2.12	950	6.09	0.577	1.41	6.06	0.571	1.60
2.23	1 000	6.41	0.639	1.56	6.38	0.633	1.76
2.45	1 100	7.05	0.773	1.87	7.02	0.766	2.14
2.67	1 200	7.70	0.920	2.20	7.66	0.911	2.53
2.90	1 300	8.34	1.08	2.56	8.30	1.07	2.94
3.12	1 400	8.98	1.25	2.95	8.93	1.24	3.40
3.34	1 500	9.62	1.44	3.37	9.57	1.42	3.91
3.56	1 600	10.3	1.64	3.82	10.2	1.62	4.45
3.79	1 700	10.9	1.85	4.29	10.8	1.83	5.00
4.01	1 800	11.5	2.07	4.79	11.5	2.05	5.58
4.23	1 900	12.2	2.31	5.31	12.1	2.29	6.19
4.46	2 000	12.8	2.56	5.86	12.8	2.53	6.84
4.90	2 200	14.1	3.09	7.02	14.0	3.06	8.26
5.35	2 400	15.4	3.68	8.31	15.3	3.65	9.80
5.79	2 600	16.7	4.32	9.70	16.6	4.28	11.47
6.24	2 800	18.0	5.01	11.20	17.9	4.96	13.3
6.68	3 000	19.2	5.75	12.8	19.1	5.70	15.2
7.13	3 200	20.5	6.55	14.5	20.4	6.48	17.3
7.58	3 400	21.8	7.39	16.4	21.7	7.32	19.5
8.02	3 600	23.1	8.28	18.4	23.0	8.20	21.9
8.47	3 800	24.4	9.23	20.5	24.3	9.14	24.4
8.91	4 000	25.7	10.2	22.6	25.5	10.1	27.0
10.03	4 500	28.9	12.9	28.5	28.7	12.8	34.0
11.1	5 000	32.1	16.0	35.1	31.9	15.8	42.0
12.3	5 500	35.3	19.3	42.5	35.1	19.1	51.0
13.4	6 000	38.5	23.0	50.5	38.3	22.8	60.5
14.5	6 500	41.7	27.0	59.1	41.5	26.7	71.0
15.6	7 000	44.9	31.3	68.3	44.7	31.0	82.0
16.7	7 500	48.1	36.0	78.1	47.9	35.6	94.0
17.8	8 000	51.3	40.9	88.6	51.1	40.5	107

IIIB-4 FRICTION LOSS FOR WATER IN FEET PER 100 FEET OF PIPE

TABLE 17

10 INCH NOMINAL		STEEL SCHEDULE 40 ID = 10.020 INCHES ϵ/D = 0.000180			ASPHALT-DIPPED CAST IRON ID = 10.00 INCHES ϵ/D = 0.000480		
DISCHARGE CFS	DISCHARGE GPM	V ft/sec	$V^2/2g$ feet	h_f feet per 100 feet of pipe	V ft/sec	$V^2/2g$ feet	h_f feet per 100 feet of pipe
0.0223	10	0.0407	0.0000257	0.000138	0.0409	0.0000259	0.000140
0.0446	20	0.0814	0.000103	0.000451	0.0817	0.000104	0.000460
0.0891	40	0.163	0.000412	0.00149	0.163	0.000415	0.00154
0.134	60	0.244	0.000926	0.00304	0.245	0.000934	0.00315
0.178	80	0.325	0.00165	0.00505	0.327	0.00166	0.00525
0.223	100	0.407	0.00257	0.00747	0.409	0.00259	0.00783
0.267	120	0.488	0.00370	0.0103	0.490	0.00373	0.01085
0.312	140	0.570	0.00504	0.0136	0.572	0.00508	0.0144
0.356	160	0.651	0.00659	0.0174	0.654	0.00664	0.0183
0.401	180	0.732	0.00834	0.0215	0.735	0.00840	0.0227
0.446	200	0.814	0.0103	0.0260	0.817	0.0104	0.0276
0.490	220	0.895	0.0125	0.0309	0.899	0.0126	0.0329
0.535	240	0.976	0.0148	0.0362	0.980	0.0149	0.0387
0.579	260	1.06	0.0174	0.0417	1.06	0.0175	0.0449
0.624	280	1.14	0.0202	0.0478	1.14	0.0203	0.0514
0.668	300	1.22	0.0232	0.0542	1.23	0.0233	0.0583
0.780	350	1.42	0.0315	0.0719	1.43	0.0318	0.0778
0.891	400	1.63	0.0412	0.0917	1.63	0.0415	0.0990
1.003	450	1.83	0.0521	0.114	1.84	0.0525	0.1235
1.11	500	2.03	0.0643	0.138	2.04	0.0648	0.151
1.23	550	2.24	0.0778	0.164	2.25	0.0785	0.181
1.34	600	2.44	0.0926	0.192	2.45	0.0934	0.214
1.45	650	2.64	0.109	0.224	2.66	0.110	0.250
1.56	700	2.85	0.126	0.256	2.86	0.127	0.288
1.67	750	3.05	0.145	0.291	3.06	0.146	0.328
1.78	800	3.25	0.165	0.328	3.27	0.166	0.370
1.89	850	3.46	0.186	0.368	3.47	0.187	0.415
2.01	900	3.66	0.208	0.410	3.68	0.210	0.462
2.12	950	3.87	0.232	0.455	3.88	0.234	0.512
2.23	1 000	4.07	0.257	0.500	4.09	0.259	0.565
2.45	1 100	4.48	0.311	0.600	4.49	0.314	0.680
2.67	1 200	4.88	0.370	0.703	4.90	0.373	0.805
2.90	1 300	5.29	0.435	0.818	5.31	0.438	0.945
3.12	1 400	5.70	0.504	0.940	5.72	0.508	1.09
3.34	1 500	6.10	0.579	1.07	6.13	0.584	1.25
3.56	1 600	6.51	0.659	1.21	6.54	0.664	1.42
3.79	1 700	6.92	0.743	1.36	6.94	0.749	1.60
4.01	1 800	7.32	0.834	1.52	7.35	0.840	1.78
4.23	1 900	7.73	0.929	1.68	7.76	0.936	1.97
4.46	2 000	8.14	1.03	1.86	8.17	1.04	2.17
4.90	2 200	8.95	1.25	2.23	8.99	1.26	2.64
5.35	2 400	9.76	1.48	2.64	9.80	1.49	3.12
5.79	2 600	10.6	1.74	3.08	10.6	1.75	3.63
6.24	2 800	11.4	2.02	3.56	11.4	2.03	4.18
6.68	3 000	12.2	2.32	4.06	12.3	2.33	4.79
7.13	3 200	13.0	2.63	4.59	13.1	2.66	5.47
7.58	3 400	13.8	2.97	5.16	13.9	3.00	6.18
8.02	3 600	14.6	3.33	5.76	14.7	3.36	6.91
8.47	3 800	15.5	3.71	6.40	15.5	3.74	7.68
8.91	4 000	16.3	4.12	7.07	16.3	4.15	8.50
10.03	4 500	18.3	5.21	8.88	18.4	5.25	10.7
11.1	5 000	20.3	6.43	10.9	20.4	6.48	13.2
12.3	5 500	22.4	7.78	13.2	22.5	7.85	15.9
13.4	6 000	24.4	9.26	15.6	24.5	9.34	18.9
14.5	6 500	26.4	10.9	18.3	26.6	11.0	22.2
15.6	7 000	28.5	12.6	21.1	28.6	12.7	25.8
16.7	7 500	30.5	14.5	24.3	30.0	14.6	29.6
17.8	8 000	32.5	16.5	27.5	32.7	16.6	33.6
18.9	8 500	34.6	18.6	30.9	34.7	18.7	37.8
20.1	9 000	36.6	20.8	34.6	36.8	21.0	42.2
21.2	9 500	38.7	23.2	38.5	38.8	23.4	46.9
22.3	10 000	40.7	25.7	42.6	40.9	25.9	51.8

IIIB-4 FRICTION LOSS FOR WATER IN FEET PER 100 FEET OF PIPE

TABLE 18

12 INCH NOMINAL		STEEL SCHEDULE 40 ID = 11.938 INCHES ϵ/D = 0.000151			ASPHALT-DIPPED CAST IRON ID = 12.00 INCHES ϵ/D = 0.000400		
DISCHARGE CFS	DISCHARGE GPM	V ft/sec	$V^2/2g$ feet	h_f feet per 100 feet of pipe	V ft/sec	$V^2/2g$ feet	h_f feet per 100 feet of pipe
0.223	100	0.287	0.00128	0.00325	0.284	0.00125	0.00320
0.267	120	0.344	0.00184	0.00448	0.340	0.00180	0.00445
0.312	140	0.401	0.00250	0.00590	0.397	0.00245	0.00589
0.356	160	0.459	0.00327	0.00747	0.454	0.00320	0.00752
0.401	180	0.516	0.00414	0.00920	0.511	0.00405	0.00932
0.446	200	0.573	0.00511	0.0111	0.567	0.00500	0.01129
0.490	220	0.631	0.00618	0.0132	0.624	0.00605	0.0135
0.535	240	0.688	0.00735	0.0155	0.681	0.00720	0.0158
0.579	260	0.745	0.00863	0.0180	0.738	0.00845	0.0182
0.624	280	0.802	0.0100	0.0206	0.794	0.00980	0.0208
0.668	300	0.860	0.0115	0.0233	0.851	0.0113	0.0236
0.780	350	1.00	0.0156	0.0306	0.993	0.0153	0.0316
0.891	400	1.15	0.0204	0.0391	1.13	0.0200	0.0404
1.00	450	1.29	0.0259	0.0485	1.28	0.0253	0.0500
1.11	500	1.43	0.0319	0.0587	1.42	0.0313	0.0604
1.23	550	1.58	0.0386	0.0698	1.56	0.0378	0.0718
1.34	600	1.72	0.0460	0.0820	1.70	0.0450	0.0845
1.45	650	1.86	0.0539	0.0950	1.84	0.0528	0.0990
1.56	700	2.01	0.0626	0.109	1.99	0.0613	0.115
1.67	750	2.15	0.0718	0.124	2.13	0.0703	0.131
1.78	800	2.29	0.0817	0.140	2.27	0.0800	0.148
1.89	850	2.44	0.0922	0.156	2.41	0.0903	0.166
2.01	900	2.58	0.103	0.173	2.55	0.101	0.184
2.12	950	2.72	0.115	0.191	2.69	0.113	0.203
2.23	1 000	2.87	0.128	0.210	2.84	0.125	0.224
2.45	1 100	3.15	0.154	0.251	3.12	0.151	0.272
2.67	1 200	3.44	0.184	0.296	3.40	0.180	0.321
2.90	1 300	3.73	0.216	0.344	3.69	0.211	0.372
3.12	1 400	4.01	0.250	0.395	3.97	0.245	0.428
3.34	1 500	4.30	0.287	0.450	4.26	0.281	0.488
3.56	1 600	4.59	0.327	0.509	4.54	0.320	0.552
3.79	1 700	4.87	0.369	0.572	4.82	0.361	0.621
4.01	1 800	5.16	0.414	0.636	5.11	0.405	0.695
4.23	1 900	5.45	0.461	0.704	5.39	0.451	0.774
4.46	2 000	5.73	0.511	0.776	5.67	0.500	0.858
4.90	2 200	6.31	0.618	0.930	6.24	0.605	1.03
5.35	2 400	6.88	0.735	1.093	6.81	0.720	1.22
5.79	2 600	7.45	0.863	1.28	7.38	0.845	1.43
6.24	2 800	8.03	1.00	1.47	7.94	0.980	1.65
6.68	3 000	8.60	1.15	1.68	8.51	1.13	1.88
7.13	3 200	9.17	1.31	1.90	9.08	1.28	2.13
7.58	3 400	9.75	1.48	2.13	9.65	1.45	2.41
8.02	3 600	10.3	1.65	2.37	10.2	1.62	2.70
8.47	3 800	10.9	1.84	2.63	10.8	1.81	3.00
8.91	4 000	11.5	2.04	2.92	11.3	2.00	3.31
10.03	4 500	12.9	2.59	3.65	12.8	2.53	4.18
11.1	5 000	14.3	3.19	4.47	14.2	3.13	5.13
12.3	5 500	15.8	3.86	5.38	15.6	3.78	6.17
13.4	6 000	17.2	4.60	6.39	17.0	4.50	7.30
14.5	6 500	18.6	5.39	7.47	18.4	5.28	8.55
15.6	7 000	20.1	6.26	8.63	19.9	6.13	9.92
16.7	7 500	21.5	7.18	9.88	21.3	7.03	11.4
17.8	8 000	22.9	8.17	11.20	22.7	8.00	13.0
18.9	8 500	24.4	9.22	12.6	24.1	9.04	14.7
20.1	9 000	25.8	10.3	14.1	25.5	10.1	16.4
21.2	9 500	27.2	11.5	15.7	26.9	11.3	18.2
22.3	10 000	28.7	12.8	17.4	28.4	12.5	20.2
24.5	11 000	31.5	15.4	21.0	31.2	15.1	24.2
26.7	12 000	34.4	18.3	24.8	34.0	18.0	28.8
29.0	13 000	37.3	21.6	28.9	36.9	21.1	34.0
31.2	14 000	40.1	25.0	33.5	39.7	24.5	39.7
33.4	15 000	43.0	28.7	38.4	42.6	28.1	45.7
35.6	16 000	45.9	32.7	43.7	45.4	32.0	51.8
37.9	17 000	48.7	36.9	49.2	48.2	36.1	58.2
40.1	18 000	51.6	41.4	55.2	51.1	40.5	65.0
42.3	19 000	54.5	46.1	61.5	53.9	45.1	72.1
44.6	20 000	57.3	51.1	68.1	56.7	50.0	79.8

IIIB-4 FRICTION LOSS FOR WATER IN FEET PER 100 FEET OF PIPE

TABLE 19

14 INCH OD		STEEL — SCHEDULE 40 ID = 13.124 INCHES ϵ/D = 0.000137			ASPHALT-DIPPED CAST IRON ID = 14.00 INCHES ϵ/D = 0.000343		
DISCHARGE CFS	DISCHARGE GPM	V ft/sec	$V^2/2g$ feet	h_f feet per 100 feet of pipe	V ft/sec	$V^2/2g$ feet	h_f feet per 100 feet of pipe
0.446	200	0.474	0.00349	0.00704	0.417	0.00270	0.00533
0.557	250	0.593	0.00546	0.0105	0.521	0.00422	0.00800
0.668	300	0.711	0.00786	0.0147	0.625	0.00608	0.0112
0.780	350	0.830	0.0107	0.0194	0.729	0.00827	0.0148
0.891	400	0.948	0.0140	0.0247	0.834	0.0108	0.0190
1.114	500	1.19	0.0218	0.0370	1.04	0.0169	0.0284
1.34	600	1.42	0.0314	0.0517	1.25	0.0243	0.0400
1.56	700	1.66	0.0428	0.0683	1.46	0.0330	0.0533
1.78	800	1.90	0.0559	0.0872	1.67	0.0432	0.0686
2.01	900	2.13	0.0708	0.108	1.88	0.0547	0.0859
2.23	1 000	2.37	0.0874	0.131	2.08	0.0675	0.1050
2.45	1 100	2.61	0.106	0.157	2.29	0.0817	0.1256
2.67	1 200	2.85	0.126	0.185	2.50	0.0972	0.148
2.90	1 300	3.08	0.148	0.215	2.71	0.114	0.172
3.12	1 400	3.32	0.171	0.247	2.92	0.132	0.198
3.34	1 500	3.56	0.197	0.281	3.13	0.152	0.225
3.56	1 600	3.79	0.224	0.317	3.33	0.173	0.254
3.79	1 700	4.03	0.252	0.355	3.54	0.195	0.285
4.01	1 800	4.27	0.283	0.395	3.75	0.219	0.318
4.23	1 900	4.50	0.315	0.438	3.96	0.244	0.353
4.46	2 000	4.74	0.349	0.483	4.17	0.270	0.390
5.57	2 500	5.93	0.546	0.738	5.21	0.422	0.600
6.68	3 000	7.11	0.786	1.04	6.25	0.608	0.855
7.80	3 500	8.30	1.07	1.40	7.29	0.827	1.16
8.91	4 000	9.48	1.40	1.81	8.34	1.08	1.50
10.02	4 500	10.7	1.77	2.27	9.38	1.37	1.88
11.1	5 000	11.9	2.18	2.78	10.4	1.69	2.30
13.4	6 000	14.2	3.14	3.95	12.5	2.43	3.31
15.6	7 000	16.6	4.28	5.32	14.6	3.30	4.50
17.8	8 000	19.0	5.59	6.90	16.7	4.32	5.87
20.1	9 000	21.3	7.08	8.7	18.8	5.47	7.42
22.3	10 000	23.7	8.74	10.7	20.8	6.75	9.15
24.5	11 000	26.1	10.6	12.9	22.9	8.17	11.05
26.7	12 000	28.5	12.6	15.2	25.0	9.72	13.0
29.0	13 000	30.8	14.8	17.8	27.1	11.4	15.2
31.2	14 000	33.2	17.1	20.7	29.2	13.2	17.6
33.4	15 000	35.6	19.7	23.7	31.3	15.2	20.2
35.6	16 000	37.9	22.4	26.8	33.3	17.3	23.0
37.9	17 000	40.3	25.2	30.2	35.4	19.5	26.0
40.1	18 000	42.7	28.3	33.9	37.5	21.9	29.2
42.3	19 000	45.0	31.5	37.8	39.6	24.4	32.6
44.6	20 000	47.4	34.9	41.9	41.7	27.0	36.2
49.0	22 000	52.2	42.3	50.6	45.9	32.7	43.5
53.5	24 000	56.9	50.3	60.0	50.0	38.9	51.9
57.9	26 000	61.6	59.1	70.3	54.2	45.6	61.2
62.4	28 000	66.4	68.5	81.4	58.4	52.9	71.2
66.8	30 000	71.1	78.6	93.4	62.5	60.8	81.6

NOTE: No allowance has been made for age, differences in diameter, or any abnormal condition of interior surface. Any factor of safety must be estimated from the local conditions and the requirements of each particular installation.

IIIB-4 FRICTION LOSS FOR WATER IN FEET PER 100 FEET OF PIPE

TABLE 20

16 INCH OD		STEEL — SCHEDULE 40 ID = 15.000 INCHES ϵ/D = 0.000120			ASPHALT-DIPPED CAST IRON ID = 16.00 INCHES ϵ/D = 0.000300		
DISCHARGE CFS	DISCHARGE GPM	V ft/sec	$V^2/2g$ feet	h_f feet per 100 feet of pipe	V ft/sec	$V^2/2g$ feet	h_f feet per 100 feet of pipe
0.668	300	0.545	0.00461	0.00769	0.479	0.00356	0.00581
0.891	400	0.726	0.00820	0.0129	0.638	0.00633	0.00980
1.114	500	0.908	0.0128	0.0193	0.798	0.00989	0.0148
1.34	600	1.09	0.0184	0.0269	0.957	0.0142	0.0207
1.56	700	1.27	0.0251	0.0356	1.12	0.0194	0.0276
1.78	800	1.45	0.0328	0.0454	1.28	0.0253	0.0354
2.01	900	1.63	0.0415	0.0563	1.44	0.0320	0.0441
2.23	1 000	1.82	0.0512	0.0683	1.60	0.0396	0.0537
2.67	1 200	2.18	0.0738	0.0953	1.91	0.0570	0.0760
3.12	1 400	2.54	0.1004	0.127	2.23	0.0775	0.101
3.56	1 600	2.90	0.131	0.163	2.55	0.101	0.130
4.01	1 800	3.27	0.166	0.203	2.87	0.128	0.163
4.46	2 000	3.63	0.205	0.248	3.19	0.158	0.200
5.57	2 500	4.54	0.320	0.377	3.99	0.247	0.307
6.68	3 000	5.45	0.461	0.535	4.79	0.356	0.435
7.80	3 500	6.35	0.627	0.718	5.58	0.485	0.584
8.91	4 000	7.26	0.820	0.921	6.38	0.633	0.754
10.02	4 500	8.17	1.04	1.15	7.18	0.801	0.948
11.1	5 000	9.08	1.28	1.41	7.98	0.989	1.17
13.4	6 000	10.9	1.84	2.01	9.57	1.42	1.66
15.6	7 000	12.7	2.51	2.69	11.2	1.94	2.26
17.8	8 000	14.5	3.28	3.49	12.8	2.53	2.96
20.1	9 000	16.3	4.15	4.38	14.4	3.20	3.73
22.3	10 000	18.2	5.12	5.38	16.0	3.96	4.57
24.5	11 000	20.0	6.20	6.49	17.6	4.79	5.50
26.7	12 000	21.8	7.38	7.69	19.1	5.70	6.52
29.0	13 000	23.6	8.66	8.99	20.7	6.69	7.63
31.2	14 000	25.4	10.04	10.4	22.3	7.75	8.81
33.4	15 000	27.2	11.5	11.9	23.9	8.90	10.1
35.6	16 000	29.0	13.1	13.5	25.5	10.1	11.5
37.9	17 000	30.9	14.8	15.3	27.1	11.4	13.0
40.1	18 000	32.7	16.6	17.2	28.7	12.8	14.6
42.3	19 000	34.5	18.5	19.2	30.3	14.3	16.3
44.6	20 000	36.3	20.5	21.2	31.9	15.8	18.1
49.0	22 000	39.9	24.8	25.5	35.1	19.1	21.8
53.5	24 000	43.6	29.5	30.2	38.3	22.8	25.9
57.9	26 000	47.2	34.6	35.4	41.5	26.7	30.4
62.4	28 000	50.8	40.2	41.0	44.7	31.0	35.3
66.8	30 000	54.5	46.1	47.0	47.9	35.6	40.5
71.3	32 000	58.1	52.4	53.5	51.1	40.5	46.0
75.8	34 000	61.7	59.2	60.2	54.3	45.7	51.9
80.2	36 000	65.4	66.4	67.2	57.4	51.3	58.1
84.7	38 000	69.0	74.0	75.0	60.6	57.1	64.7
89.1	40 000	72.6	82.0	83.0	63.8	63.3	71.7
93.6	42 000	76.2	90.4	91.5	67.0	69.8	79.1
98.0	44 000	79.9	99.2	101	70.2	76.6	86.9
102.5	46 000	83.5	108.4	110	73.4	83.7	95.0
107	48 000	87.1	118	118	76.6	91.2	103
111	50 000	90.8	128	128	79.8	98.9	112

NOTE: No allowance has been made for age, differences in diameter, or any abnormal condition of interior surface. Any factor of safety must be estimated from the local conditions and the requirements of each particular installation.

Fluid Flow

Friction Loss - Water

IIIB-4 FRICTION LOSS FOR WATER IN FEET PER 100 FEET OF PIPE

TABLE 21

18 INCH OD		STEEL — SCHEDULE 40 ID = 16.876 INCHES ϵ/D = 0.000107			ASPHALT-DIPPED CAST IRON ID = 18.00 INCHES ϵ/D = 0.000267		
DISCHARGE CFS	DISCHARGE GPM	V ft/sec	$V^2/2g$ feet	h_f feet per 100 feet of pipe	V ft/sec	$V^2/2g$ feet	h_f feet per 100 feet of pipe
0.668	300	0.430	0.00288	0.00437	0.378	0.00222	0.00328
0.891	400	0.574	0.00512	0.00730	0.504	0.00359	0.00554
1.114	500	0.717	0.00799	0.0109	0.630	0.00618	0.00832
1.34	600	0.861	0.0115	0.0152	0.756	0.00889	0.01162
1.56	700	1.00	0.0157	0.0201	0.883	0.0121	0.0154
1.78	800	1.15	0.0205	0.0256	1.01	0.0158	0.0197
2.01	900	1.29	0.0259	0.0318	1.13	0.0200	0.0245
2.23	1 000	1.43	0.0320	0.0386	1.26	0.0247	0.0298
2.67	1 200	1.72	0.0460	0.0541	1.51	0.0356	0.0420
3.12	1 400	2.01	0.0627	0.0719	1.77	0.0484	0.0560
3.56	1 600	2.30	0.0819	0.092	2.02	0.0632	0.0728
4.01	1 800	2.58	0.1036	0.114	2.27	0.0800	0.0910
4.46	2 000	2.87	0.128	0.139	2.52	0.0988	0.110
5.57	2 500	3.59	0.200	0.211	3.15	0.154	0.170
6.68	3 000	4.30	0.288	0.297	3.78	0.222	0.240
7.80	3 500	5.02	0.392	0.397	4.41	0.303	0.320
8.91	4 000	5.74	0.512	0.511	5.04	0.395	0.415
10.02	4 500	6.45	0.647	0.639	5.67	0.500	0.525
11.1	5 000	7.17	0.799	0.781	6.30	0.618	0.645
13.4	6 000	8.61	1.15	1.11	7.56	0.889	0.920
15.6	7 000	10.0	1.57	1.49	8.83	1.21	1.24
17.8	8 000	11.5	2.05	1.93	10.09	1.58	1.61
20.1	9 000	12.9	2.59	2.42	11.3	2.00	2.02
22.3	10 000	14.3	3.20	2.97	12.6	2.47	2.48
26.7	12 000	17.2	4.60	4.21	15.1	3.56	3.56
31.2	14 000	20.1	6.27	5.69	17.7	4.84	4.85
35.6	16 000	22.9	8.19	7.41	20.2	6.32	6.34
40.1	18 000	25.8	10.36	9.33	22.7	8.00	8.02
44.6	20 000	28.7	12.8	11.5	25.2	9.88	9.88
49.0	22 000	31.6	15.5	13.9	27.7	12.0	11.90
53.5	24 000	34.4	18.4	16.5	30.3	14.2	14.10
57.9	26 000	37.3	21.6	19.2	32.8	16.7	16.50
62.4	28 000	40.2	25.1	22.2	35.3	19.4	19.1
66.8	30 000	43.0	28.8	25.5	37.8	22.9	21.9
71.3	32 000	45.9	32.7	29.0	40.3	25.3	24.9
75.8	34 000	48.8	37.0	32.8	42.9	28.6	28.1
80.2	36 000	51.6	41.4	36.8	45.4	32.0	31.5
84.7	38 000	54.5	46.2	40.8	47.9	35.7	35.1
89.1	40 000	57.4	51.2	45.0	50.4	39.5	38.9
93.6	42 000	60.2	56.4	49.7	53.0	43.6	42.9
98	44 000	63.1	61.9	54.5	55.5	47.8	47.0
102	46 000	66.0	67.7	59.5	58.0	52.3	51.3
107	48 000	68.9	73.7	64.8	60.5	56.9	55.8
111	50 000	71.7	79.9	70.2	63.0	61.8	60.5
123	55 000	78.9	96.7	84.8	69.3	74.7	73.0
134	60 000	86.1	115	101	75.6	88.9	86.7
145	65 000	93.2	135	118	82.0	104.4	101.8
156	70 000	100.4	157	136	88.3	121	118

NOTE: No allowance has been made for age, differences in diameter, or any abnormal condition of interior surface. Any factor of safety must be estimated from the local conditions and the requirements of each particular installation.

IIIB-4 FRICTION LOSS FOR WATER IN FEET PER 100 FEET OF PIPE

TABLE 22

20 INCH OD		STEEL — SCHEDULE 40 ID = 18.812 INCHES ϵ/D = 0.0000957			ASPHALT-DIPPED CAST IRON ID = 20.00 INCHES ϵ/D = 0.000240		
DISCHARGE CFS	GPM	V ft/sec	$V^2/2g$ feet	h_f feet per 100 feet of pipe	V ft/sec	$V^2/2g$ feet	h_f feet per 100 feet of pipe
0.668	300	0.346	0.00186	0.00258	0.306	0.00146	0.00197
0.891	400	0.462	0.00331	0.00432	0.408	0.00259	0.00332
1.114	500	0.577	0.00517	0.00645	0.511	0.00405	0.00496
1.34	600	0.692	0.00745	0.00897	0.613	0.00583	0.00691
1.56	700	0.808	0.0101	0.01186	0.715	0.00794	0.00918
1.78	800	0.923	0.0132	0.0152	0.817	0.0104	0.0117
2.01	900	1.039	0.0168	0.0188	0.919	0.0131	0.0146
2.23	1 000	1.15	0.0207	0.0227	1.02	0.0162	0.0177
2.67	1 200	1.38	0.0298	0.0318	1.23	0.0233	0.0249
3.12	1 400	1.62	0.0406	0.0422	1.43	0.0318	0.0332
3.56	1 600	1.85	0.0530	0.0538	1.63	0.0415	0.0427
4.01	1 800	2.08	0.0671	0.0669	1.84	0.0525	0.0533
4.46	2 000	2.31	0.0828	0.0812	2.04	0.0648	0.0650
5.57	2 500	2.89	0.129	0.123	2.55	0.1013	0.0998
6.68	3 000	3.46	0.186	0.174	3.06	0.146	0.140
7.80	3 500	4.04	0.254	0.232	3.57	0.198	0.188
8.91	4 000	4.62	0.331	0.298	4.08	0.259	0.243
10.02	4 500	5.19	0.419	0.372	4.59	0.328	0.306
11.1	5 000	5.77	0.517	0.455	5.11	0.405	0.376
13.4	6 000	6.92	0.745	0.645	6.13	0.583	0.533
15.6	7 000	8.08	1.014	0.862	7.15	0.794	0.721
17.8	8 000	9.23	1.32	1.11	8.17	1.04	0.935
20.1	9 000	10.39	1.68	1.39	9.19	1.31	1.18
22.3	10 000	11.5	2.07	1.70	10.2	1.62	1.45
26.7	12 000	13.8	2.98	2.44	12.3	2.33	2.07
31.2	14 000	16.2	4.06	3.29	14.3	3.18	2.80
35.6	16 000	18.5	5.30	4.26	16.3	4.15	3.66
40.1	18 000	20.8	6.71	5.35	18.4	5.25	4.62
44.6	20 000	23.1	8.28	6.56	20.4	6.48	5.67
49.0	22 000	25.4	10.02	7.91	22.5	7.84	6.85
53.5	24 000	27.7	11.9	9.39	24.5	9.33	8.13
57.9	26 000	30.0	14.0	11.0	26.5	10.95	9.54
62.4	28 000	32.3	16.2	12.7	28.6	12.7	11.1
66.8	30 000	34.6	18.6	14.6	30.6	14.6	12.7
71.3	32 000	36.9	21.2	16.6	32.7	16.6	14.4
75.8	34 000	39.2	23.9	18.7	34.7	18.7	16.3
80.2	36 000	41.5	26.8	20.9	36.8	21.0	18.2
84.7	38 000	43.9	29.9	23.2	38.8	23.4	20.2
89.1	40 000	46.2	33.1	25.7	40.8	25.9	22.4
93.6	42 000	48.5	36.5	28.4	42.9	28.6	24.7
98	44 000	50.8	40.1	31.3	44.9	31.4	27.1
102	46 000	53.1	43.8	34.2	47.0	34.3	29.7
107	48 000	55.4	47.7	37.1	49.0	37.3	32.4
111	50 000	57.7	51.7	40.0	51.1	40.5	35.2
123	55 000	63.5	62.6	48.3	56.2	49.0	42.4
134	60 000	69.2	74.5	57.4	61.3	58.3	50.4
145	65 000	75.0	87.4	67.2	66.4	68.5	59.0
156	70 000	80.8	101.4	77.8	71.5	79.4	68.4
167	75 000	86.6	116	89.3	76.6	91.1	78.6
178	80 000	92.3	132	102	81.7	103.7	89.5
189	85 000	98.1	150	115	86.8	117	101
201	90 000	103.9	168	129	91.9	131	113
212	95 000	109.6	187	143	97.0	146	126
223	100 000	115.4	207	158	102.1	162	139

IIIB-4 FRICTION LOSS FOR WATER IN FEET PER 100 FEET OF PIPE

TABLE 23

24 INCH OD		STEEL — SCHEDULE 40 ID = 22.624 INCHES ϵ/D = 0.0000796			ASPHALT-DIPPED CAST IRON ID = 24.00 INCHES ϵ/D = 0.000200		
DISCHARGE CFS	GPM	V ft/sec	$V^2/2g$ feet	h_f feet per 100 feet of pipe	V ft/sec	$V^2/2g$ feet	h_f feet per 100 feet of pipe
0.668	300	0.239	0.000891	0.00107	0.213	0.000703	0.000821
0.891	400	0.319	0.00158	0.00178	0.284	0.00125	0.00137
1.114	500	0.399	0.00247	0.00267	0.355	0.00195	0.00205
1.34	600	0.479	0.00356	0.00371	0.426	0.00281	0.00284
1.56	700	0.559	0.00485	0.00490	0.496	0.00383	0.00376
1.78	800	0.638	0.00633	0.00621	0.567	0.00500	0.00480
2.01	900	0.718	0.00801	0.00767	0.638	0.00633	0.00597
2.23	1 000	0.798	0.00989	0.00928	0.709	0.00782	0.00724
2.67	1 200	0.958	0.0142	0.0129	0.851	0.01126	0.0102
3.12	1 400	1.12	0.0194	0.0171	0.993	0.0153	0.0135
3.56	1 600	1.28	0.0253	0.0219	1.135	0.0200	0.0173
4.01	1 800	1.44	0.0321	0.0272	1.276	0.0253	0.0216
4.46	2 000	1.60	0.0396	0.0330	1.42	0.0313	0.0262
5.57	2 500	1.99	0.0618	0.0499	1.77	0.0489	0.0398
6.68	3 000	2.39	0.0891	0.0700	2.13	0.0703	0.0563
7.80	3 500	2.79	0.121	0.0934	2.48	0.0957	0.0759
8.91	4 000	3.19	0.158	0.120	2.84	0.125	0.098
10.02	4 500	3.59	0.200	0.149	3.19	0.158	0.122
11.1	5 000	3.99	0.247	0.181	3.55	0.195	0.149
13.4	6 000	4.79	0.356	0.257	4.26	0.281	0.211
15.6	7 000	5.59	0.485	0.343	4.96	0.383	0.284
17.8	8 000	6.38	0.633	0.441	5.67	0.500	0.368
20.1	9 000	7.18	0.801	0.551	6.38	0.633	0.464
22.3	10 000	7.98	0.989	0.671	7.09	0.782	0.571
26.7	12 000	9.58	1.42	0.959	8.51	1.126	0.816
31.2	14 000	11.2	1.94	1.29	9.93	1.53	1.11
35.6	16 000	12.8	2.53	1.67	11.35	2.00	1.43
40.1	18 000	14.4	3.21	2.10	12.76	2.53	1.80
44.6	20 000	16.0	3.96	2.58	14.2	3.13	2.21
49.0	22 000	17.6	4.79	3.10	15.6	3.78	2.67
53.5	24 000	19.2	5.70	3.67	17.0	4.50	3.16
57.9	26 000	20.7	6.69	4.29	18.4	5.28	3.71
62.4	28 000	22.3	7.76	4.96	19.9	6.13	4.32
66.8	30 000	23.9	8.91	5.68	21.3	7.03	4.97
71.3	32 000	25.5	10.13	6.42	22.7	8.00	5.65
75.8	34 000	27.1	11.4	7.22	24.1	9.04	6.35
80.2	36 000	28.7	12.8	8.08	25.5	10.13	7.10
84.7	38 000	30.3	14.3	9.00	26.9	11.3	7.90
89.1	40 000	31.9	15.8	9.98	28.4	12.5	8.75
93.6	42 000	33.5	17.5	11.0	29.8	13.8	9.63
98.0	44 000	35.1	19.2	12.1	31.2	15.1	10.5
102	46 000	36.7	20.9	13.2	32.6	16.5	11.5
107	48 000	38.3	22.8	14.3	34.0	18.0	12.5
111	50 000	39.9	24.7	15.5	35.5	19.5	13.6
123	55 000	43.9	29.9	18.7	39.0	23.6	16.4
134	60 000	47.9	35.6	22.3	42.6	28.1	19.5
145	65 000	51.9	41.8	26.2	46.1	33.0	22.9
156	70 000	55.9	48.5	30.4	49.6	38.3	26.5
167	75 000	59.8	55.7	34.8	53.2	44.0	30.5
178	80 000	63.8	63.3	39.4	56.7	50.0	34.7
189	85 000	67.8	71.5	44.4	60.3	56.5	39.2
201	90 000	71.8	80.1	49.7	63.8	63.3	43.9
212	95 000	75.8	89.3	55.5	67.4	70.5	48.9
223	100 000	79.8	98.9	61.5	70.9	78.2	54.2
245	110 000	87.8	120	74.0	78.0	94.6	65.5
267	120 000	95.8	142	88.0	85.1	113	78.0
290	130 000	103.7	167	103	92.2	132	91.5
312	140 000	112	194	119	99.3	153	106
334	150 000	120	223	137	106	200	121

IIIB-4 FRICTION LOSS FOR WATER IN FEET PER 100 FEET OF PIPE

TABLE 24

30 INCH OD		STEEL — SCHEDULE 20 ID = 29.000 INCHES ϵ/D = 0.0000621			ASPHALT-DIPPED CAST IRON ID = 30.00 INCHES ϵ/D = 0.000160		
DISCHARGE CFS	DISCHARGE GPM	V ft/sec	$V^2/2g$ feet	h_f feet per 100 feet of pipe	V ft/sec	$V^2/2g$ feet	h_f feet per 100 feet of pipe
0.891	400	0.194	0.000587	0.000540	0.182	0.000512	0.000466
1.114	500	0.243	0.000917	0.000805	0.227	0.000800	0.000695
1.34	600	0.291	0.00132	0.001115	0.272	0.00115	0.000964
1.56	700	0.340	0.00180	0.00147	0.318	0.00157	0.00128
1.78	800	0.389	0.00235	0.00187	0.363	0.00205	0.00163
2.01	900	0.437	0.00297	0.00231	0.408	0.00259	0.00202
2.23	1 000	0.486	0.00367	0.00280	0.454	0.00320	0.00244
2.67	1 200	0.583	0.00528	0.00390	0.545	0.00461	0.00343
3.12	1 400	0.680	0.00719	0.00514	0.635	0.00627	0.00452
3.56	1 600	0.777	0.00939	0.00652	0.726	0.00819	0.00577
4.01	1 800	0.874	0.0119	0.00814	0.817	0.0104	0.00720
4.46	2 000	0.971	0.0147	0.00986	0.908	0.0128	0.00876
5.57	2 500	1.21	0.0229	0.0148	1.13	0.0200	0.0132
6.68	3 000	1.46	0.0330	0.0206	1.36	0.0288	0.0186
7.80	3 500	1.70	0.0449	0.0276	1.59	0.0392	0.0248
8.91	4 000	1.94	0.0587	0.0354	1.82	0.0512	0.0320
10.02	4 500	2.19	0.0742	0.0440	2.04	0.0648	0.0400
11.14	5 000	2.43	0.0917	0.0535	2.27	0.0800	0.0488
13.4	6 000	2.91	0.132	0.0750	2.72	0.115	0.0690
15.6	7 000	3.40	0.180	0.100	3.18	0.157	0.0923
17.8	8 000	3.89	0.235	0.129	3.63	0.205	0.119
20.1	9 000	4.37	0.297	0.161	4.08	0.259	0.149
22.3	10 000	4.86	0.367	0.196	4.54	0.320	0.183
26.7	12 000	5.83	0.528	0.277	5.45	0.461	0.260
31.2	14 000	6.80	0.719	0.371	6.35	0.627	0.351
35.6	16 000	7.77	0.939	0.478	7.26	0.819	0.455
40.1	18 000	8.74	1.19	0.598	8.17	1.04	0.572
44.6	20 000	9.71	1.47	0.732	9.08	1.28	0.703
55.7	25 000	12.1	2.29	1.13	11.3	2.00	1.09
66.8	30 000	14.6	3.30	1.61	13.6	2.88	1.57
78.0	35 000	17.0	4.49	2.17	15.9	3.92	2.13
89.1	40 000	19.4	5.87	2.83	18.2	5.12	2.77
100	45 000	21.9	7.42	3.56	20.4	6.48	3.50
111	50 000	24.3	9.17	4.38	22.7	8.00	4.30
134	60 000	29.1	13.2	6.23	27.2	11.5	6.19
156	70 000	34.0	18.0	8.43	31.8	15.7	8.39
178	80 000	38.9	23.5	11.0	36.3	20.5	10.9
201	90 000	43.7	29.7	13.8	40.8	25.9	13.8
223	100 000	48.6	36.7	17.0	45.4	32.0	17.0
245	110 000	53.4	44.4	20.6	49.9	38.7	20.5
267	120 000	58.3	52.8	24.5	54.5	46.1	24.4
290	130 000	63.1	62.0	28.7	59.0	54.1	28.6
312	140 000	68.0	71.9	33.3	63.5	62.7	33.1
334	150 000	72.9	82.5	38.2	68.1	72.0	38.0
356	160 000	77.7	93.9	43.3	72.6	81.9	43.2
379	170 000	82.6	106	48.8	77.2	92.5	48.7
401	180 000	87.4	119	54.7	81.7	104	54.7
423	190 000	92.3	132	60.8	86.2	116	61.0
446	200 000	97.1	147	67.1	90.8	128	67.6
468	210 000	102	162	73.8	95.3	141	74.5
490	220 000	107	177	81.0	99.8	155	81.7
512	230 000	112	194	88.6	104	169	89.2
535	240 000	117	211	96.7	109	184	97.0
557	250 000	121	229	106	113	200	105

Fluid Flow
Friction Loss - Water

IIIB-4 FRICTION LOSS FOR WATER IN FEET PER 100 FEET OF PIPE

TABLE 25

36 INCH ID				STEEL $\epsilon/D = 0.0000500$	CAST IRON $\epsilon/D = 0.000133$
DISCHARGE CFS	GPM	V ft/sec	$V^2/2g$ feet	h_f feet per 100 feet of pipe	h_f feet per 100 feet of pipe
2.23	1 000	0.315	0.00154	0.000988	0.00101
2.67	1 200	0.378	0.00222	0.00137	0.00140
3.12	1 400	0.441	0.00303	0.00181	0.00186
3.56	1 600	0.504	0.00395	0.00231	0.00237
4.01	1 800	0.567	0.00500	0.00285	0.00295
4.46	2 000	0.630	0.00618	0.00344	0.00357
5.57	2 500	0.788	0.00965	0.00517	0.00538
6.68	3 000	0.946	0.0139	0.00721	0.00751
7.80	3 500	1.103	0.0189	0.00957	0.0101
8.91	4 000	1.26	0.0247	0.0122	0.0129
10.02	4 500	1.41	0.0313	0.0152	0.0161
11.14	5 000	1.58	0.0386	0.0185	0.0196
13.4	6 000	1.89	0.0556	0.0260	0.0276
15.6	7 000	2.21	0.0756	0.0345	0.0369
17.8	8 000	2.52	0.0988	0.0442	0.0475
20.1	9 000	2.84	0.125	0.0551	0.0593
22.3	10 000	3.15	0.154	0.0670	0.0724
26.7	12 000	3.78	0.222	0.0942	0.103
31.2	14 000	4.41	0.303	0.126	0.139
35.6	16 000	5.04	0.395	0.162	0.180
40.1	18 000	5.67	0.500	0.203	0.227
44.6	20 000	6.30	0.618	0.248	0.279
55.7	25 000	7.88	0.965	0.378	0.430
66.8	30 000	9.46	1.39	0.540	0.617
78.0	35 000	11.03	1.89	0.724	0.832
89.1	40 000	12.6	2.47	0.941	1.08
100	45 000	14.1	3.13	1.18	1.36
111	50 000	15.8	3.86	1.45	1.68
134	60 000	18.9	5.56	2.07	2.40
156	70 000	22.1	7.56	2.81	3.25
178	80 000	25.2	9.88	3.66	4.23
201	90 000	28.4	12.5	4.59	5.34
223	100 000	31.5	15.4	5.64	6.58
267	120 000	37.8	22.2	8.05	9.50
312	140 000	44.1	30.3	10.9	12.9
356	160 000	50.4	39.5	14.2	16.8
401	180 000	56.7	50.0	17.9	21.3
446	200 000	63.0	61.8	22.1	26.3
557	250 000	78.8	96.5	34.4	41.0
668	300 000	94.6	139	49.4	58.8
780	350 000	110	189	67.0	80.0
891	400 000	126	247	87.3	105

NOTE: No allowance has been made for age, differences in diameter, or any abnormal condition of interior surface. Any factor of safety must be estimated from the local conditions and the requirements of each particular installation.

IIIB-4 FRICTION LOSS FOR WATER IN FEET PER 100 FEET OF PIPE

TABLE 26

42 INCH ID				STEEL $\epsilon/D = 0.0000429$	CAST IRON $\epsilon/D = 0.000114$
DISCHARGE CFS	GPM	V ft/sec	$V^2/2g$ feet	h_f feet per 100 feet of pipe	h_f feet per 100 feet of pipe
2.23	1 000	0.232	0.000833	0.000471	0.000481
3.34	1 500	0.347	0.00187	0.000977	0.000997
4.46	2 000	0.463	0.00333	0.00164	0.00168
5.57	2 500	0.579	0.00521	0.00246	0.00252
6.68	3 000	0.695	0.00750	0.00343	0.00353
7.80	3 500	0.811	0.0102	0.00454	0.00470
8.91	4 000	0.926	0.0133	0.00580	0.00602
10.02	4 500	1.042	0.0169	0.00720	0.00750
11.14	5 000	1.16	0.0208	0.00874	0.00915
13.4	6 000	1.39	0.0300	0.0122	0.0128
15.6	7 000	1.62	0.0408	0.0162	0.0172
17.8	8 000	1.85	0.0533	0.0208	0.0222
20.1	9 000	2.08	0.0675	0.0258	0.0276
22.3	10 000	2.32	0.0833	0.0314	0.0337
26.7	12 000	2.78	0.120	0.0441	0.0477
31.2	14 000	3.24	0.163	0.0591	0.0641
35.6	16 000	3.71	0.213	0.0758	0.0829
40.1	18 000	4.17	0.270	0.0944	0.104
44.6	20 000	4.63	0.333	0.115	0.127
55.7	25 000	5.79	0.521	0.176	0.196
66.8	30 000	6.95	0.750	0.250	0.279
78.0	35 000	8.11	1.02	0.334	0.377
89.1	40 000	9.26	1.33	0.433	0.490
100	45 000	10.42	1.69	0.545	0.619
111	50 000	11.6	2.08	0.668	0.760
134	60 000	13.9	3.00	0.946	1.09
156	70 000	16.2	4.08	1.27	1.48
178	80 000	18.5	5.33	1.66	1.92
201	90 000	20.8	6.75	2.08	2.42
223	100 000	23.2	8.33	2.57	2.98
267	120 000	27.8	12.0	3.67	4.30
312	140 000	32.4	16.3	4.98	5.82
356	160 000	37.1	21.3	6.46	7.58
401	180 000	41.7	27.0	8.12	9.58
446	200 000	46.3	33.3	10.00	11.8
557	250 000	57.9	52.1	15.6	18.4
668	300 000	69.5	75.0	22.3	26.5
780	350 000	81.1	102	30.4	36.1
891	400 000	92.6	133	39.6	47.2
1002	450 000	104.2	169	50.1	59.7
1114	500 000	116	208	67.7	73.6

NOTE: No allowance has been made for age, differences in diameter, or any abnormal condition of interior surface. Any factor of safety must be estimated from the local conditions and the requirements of each particular installation.

IIIB-4 FRICTION LOSS FOR WATER IN FEET PER 100 FEET OF PIPE

TABLE 27

48 INCH ID				STEEL $\epsilon/D = 0.0000375$	CAST IRON $\epsilon/D = 0.000100$
DISCHARGE CFS	DISCHARGE GPM	V ft/sec	$V^2/2g$ feet	h_f feet per 100 feet of pipe	h_f feet per 100 feet of pipe
3.34	1 500	0.266	0.00110	0.000508	0.000521
4.46	2 000	0.355	0.00195	0.000855	0.000883
5.57	2 500	0.443	0.00305	0.00129	0.00133
6.68	3 000	0.532	0.00440	0.00180	0.00185
7.80	3 500	0.621	0.00598	0.00238	0.00245
8.91	4 000	0.709	0.00782	0.00304	0.00314
10.02	4 500	0.798	0.00989	0.00378	0.00391
11.14	5 000	0.887	0.01221	0.00458	0.00474
13.4	6 000	1.064	0.0176	0.00636	0.00667
15.6	7 000	1.24	0.0239	0.00844	0.00890
17.8	8 000	1.42	0.0313	0.0108	0.0114
20.1	9 000	1.60	0.0396	0.0134	0.0142
22.3	10 000	1.77	0.0489	0.0163	0.0173
26.7	12 000	2.13	0.0703	0.0229	0.0244
31.2	14 000	2.48	0.0957	0.0305	0.0327
35.6	16 000	2.84	0.125	0.0391	0.0422
40.1	18 000	3.19	0.158	0.0488	0.0529
44.6	20 000	3.55	0.195	0.0598	0.0648
55.7	25 000	4.43	0.305	0.0910	0.0996
66.8	30 000	5.32	0.440	0.128	0.142
78.0	35 000	6.21	0.598	0.172	0.192
89.1	40 000	7.09	0.782	0.222	0.248
100.2	45 000	7.98	0.989	0.278	0.314
111.4	50 000	8.87	1.221	0.341	0.384
134	60 000	10.64	1.76	0.484	0.548
156	70 000	12.4	2.39	0.652	0.742
178	80 000	14.2	3.13	0.849	0.968
201	90 000	16.0	3.96	1.06	1.22
223	100 000	17.7	4.89	1.30	1.50
267	120 000	21.3	7.03	1.87	2.15
312	140 000	24.8	9.57	2.51	2.92
356	160 000	28.4	12.5	3.26	3.81
401	180 000	31.9	15.8	4.11	4.83
446	200 000	35.5	19.5	5.05	5.97
557	250 000	44.3	30.5	7.88	9.28
668	300 000	53.2	44.0	11.3	13.4
780	350 000	62.1	59.8	15.3	18.2
891	400 000	70.9	78.2	20.0	23.7
1002	450 000	79.8	98.9	25.2	29.9
1114	500 000	88.7	122.1	31.1	36.8
1225	550 000	97.5	148	37.6	44.5
1337	600 000	106.4	176	44.7	53.0

NOTE: No allowance has been made for age, differences in diameter, or any abnormal condition of interior surface. Any factor of safety must be estimated from the local conditions and the requirements of each particular installation.

IIIB-4 FRICTION LOSS FOR WATER IN FEET PER 100 FEET OF PIPE

TABLE 28

54 INCH ID				STEEL $\epsilon/D = 0.0000333$	CAST IRON $\epsilon/D = 0.0000889$
DISCHARGE CFS	DISCHARGE GPM	V ft/sec	$V^2/2g$ feet	h_f feet per 100 feet of pipe	h_f feet per 100 feet of pipe
4.46	2 000	0.280	0.00122	0.000488	0.000498
5.57	2 500	0.350	0.00191	0.000733	0.000748
6.68	3 000	0.420	0.00274	0.00102	0.00104
7.80	3 500	0.490	0.00374	0.00134	0.00138
8.91	4 000	0.560	0.00488	0.00172	0.00177
10.02	4 500	0.630	0.00618	0.00213	0.00220
11.14	5 000	0.700	0.00762	0.00257	0.00266
13.4	6 000	0.840	0.01098	0.00358	0.00373
15.6	7 000	0.981	0.0149	0.00476	0.00499
17.8	8 000	1.121	0.0195	0.00610	0.00642
20.1	9 000	1.26	0.0247	0.00760	0.00800
22.3	10 000	1.40	0.0305	0.00920	0.00970
26.7	12 000	1.68	0.0439	0.0129	0.0137
31.2	14 000	1.96	0.0598	0.0171	0.0182
35.6	16 000	2.24	0.0781	0.0219	0.0234
40.1	18 000	2.52	0.0988	0.0273	0.0294
44.6	20 000	2.80	0.122	0.0333	0.0360
55.7	25 000	3.50	0.191	0.0504	0.0550
66.8	30 000	4.20	0.274	0.0713	0.0782
78.0	35 000	4.90	0.374	0.0958	0.106
89.1	40 000	5.60	0.488	0.124	0.137
100.2	45 000	6.30	0.618	0.155	0.172
111.4	50 000	7.00	0.762	0.189	0.211
134	60 000	8.40	1.098	0.267	0.301
156	70 000	9.81	1.49	0.358	0.408
178	80 000	11.21	1.95	0.465	0.530
201	90 000	12.6	2.47	0.586	0.668
223	100 000	14.0	3.05	0.715	0.820
267	120 000	16.8	4.39	1.02	1.18
312	140 000	19.6	5.98	1.38	1.59
356	160 000	22.4	7.81	1.80	2.07
401	180 000	25.2	9.88	2.26	2.62
446	200 000	28.0	12.2	2.77	3.22
557	250 000	35.0	19.1	4.32	5.01
668	300 000	42.0	27.4	6.19	7.23
780	350 000	49.0	37.4	8.40	9.82
891	400 000	56.0	48.8	11.00	12.8
1002	450 000	63.0	61.8	13.9	16.2
1114	500 000	70.0	76.2	17.0	20.0
1225	550 000	77.0	92.3	20.6	24.2
1337	600 000	84.0	109.8	24.5	28.8
1448	650 000	98.1	129	28.7	33.7
1560	700 000	112.1	149	33.2	39.0

NOTE: No allowance has been made for age, differences in diameter, or any abnormal condition of interior surface. Any factor of safety must be estimated from the local conditions and the requirements of each particular installation.

IIIB-4 FRICTION LOSS FOR WATER IN FEET PER 100 FEET OF PIPE

TABLE 29

60 INCH ID				STEEL $\epsilon/D = 0.0000300$	CAST IRON $\epsilon/D = 0.0000800$
DISCHARGE CFS	DISCHARGE GPM	V ft/sec	$V^2/2g$ feet	h_f feet per 100 feet of pipe	h_f feet per 100 feet of pipe
4.46	2 000	0.227	0.000800	0.000293	0.000298
5.57	2 500	0.284	0.00125	0.000440	0.000446
6.68	3 000	0.340	0.00180	0.000612	0.000621
7.80	3 500	0.397	0.00245	0.000810	0.000824
8.91	4 000	0.454	0.00320	0.00103	0.00105
10.02	4 500	0.511	0.00405	0.00128	0.00131
11.14	5 000	0.567	0.00500	0.00155	0.00159
13.4	6 000	0.681	0.00720	0.00216	0.00223
15.6	7 000	0.794	0.00980	0.00285	0.00297
17.8	8 000	0.908	0.0128	0.00365	0.00382
20.1	9 000	1.021	0.0162	0.00454	0.00476
22.3	10 000	1.13	0.0200	0.00550	0.00579
26.7	12 000	1.36	0.0288	0.00766	0.00815
31.2	14 000	1.59	0.0392	0.0102	0.0108
35.6	16 000	1.82	0.0512	0.0131	0.0140
40.1	18 000	2.04	0.0648	0.0163	0.0174
44.6	20 000	2.27	0.0800	0.0198	0.0212
55.7	25 000	2.84	0.125	0.0301	0.0325
66.8	30 000	3.40	0.180	0.0424	0.0460
78.0	35 000	3.97	0.245	0.0567	0.0618
89.1	40 000	4.54	0.320	0.0730	0.0800
100.2	45 000	5.11	0.405	0.0916	0.100
111.4	50 000	5.67	0.500	0.112	0.124
134	60 000	6.81	0.720	0.158	0.176
156	70 000	7.94	0.980	0.213	0.237
178	80 000	9.08	1.28	0.275	0.307
201	90 000	10.21	1.62	0.344	0.387
223	100 000	11.3	2.00	0.420	0.478
267	120 000	13.6	2.88	0.600	0.688
312	140 000	15.9	3.92	0.806	0.930
356	160 000	18.2	5.12	1.04	1.20
401	180 000	20.4	6.48	1.32	1.52
446	200 000	22.7	8.00	1.62	1.87
557	250 000	28.4	12.5	2.52	2.92
668	300 000	34.0	18.0	3.60	4.20
780	350 000	39.7	24.5	4.88	5.71
891	400 000	45.4	32.0	6.34	7.42
1002	450 000	51.1	40.5	8.01	9.40
1114	500 000	56.7	50.0	9.87	11.6
1225	550 000	62.4	60.5	11.9	14.0
1337	600 000	68.1	72.0	14.1	16.7
1448	650 000	73.8	84.5	16.6	19.6
1560	700 000	79.4	98.0	19.2	22.7
1671	750 000	85.1	112.6	22.0	26.0
1782	800 000	90.8	128	25.0	29.6

NOTE: No allowance has been made for age, differences in diameter, or any abnormal condition of interior surface. Any factor of safety must be estimated from the local conditions and the requirements of each particular installation.

IIIB-4 FRICTION LOSS FOR WATER IN FEET PER 100 FEET OF PIPE

TABLE 30

72 INCH ID				STEEL $\epsilon/D = 0.0000250$	CAST IRON $\epsilon/D = 0.000066$
DISCHARGE CFS	DISCHARGE GPM	V ft/sec	$V^2/2g$ feet	h_f feet per 100 feet of pipe	h_f feet per 100 feet of pipe
4.46	2 000	0.158	0.000387	0.000123	0.000125
5.57	2 500	0.197	0.000604	0.000183	0.000186
6.68	3 000	0.237	0.000870	0.000254	0.000258
7.80	3 500	0.276	0.00118	0.000336	0.000342
8.91	4 000	0.316	0.00155	0.000427	0.000435
10.02	4 500	0.355	0.00196	0.000530	0.000540
11.14	5 000	0.394	0.00242	0.000640	0.000656
13.4	6 000	0.473	0.00348	0.000890	0.000913
15.6	7 000	0.552	0.00474	0.00118	0.00121
17.8	8 000	0.631	0.00619	0.00150	0.00155
20.1	9 000	0.710	0.00783	0.00186	0.00193
22.3	10 000	0.789	0.00967	0.00227	0.00235
26.7	12 000	0.947	0.0139	0.00313	0.00329
31.2	14 000	1.104	0.0189	0.00418	0.00438
35.6	16 000	1.26	0.0248	0.00538	0.00563
40.1	18 000	1.42	0.0313	0.00673	0.00703
44.6	20 000	1.58	0.0387	0.00822	0.00859
55.7	25 000	1.97	0.0604	0.0124	0.0131
66.8	30 000	2.37	0.0870	0.0173	0.0184
78.0	35 000	2.76	0.118	0.0231	0.0248
89.1	40 000	3.16	0.155	0.0297	0.0320
100.2	45 000	3.55	0.196	0.0370	0.0402
111.4	50 000	3.94	0.242	0.0450	0.0493
134	60 000	4.73	0.348	0.0637	0.0700
156	70 000	5.52	0.474	0.0850	0.0940
178	80 000	6.31	0.619	0.110	0.122
201	90 000	7.10	0.783	0.138	0.154
223	100 000	7.89	0.967	0.168	0.189
267	120 000	9.47	1.39	0.237	0.271
312	140 000	11.04	1.89	0.321	0.365
356	160 000	12.6	2.48	0.414	0.475
401	180 000	14.2	3.13	0.522	0.600
446	200 000	15.8	3.87	0.642	0.736
557	250 000	19.7	6.04	1.00	1.14
668	300 000	23.7	8.70	1.42	1.64
780	350 000	27.6	11.8	1.92	2.23
891	400 000	31.6	15.5	2.50	2.90
1002	450 000	35.5	19.6	3.16	3.66
1114	500 000	39.4	24.2	3.88	4.52
1225	550 000	43.4	29.2	4.69	5.47
1337	600 000	47.3	34.8	5.56	6.50
1448	650 000	51.3	40.8	6.52	7.61
1560	700 000	55.2	47.4	7.56	8.81
1671	750 000	59.2	54.4	8.67	10.1
1782	800 000	63.1	61.9	9.83	11.5
1894	850 000	67.0	69.9	11.1	13.0
2005	900 000	71.0	78.3	12.4	14.6

NOTE: No allowance has been made for age, differences in diameter, or any abnormal condition of interior surface. Any factor of safety must be estimated from the local conditions and the requirements of each particular installation.

IIIB-4 FRICTION LOSS FOR WATER IN FEET PER 100 FEET OF PIPE

TABLE 31

84 INCH ID				STEEL $\epsilon/D = 0.0000214$	CAST IRON $\epsilon/D = 0.0000571$
DISCHARGE CFS	GPM	V ft/sec	$V^2/2g$ feet	h_f feet per 100 feet of pipe	h_f feet per 100 feet of pipe
6.68	3 000	0.174	0.000469	0.000121	0.000122
8.91	4 000	0.232	0.000833	0.000203	0.000206
11.14	5 000	0.289	0.00130	0.000306	0.000309
13.4	6 000	0.347	0.00188	0.000425	0.000432
15.6	7 000	0.405	0.00255	0.000562	0.000573
17.8	8 000	0.463	0.00333	0.000717	0.000731
20.1	9 000	0.521	0.00422	0.000891	0.000910
22.3	10 000	0.579	0.00521	0.00108	0.00110
26.7	12 000	0.695	0.00750	0.00150	0.00154
31.2	14 000	0.811	0.01021	0.00199	0.00205
35.6	16 000	0.926	0.0133	0.00255	0.00262
40.1	18 000	1.042	0.0169	0.00316	0.00327
44.6	20 000	1.16	0.0208	0.00384	0.00400
55.7	25 000	1.45	0.0326	0.00579	0.00606
66.8	30 000	1.74	0.0469	0.00810	0.00858
78.0	35 000	2.03	0.0638	0.0108	0.0115
89.1	40 000	2.32	0.0833	0.0139	0.0148
100.2	45 000	2.61	0.105	0.0174	0.0185
111.4	50 000	2.89	0.130	0.0212	0.0226
134	60 000	3.47	0.180	0.0298	0.0321
156	70 000	4.05	0.255	0.0398	0.0431
178	80 000	4.63	0.333	0.0513	0.0558
201	90 000	5.21	0.422	0.0640	0.0700
223	100 000	5.79	0.521	0.0781	0.0660
267	120 000	6.95	0.750	0.111	0.122
312	140 000	8.11	1.021	0.149	0.166
356	160 000	9.26	1.33	0.193	0.216
401	180 000	10.42	1.69	0.242	0.272
446	200 000	11.6	2.08	0.297	0.334
557	250 000	14.5	3.26	0.458	0.516
668	300 000	17.4	4.69	0.649	0.740
780	350 000	20.3	6.38	0.880	1.00
891	400 000	23.2	8.33	1.14	1.30
1002	450 000	26.1	10.5	1.44	1.65
1114	500 000	28.9	13.0	1.78	2.04
1225	550 000	31.8	15.8	2.14	2.47
1337	600 000	34.7	18.0	2.54	2.94
1448	650 000	37.6	22.0	2.97	3.45
1560	700 000	40.5	25.5	3.43	4.00
1671	750 000	43.4	29.3	3.93	4.58
1782	800 000	46.3	33.3	4.47	5.20
1894	850 000	49.2	37.6	5.04	5.87
2005	900 000	52.1	42.2	5.64	6.58
2117	950 000	55.0	47.0	6.29	7.32
2228	1 000 000	57.9	52.1	6.95	8.10

NOTE: No allowance has been made for age, differences in diameter, or any abnormal condition of interior surface. Any factor of safety must be estimated from the local conditions and the requirements of each particular installation.

IIIB-5 TYPICAL RESISTANCE COEFFICIENTS FOR VALVES AND FITTINGS

TABLE 32(a)

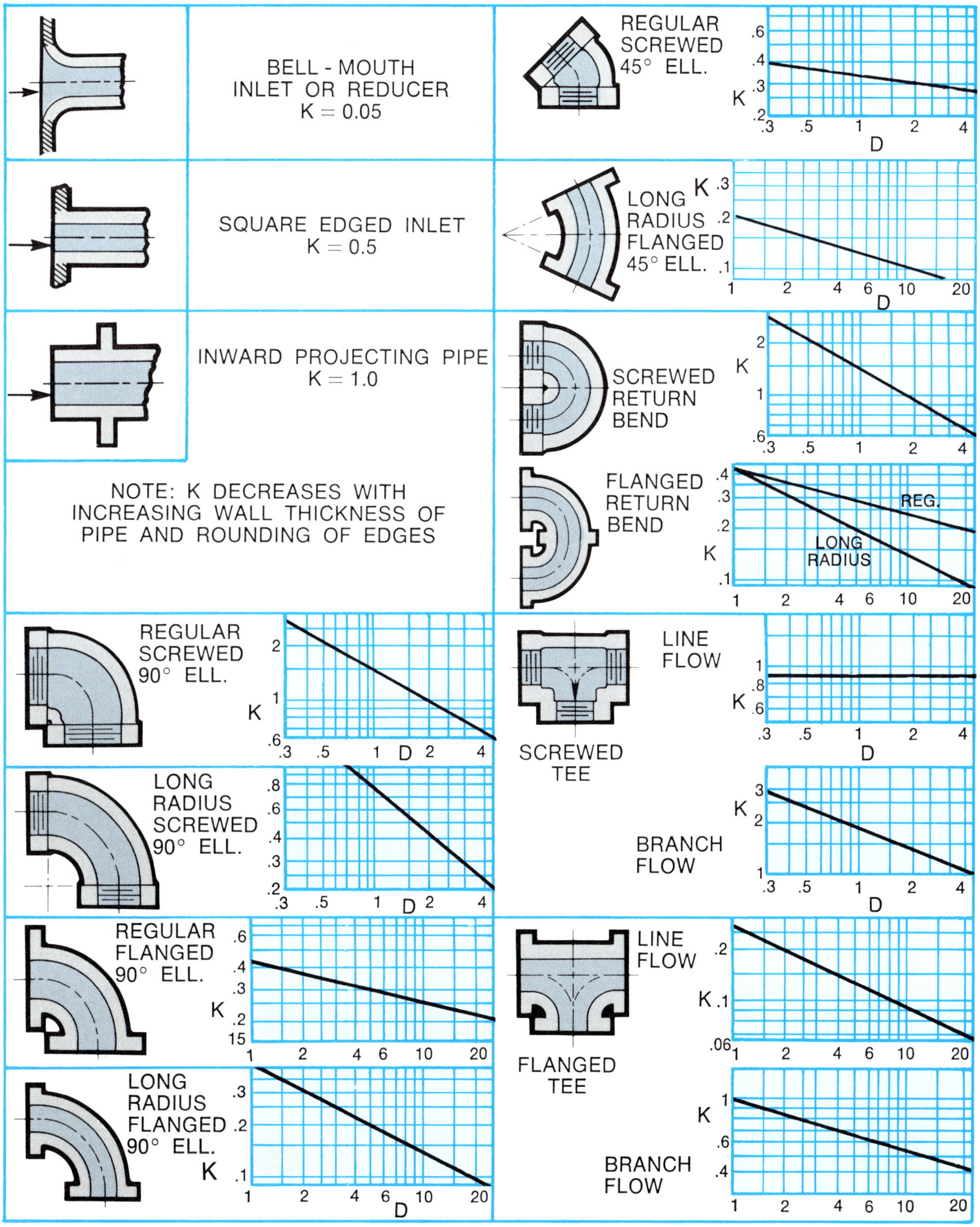

$$h = K \frac{V^2}{2g} \text{ FEET OF FLUID}$$

IIIB-5 TYPICAL RESISTANCE COEFFICIENTS FOR VALVES AND FITTINGS

TABLE 32(b)

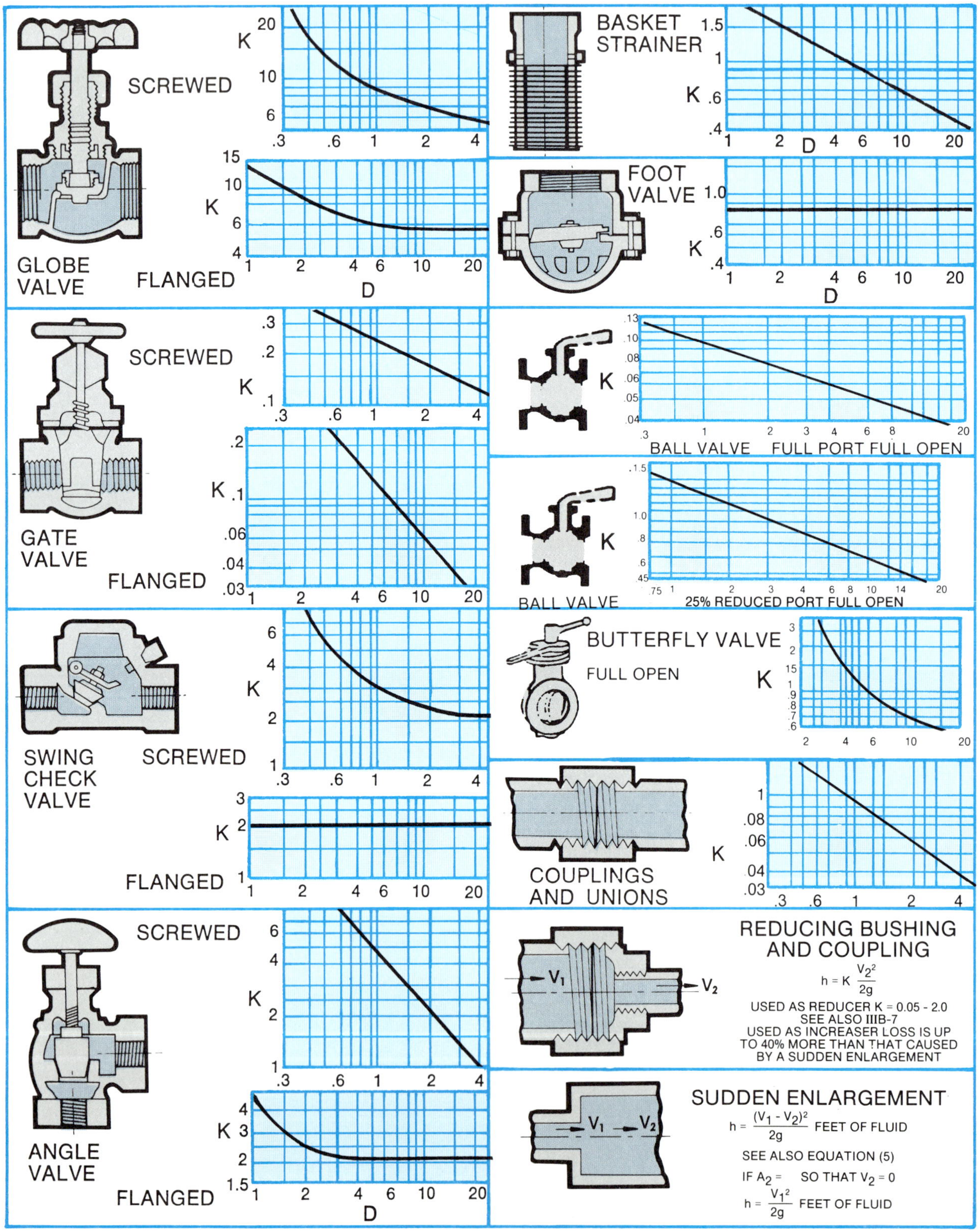

$$h = K \frac{V^2}{2g} \text{ FEET OF FLUID}$$

IIIB-5 RESISTANCE COEFFICIENTS FOR VALVES AND FITTINGS

TABLE 32(c)

Approximate Range of Variation For K		
Fitting		**Range of Variation**
90 Deg. Elbow	Regular Screwed	± 20 per cent above 2 inch size
	Regular Screwed	± 40 per cent below 2 inch size
	Long Radius, Screwed	± 25 per cent
	Regular Flanged	± 35 per cent
	Long Radius, Flanged	± 30 per cent
45 Deg. Elbow	Regular Screwed	± 10 per cent
	Long Radius, Flanged	± 10 per cent
180 Deg. Bend	Regular Screwed	± 25 per cent
	Regular Flanged	± 35 per cent
	Long Radius, Flanged	± 30 per cent
Tee	Screwed, Line or Branch Flow	± 25 per cent
	Flanged, Line or Branch Flow	± 35 per cent
Globe Valve	Screwed	± 25 per cent
	Flanged	± 25 per cent
Gate Valve	Screwed	± 25 per cent
	Flanged	± 50 per cent
Check Valve	Screwed	± 30 per cent
	Flanged	+ 200 per cent − 80 per cent
Sleeve Check Valve		Multiply flanged values by .2 to .5
Tilting Check Valve		Multiply flanged values by .13 to .19
Drainage Gate Check		Multiply flanged values by .03 to .07
Angle Valve	Screwed	± 20 per cent
	Flanged	± 50 per cent
Basket Strainer		± 50 per cent
Foot Valve		± 50 per cent
Couplings		± 50 per cent
Unions		± 50 per cent
Reducers		± 50 per cent

1. The value of D given in the charts is nominal IPS (Iron Pipe Size).
2. For velocities below 15 feet per second, check valves and foot valves will be only partially open and will exhibit higher values of K than that shown in the charts.

IIIB-5 RESISTANCE COEFFICIENTS FOR FITTINGS

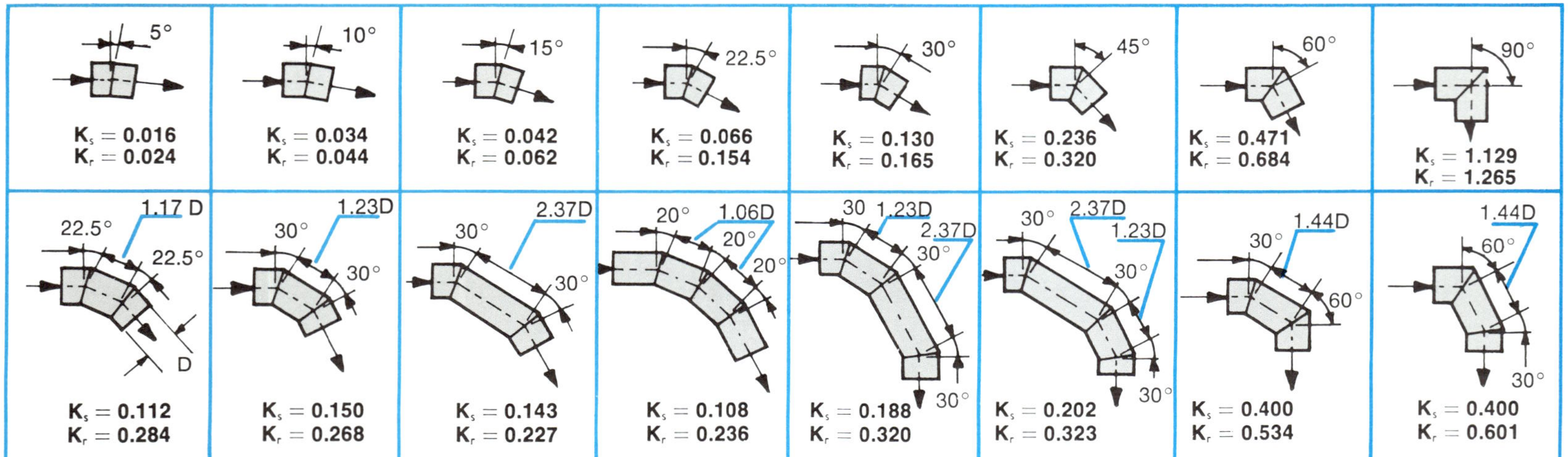

5°	10°	15°	22.5°	30°	45°	60°	90°
$K_s = 0.016$ $K_r = 0.024$	$K_s = 0.034$ $K_r = 0.044$	$K_s = 0.042$ $K_r = 0.062$	$K_s = 0.066$ $K_r = 0.154$	$K_s = 0.130$ $K_r = 0.165$	$K_s = 0.236$ $K_r = 0.320$	$K_s = 0.471$ $K_r = 0.684$	$K_s = 1.129$ $K_r = 1.265$
22.5°, 22.5°, 1.17 D $K_s = 0.112$ $K_r = 0.284$	30°, 30°, 1.23D $K_s = 0.150$ $K_r = 0.268$	30°, 30°, 2.37D $K_s = 0.143$ $K_r = 0.227$	20°, 20°, 20°, 1.06D $K_s = 0.108$ $K_r = 0.236$	30, 30°, 30°, 1.23D, 2.37D $K_s = 0.188$ $K_r = 0.320$	30°, 30°, 30°, 2.37D, 1.23D $K_s = 0.202$ $K_r = 0.323$	30°, 60°, 1.44D $K_s = 0.400$ $K_r = 0.534$	60°, 30°, 1.44D $K_s = 0.400$ $K_r = 0.601$

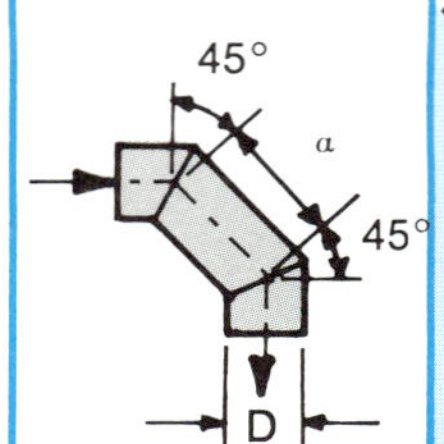

a/D	K_s	K_r
0.71	0.507	0.510
0.943	0.350	0.415
1.174	0.333	0.384
1.42	0.261	0.377
1.50*	0.280	0.376
1.86	0.289	0.390
2.56	0.356	0.429
3.14	0.346	0.426
3.72	0.356	0.460
4.89	0.389	0.455
5.59	0.392	0.444
6.28	0.399	0.444

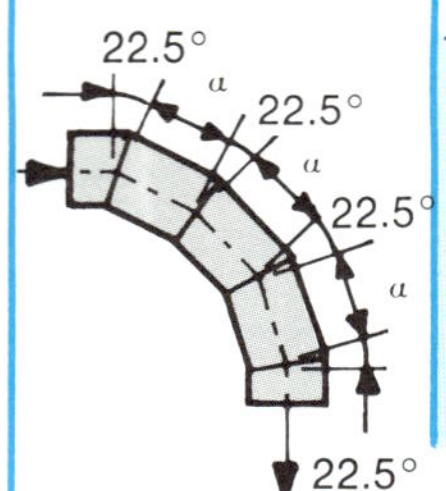

a/D	K_s	K_r
1.186	0.120	0.294
1.40	0.125	0.252
1.50*	—	0.250
1.63	0.124	0.266
1.86	0.117	0.272
2.325	0.096	0.317
2.40*	0.095	—
2.91	0.108	0.317
3.49	0.130	0.318
4.65	0.148	0.310
6.05	0.142	0.313

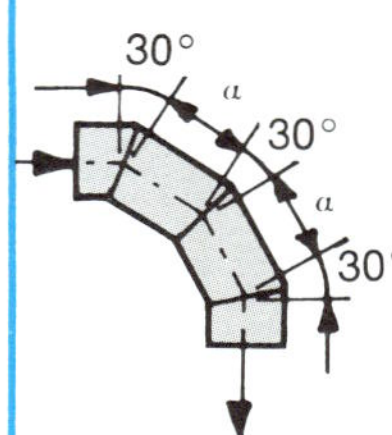

a/D	K_s	K_r
1.23	0.195	0.347
1.44	0.196	0.320
1.67	0.150	0.300
1.70*	0.149	0.299
1.91	0.154	0.312
2.37	0.167	0.337
2.96	0.172	0.342
4.11	0.190	0.354
4.70	0.192	0.360
6.10	0.201	0.360

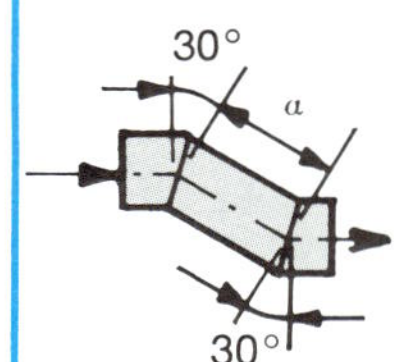

a/D	K_s	K_r
1.23	0.157	0.300
1.67	0.156	0.378
2.37	0.143	0.264
3.77	0.160	0.242

K_s = RESISTANCE COEFFICIENT FOR SMOOTH SURFACE
K_r = RESISTANCE COEFFICIENT FOR ROUGH SURFACE, $\frac{\epsilon}{D} \cong 0.0022$

*OPTIMUM VALUE OF a INTERPOLATED

TABLE 33. RESISTANCE COEFFICIENTS FOR MITER BENDS AT REYNOLDS NUMBER $\cong 2.25 \times 10^5$

IIIB-5A RESISTANCE COEFFICIENTS FOR 90 DEGREE BENDS OF UNIFORM DIAMETER

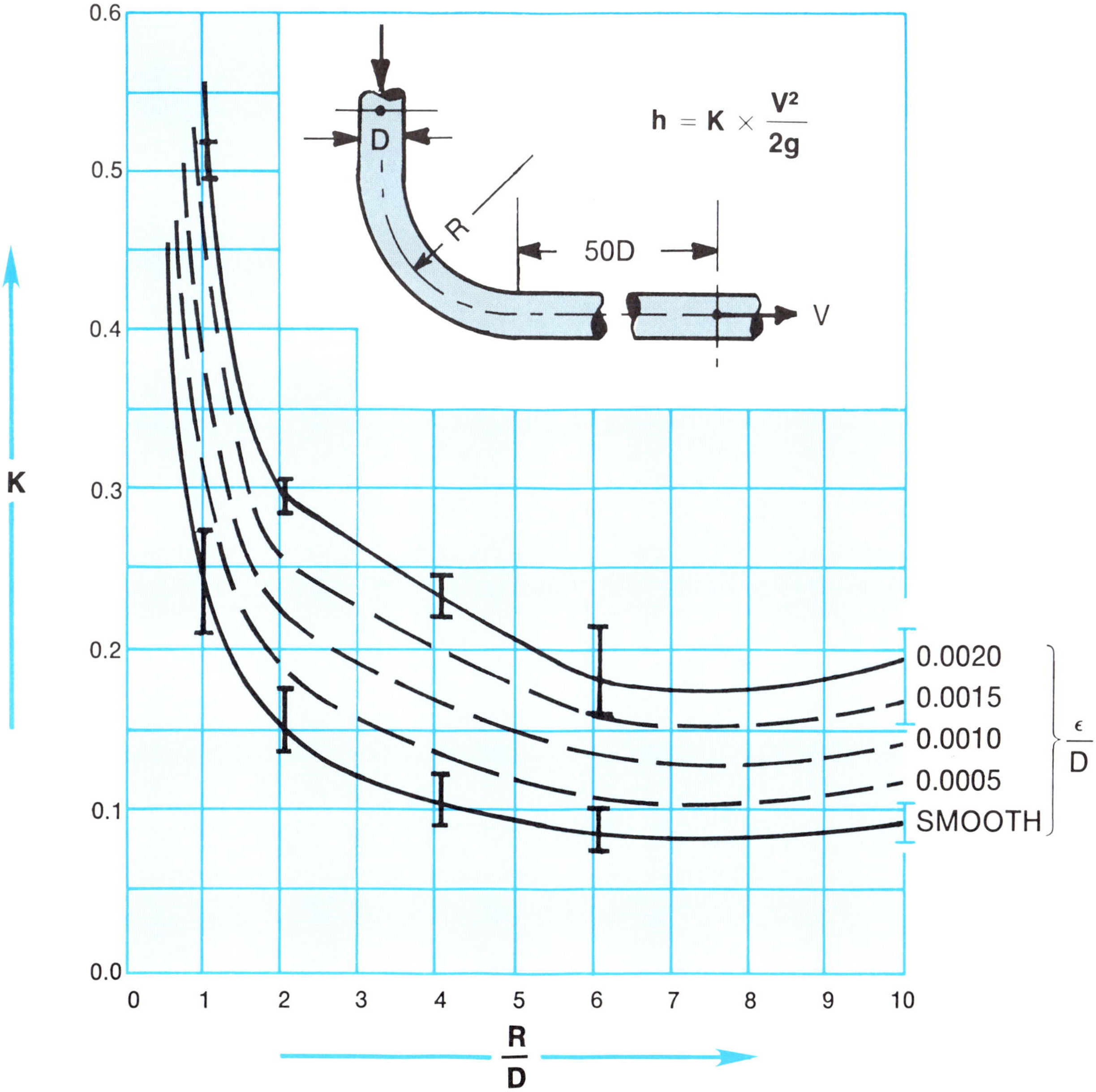

IIIB-5B RESISTANCE COEFFICIENTS FOR BENDS OF UNIFORM DIAMETER AND SMOOTH SURFACE AT REYNOLDS No. $\cong 2.25 \times 10^5$

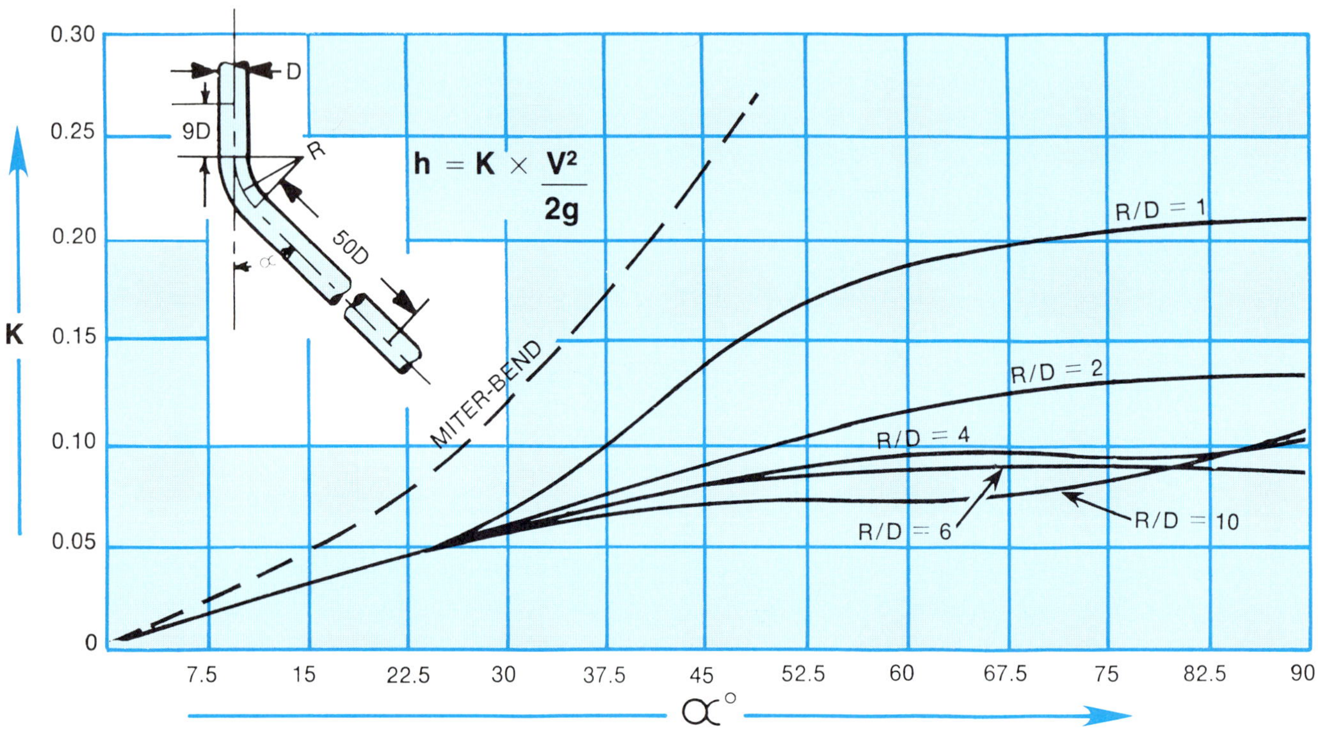

Fluid Flow

Friction Loss - Water

IIIB-6 RESISTANCE COEFFICIENTS FOR INCREASERS AND DIFFUSERS

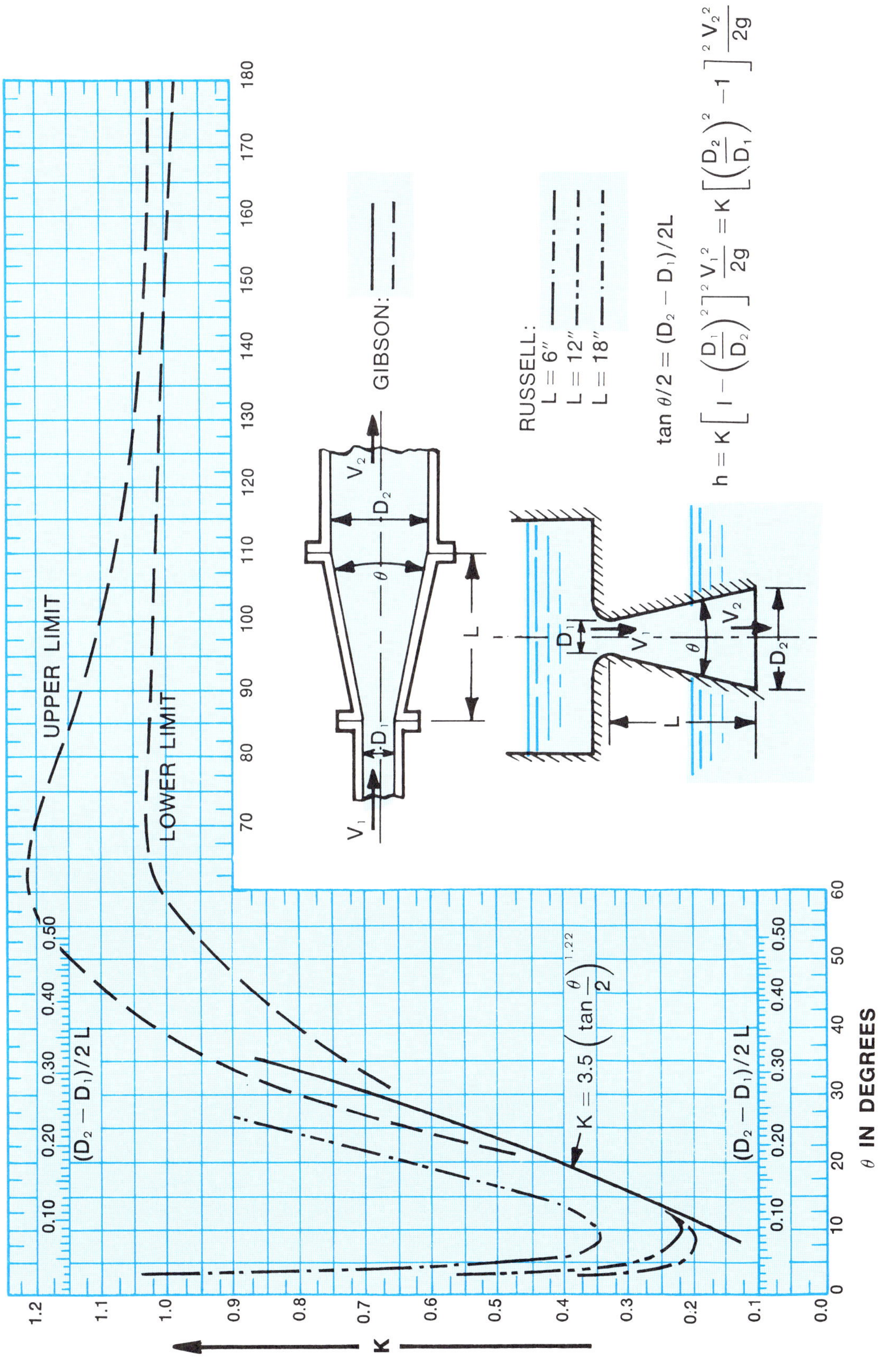

IIIB-7 RESISTANCE COEFFICIENTS FOR REDUCERS

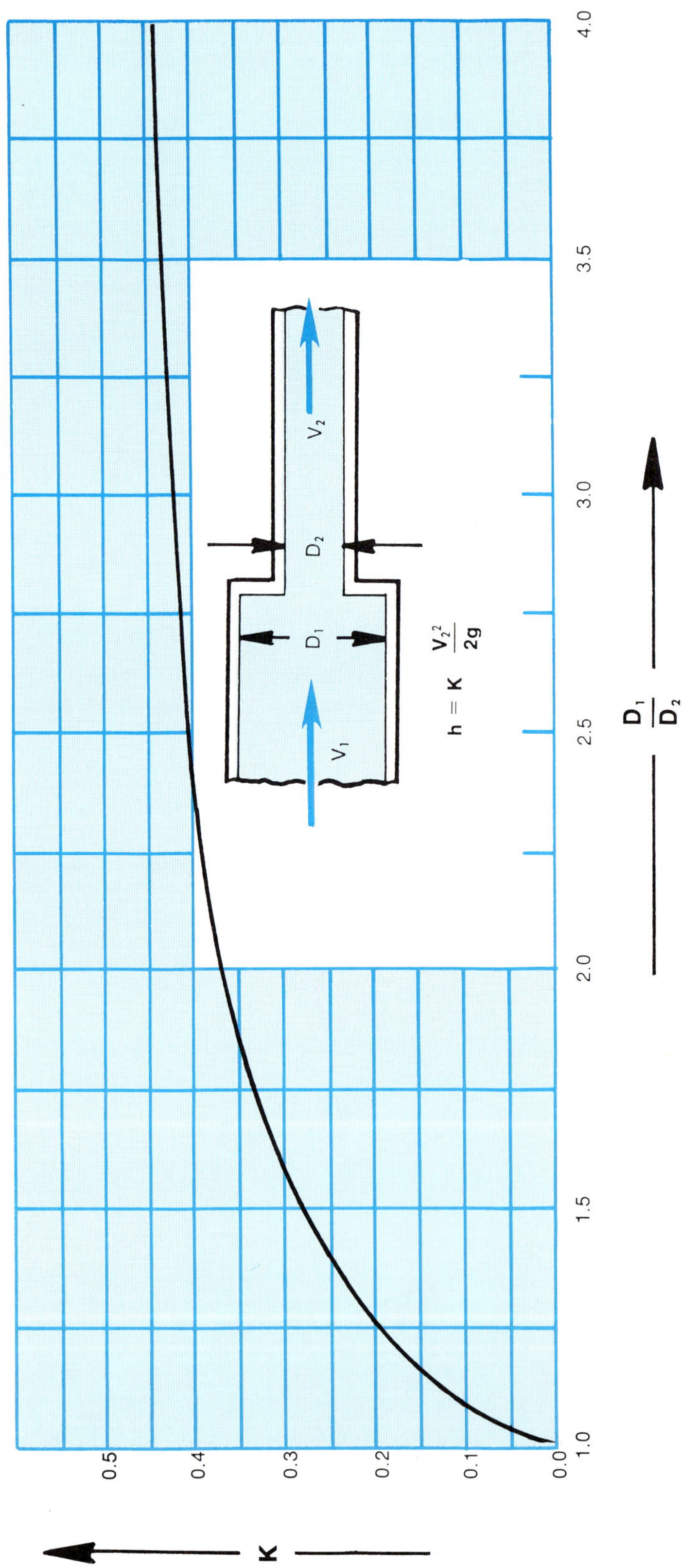

Fluid Flow

Friction Loss - Other Fluids

IIIC —CALCULATION OF FRICTION LOSS FOR ANY FLUID IN STEEL OR WROUGHT IRON PIPES

The pipe friction charts, Figs. IIIC-2 thru IIIC-13 incl., show friction-loss moduli for the incompressible flow of viscous fluids, including water, in several sizes of new clean steel or wrought iron pipes having Schedule 40 wall thickness. Each chart covers the losses for a single size of pipe based on the kinematic viscosity in centistokes (cs). As viscosity is frequently given in Seconds Saybolt Universal, (SSu), corresponding rounded values at 100° F are shown except in the case of low viscositites where no accurate SSU equivalents exist. Further information on viscosity and commonly accepted viscosity values for liquids will be found in Section II-C. No allowance has been made for abnormal conditions of interior surface or installation nor for the deterioration with age. (See Section IIIA.)

Friction loss moduli for laminar flow are shown by the 45-degree lines in the upper left hand portion of each chart. Moduli for turbulent flow are shown by the steeper curves in the lower right hand portion. Both of these regions represent stable states of flow. A diagonal line separates the regions of laminar and turbulent flow and represents the critical zone, a region in which it is difficult to predict the state of flow and hence, the friction loss. The critical zone usually represents a region of unstable flow. The critical zone line gives approximate moduli on the high side for this region of unstable flow.

The bottom scale of each chart represents flow in gallons per minute, gpm. An auxiliary top scale shows the average velocity in the pipe in feet per second. Read vertically from the gpm scale to find the corresponding velocity in feet per second. The vertical scales, labeled "Friction Loss Modulus for 100 Feet of Pipe", represent values of the ratio

$$\mathbf{M} = \frac{\Delta \mathbf{p}}{\mathbf{SG}} \qquad (10)$$

wherein

M = Friction loss modulus for 100 feet of pipe

Δp = Pressure loss in pounds per square inch per 100 feet of pipe

SG = Specific gravity of fluid at 60° F

The loss due to pipe friction may be obtained as follows:

$$\Delta \mathbf{p} = \mathbf{M} \times \mathbf{SG} \qquad (11)$$

and

$$\mathbf{h_f} = \mathbf{2.31\ M} \qquad (12)$$

wherein

h_f = Friction head loss in feet of fluid per 100 feet of pipe

The other quantities are as listed under Equation (10).

To use the charts, proceed as follows:

(a) Select the chart for the size of pipe in question.

(b) Follow the vertical line representing the flow in gpm to its intersection with the desired viscosity curve, and read the modulus at the left.

(c) If the vertical line representing the flow in gpm does not intersect the viscosity line in either turbulent or laminar flow, use the intersection with the critical zone line.

(d) Compute the friction loss in pressure drop or head, as desired, from Equations (11) or (12) above. These equations are repeated on each chart.

The chart for ½-inch pipe, Fig. IIIC-2, may be used to illustrate the method for finding the friction modulus for a flow of 10 gpm at viscosities of 50 SSU, 150 SSU and 300 SSU. For 50 SSU, the modulus value is 56. For 300 SSU, the modulus value is 119. The 10 gpm line does not intersect the 150 SSU line in either the laminar or turbulent region. Therefore, using the intersection of the 10 gpm line with the critical zone line, the modulus value is 68.

The range of the *laminar flow region* of each chart may be indefinitely increased since the friction loss modulus varies directly with the viscosity (above 1000 SSU) and the rate of flow. This does *not* apply to the turbulent flow region. For example, let it be required to find the friction modulus for the flow of a fluid of 20,000 SSU viscosity at a rate of 0.10 gpm in new ½-inch steel pipe, Fig. IIIC-1. The modulus for a flow of 1.0 gpm and 10,000 SSU is 395. The required modulus will be

$$M = 395 \times \frac{20{,}000}{10{,}000} \times \frac{0.10}{1.0} = 79$$

An example illustrating the use of the charts for

the solution of a practical problem follows:

Example 4. Use of Charts for Friction Loss of Any Fluid. A pump is required to take oil from a vented tank and deliver it at a rate of 50 gpm to a distributing manifold as shown in Fig. IIIC-14. The suction line is 20 feet long including fittings. The discharge pipe is 1½-inch new steel, Schedule 40, 300 feet long including fittings. The manifold is at the same elevation as the centerline of the pump and its pressure must be maintained at 250 psi. The oil has a viscosity of 1000 SSU and a gravity of 25 degrees API. The minimum desired suction pressure at the pump inlet is 15 inches of mercury vacuum and it is required that new steel, Schedule 40 pipe be used, preferably size 1½-inch. It is required to find the correct size for the suction pipe so as not to exceed 15 inches of mercury vacuum at the pump inlet and the pressure in pounds per square inch at the discharge flange of the pump.

SOLUTION: **a. Suction Line.** The specific gravity of the oil may be obtained from Section IIB and is equal to 0.9042.

Mercury weighs 0.4898 lb/cu in. at 60° F so that the maximum desired suction lift is $15 \times 0.4898 = 7.35$ psi. The static lift at the minimum tank level corresponds to

$$10\left[\frac{0.904 \times 62.34}{144}\right] = 3.91 \text{ psi}$$

The maximum allowable pressure drop in the suction pipe will be $7.35 - 3.91 = 3.44$ psi. The required friction loss modulus, by Equation (11) and neglecting the pipe fittings, will be

$$\mathbf{M} = \frac{\Delta \mathbf{p}}{\mathbf{SG}} \times \frac{\mathbf{100}}{\textbf{Actual Length of Pipe}}$$

$$= \frac{3.44}{0.904} \times \frac{100}{20}$$

$$= 19.0 \text{ for 100 feet of pipe}$$

The size of the suction pipe may be determined by selecting a size from the charts (Figs. IIIC-1–13 incl.) such that the intersection of the 50 gpm line with the 1000 SSU curve is at a modulus value equal to or less than 19.0. The intersection on Fig. III-5 for 1½-inch pipe is at a modulus value 44 which is too high. The intersection on Fig. IIIC-6 for 2-inch pipe is at a modulus value 16 which is satisfactory. Furthermore, the intersection is in the upper left hand portion of the chart showing that the flow will be laminar. A 2-inch pipe will be satisfactory for the suction line, however, judgment will have to be used to decide if local conditions require an additional allowance for aging of the pipe, valves and fittings.

An alternative solution is to use Table 34 to solve directly for pipe size by entering the table on page 99 on the line for 50 gal/min. and moving horizontally to the proper viscosity column. Under the column headed 1000 SSU it can be seen that the 1½-inch pipe which has a modulus of 43.8 is too small, but the modulus for the 2-inch pipe is 16.1, which is satisfactory. It can also be seen that if additional allowance for the age of the pipe must be made so that a modulus of 15.0 or less is required, it would be advisable to use a 2½-inch suction line, which has a modulus of only 7.9 and is satisfactory.

b. Discharge Line. The friction loss modulus for the 1½-inch pipe (Fig. IIIC-5) is 44 and the intersection shows the flow to be laminar. No additional allowance for the fittings will be made here. The pressure loss in the discharge pipe will be

$$\Delta p = 44 \times 0.9042 \times \frac{300}{100} \cong 119 \text{ psi}$$

The pressure loss in the discharge line could also have been obtained from Table 34. The modulus for 50 gal/min. of oil with a viscosity of 1000 SSU in an 1½-inch pipe is 43.8, which is approximately the same as obtained from Figure IIIC-5.

Therefore, the pressure at the pump discharge flange will be

$$P = 119 + 250 = 369 \text{ psi}$$

IIIC-14 INSTALLATION FOR EXAMPLE No. 4

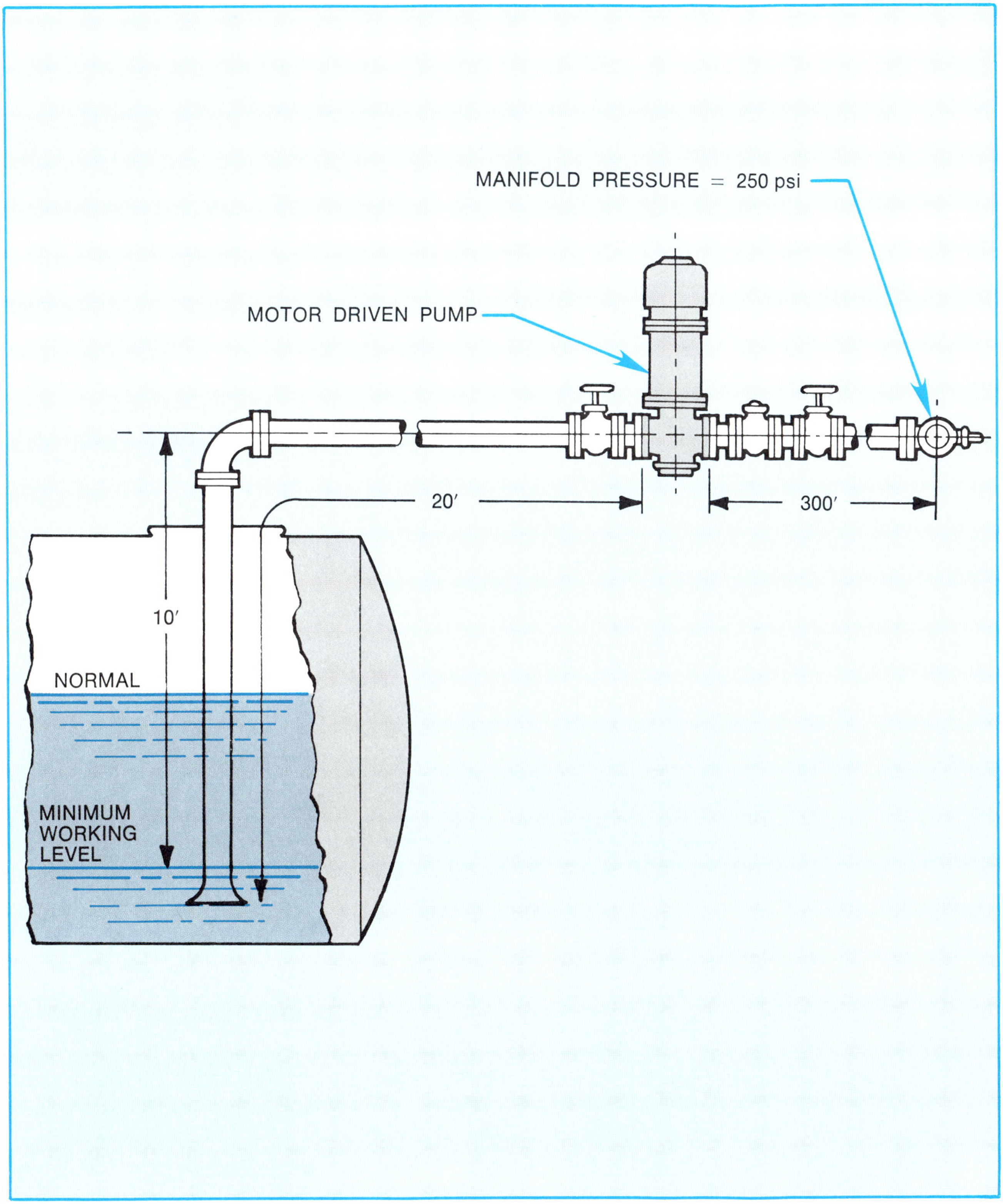

Fluid Flow
Friction Loss - Other Fluids

IIIC-1 FRICTION LOSS MODULUS FOR 100 FEET OF PIPE

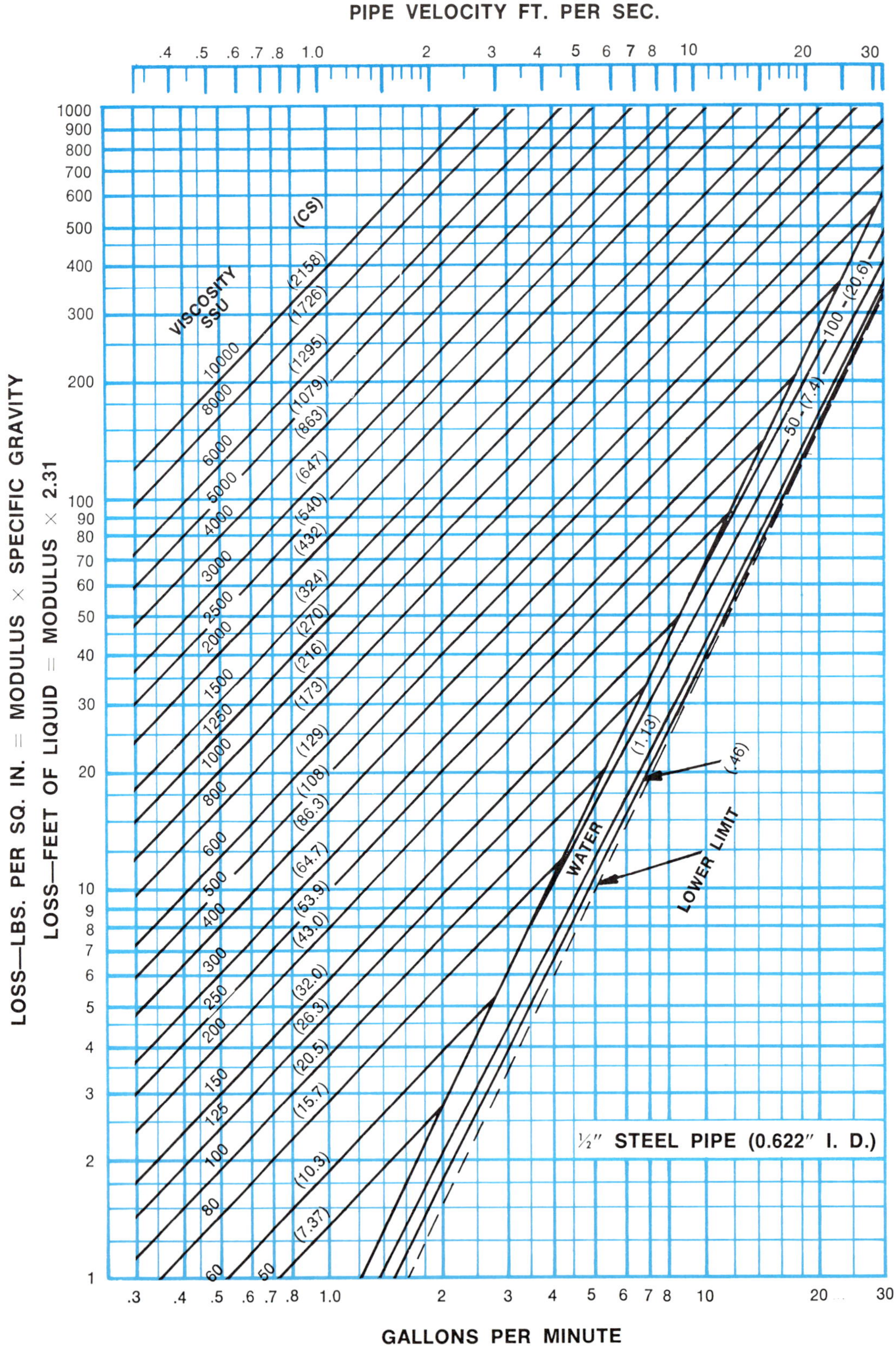

IIIC-2 FRICTION LOSS MODULUS FOR 100 FEET OF PIPE

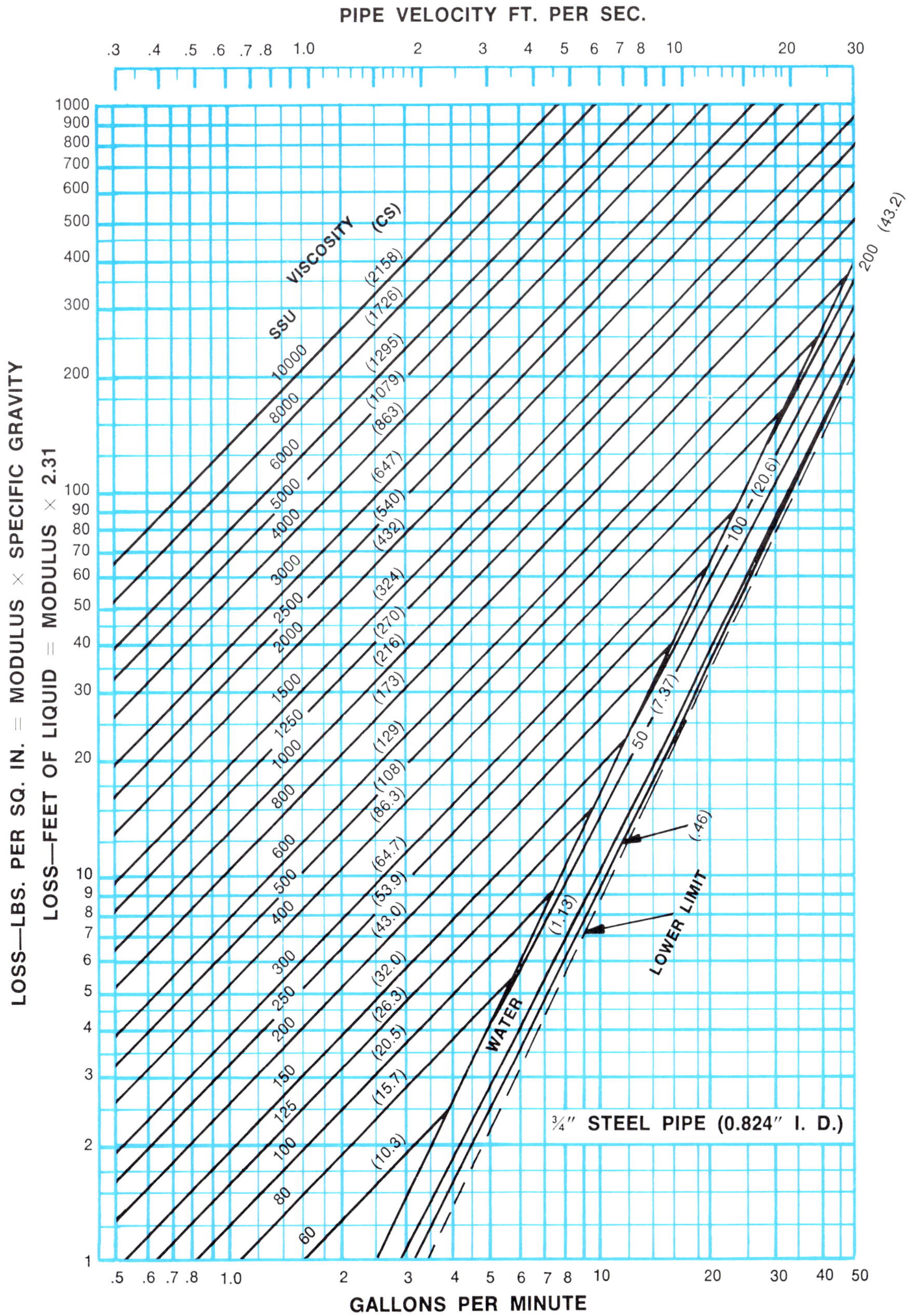

IIIC-3 FRICTION LOSS MODULUS FOR 100 FEET OF PIPE

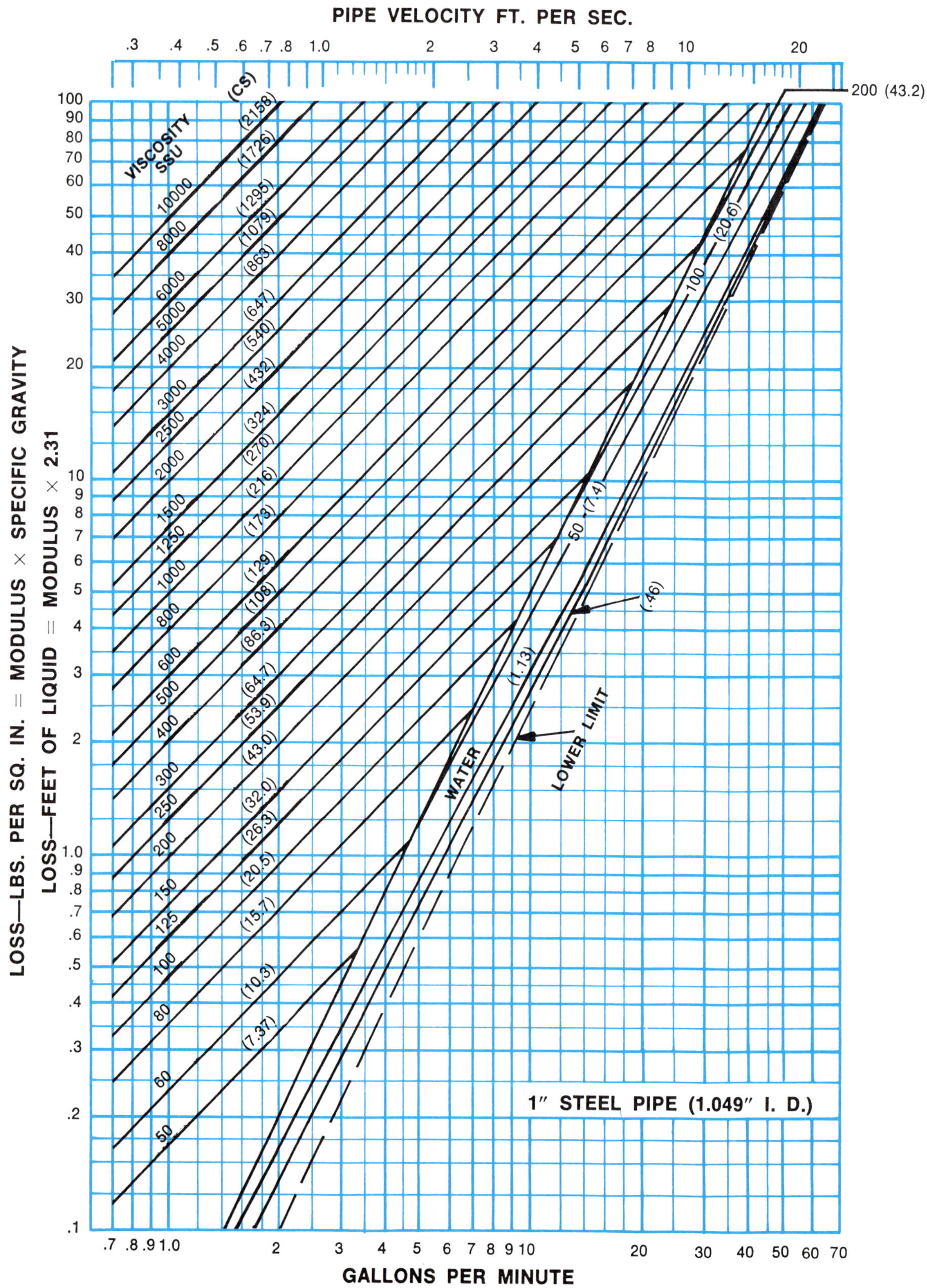

IIIC-4 FRICTION LOSS MODULUS FOR 100 FEET OF PIPE

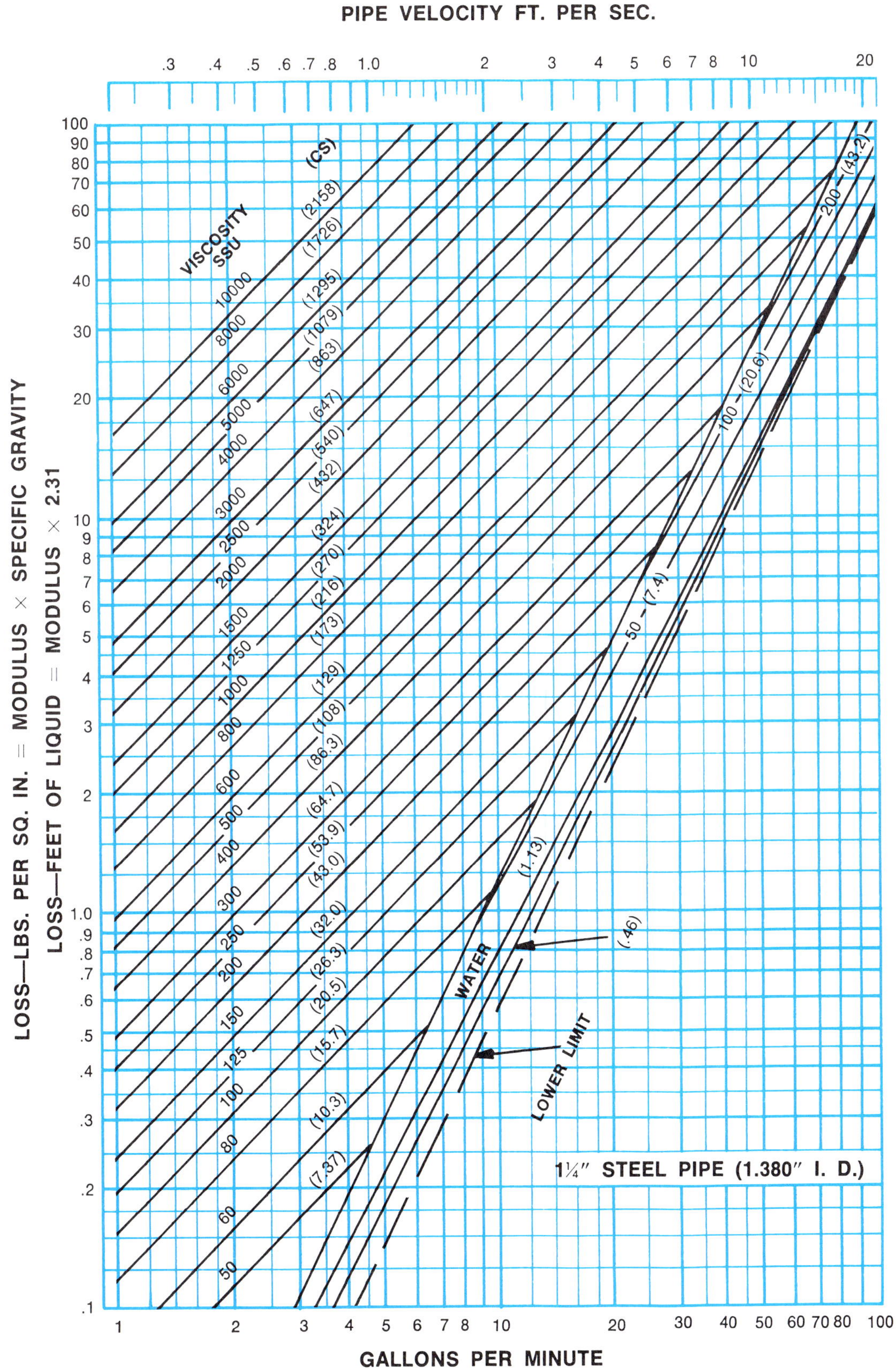

IIIC-5 FRICTION LOSS MODULUS FOR 100 FEET OF PIPE

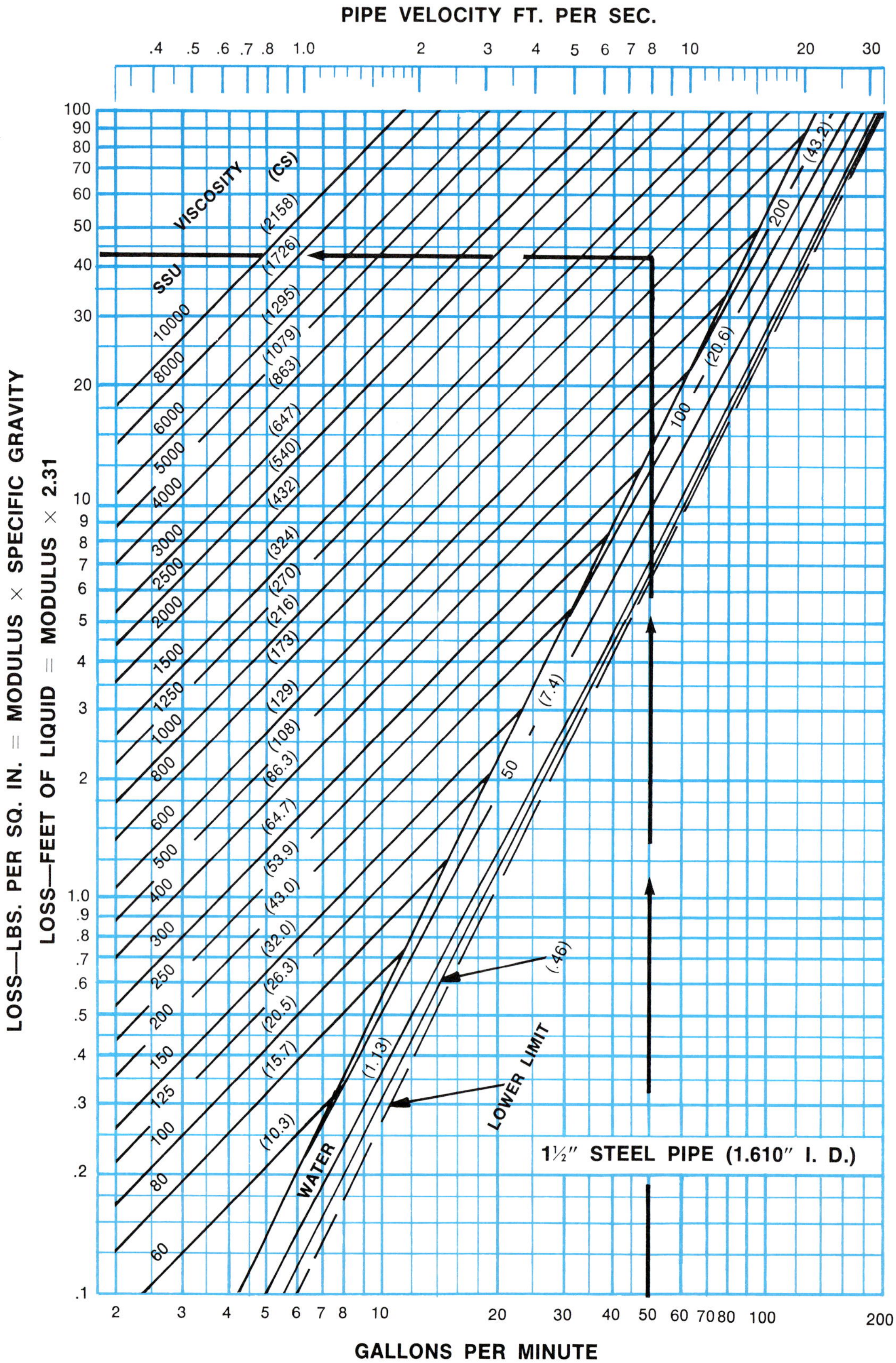

IIIC-6 FRICTION LOSS MODULUS FOR 100 FEET OF PIPE

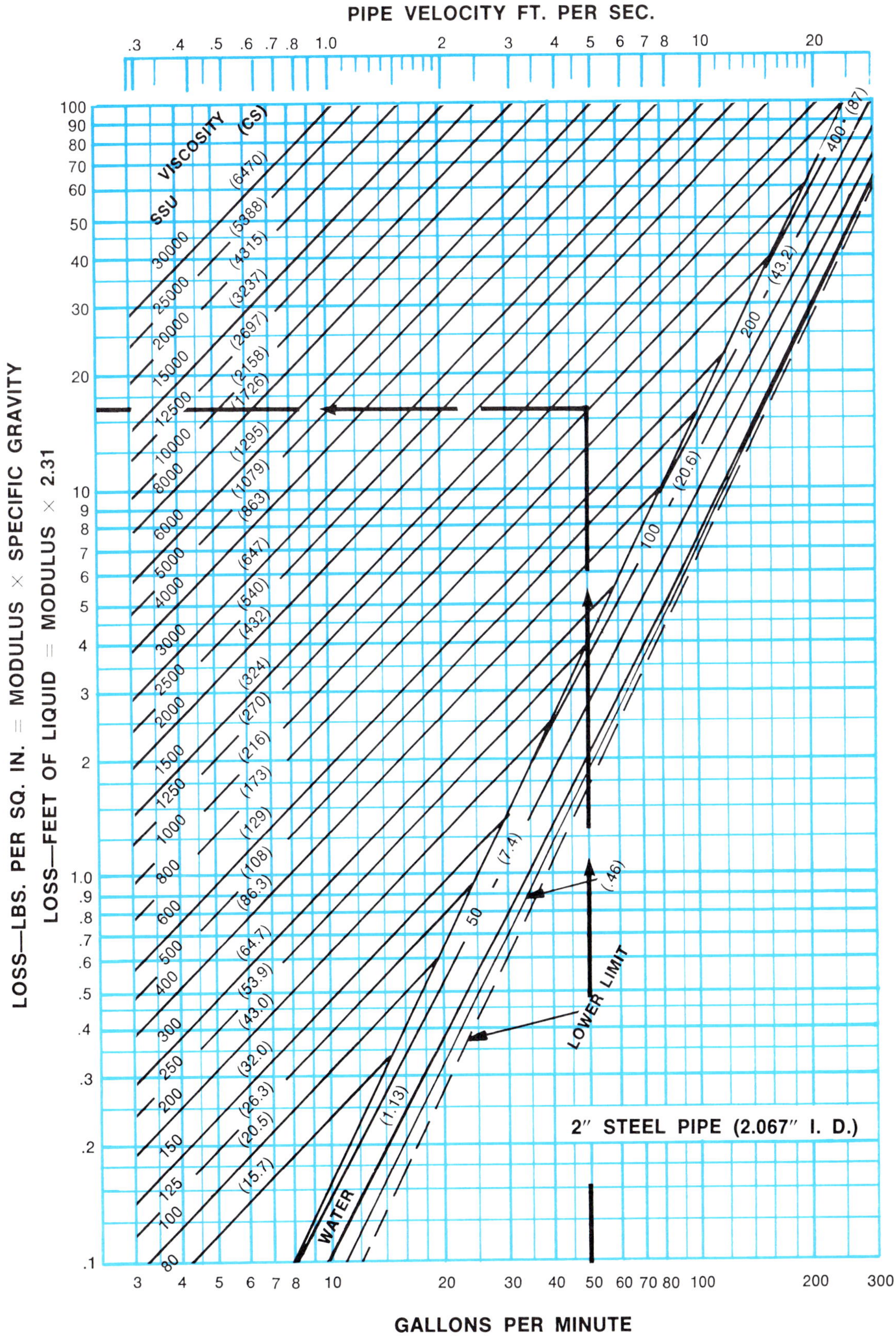

IIIC-7 FRICTION LOSS MODULUS FOR 100 FEET OF PIPE

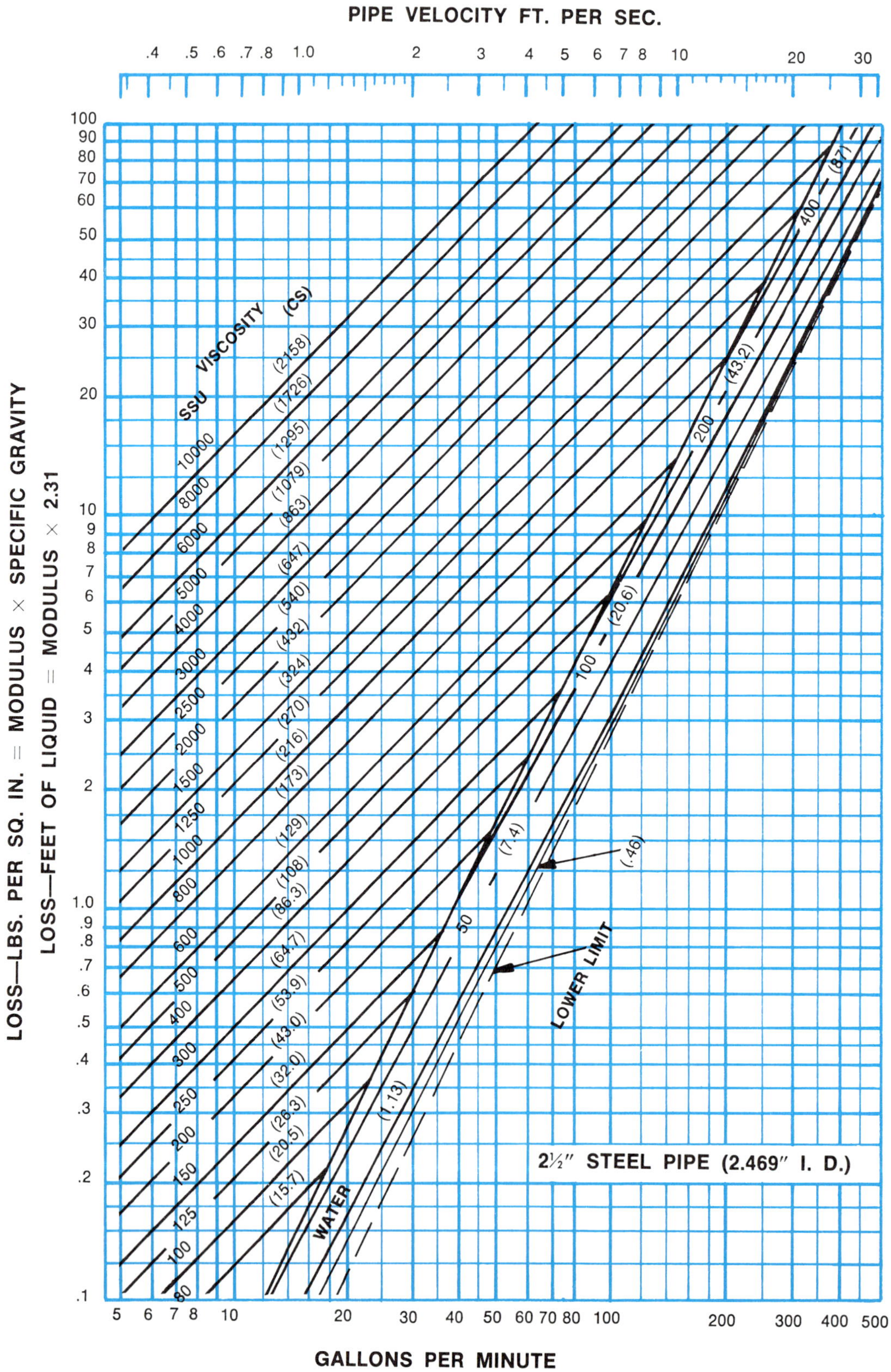

IIIC-8 FRICTION LOSS MODULUS FOR 100 FEET OF PIPE

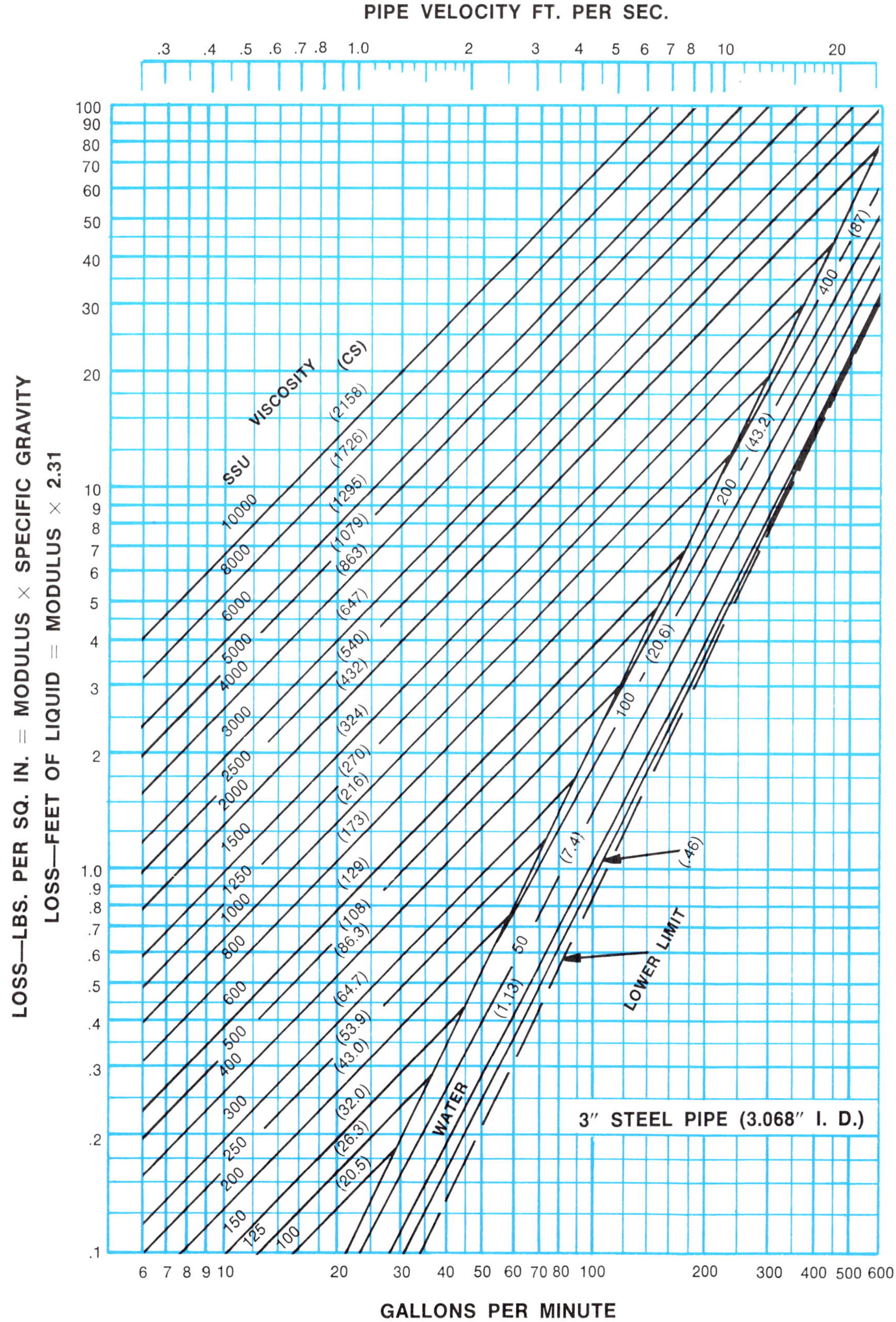

IIIC-9 FRICTION LOSS MODULUS FOR 100 FEET OF PIPE

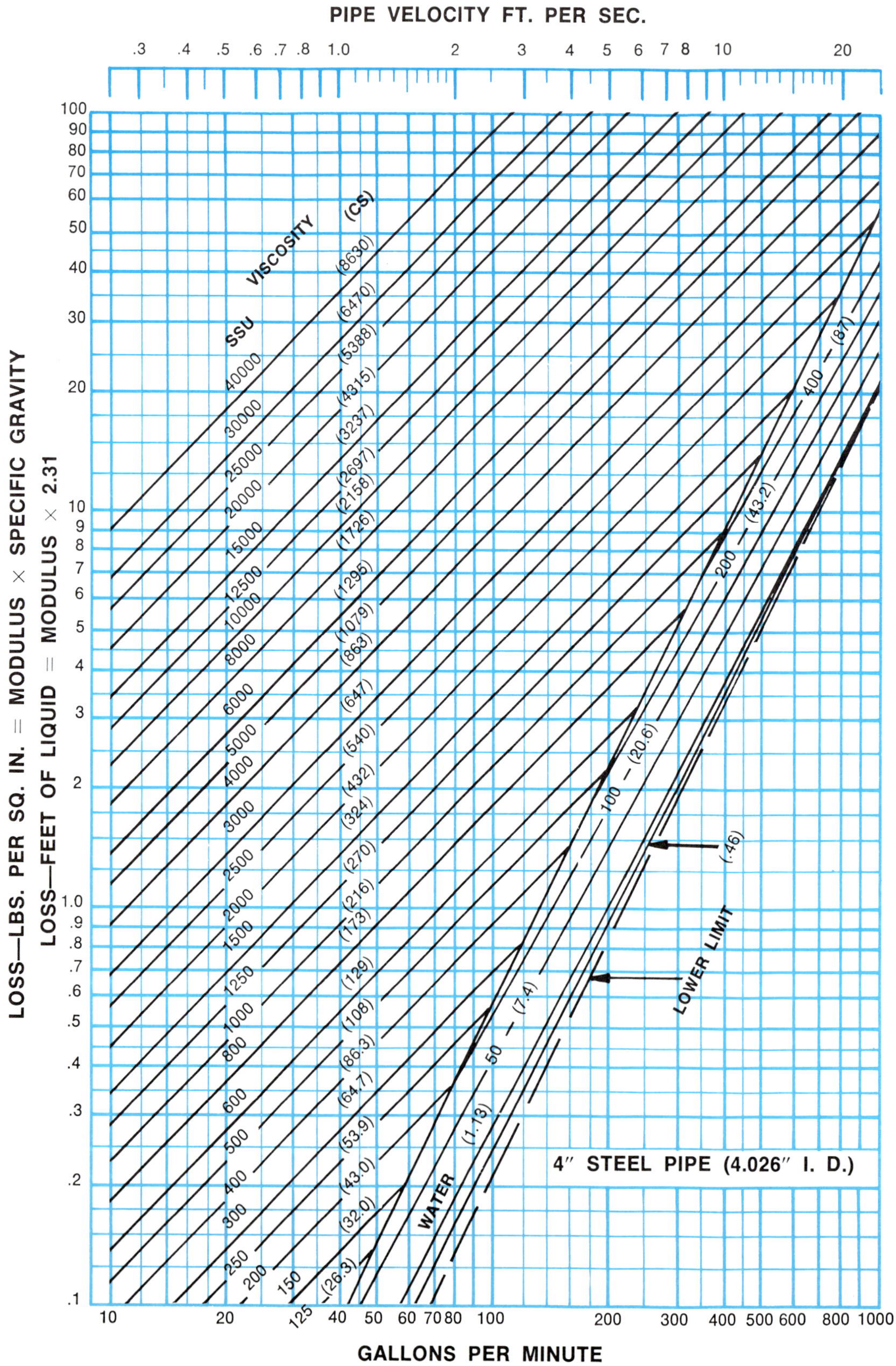

IIIC-10 FRICTION LOSS MODULUS FOR 100 FEET OF PIPE

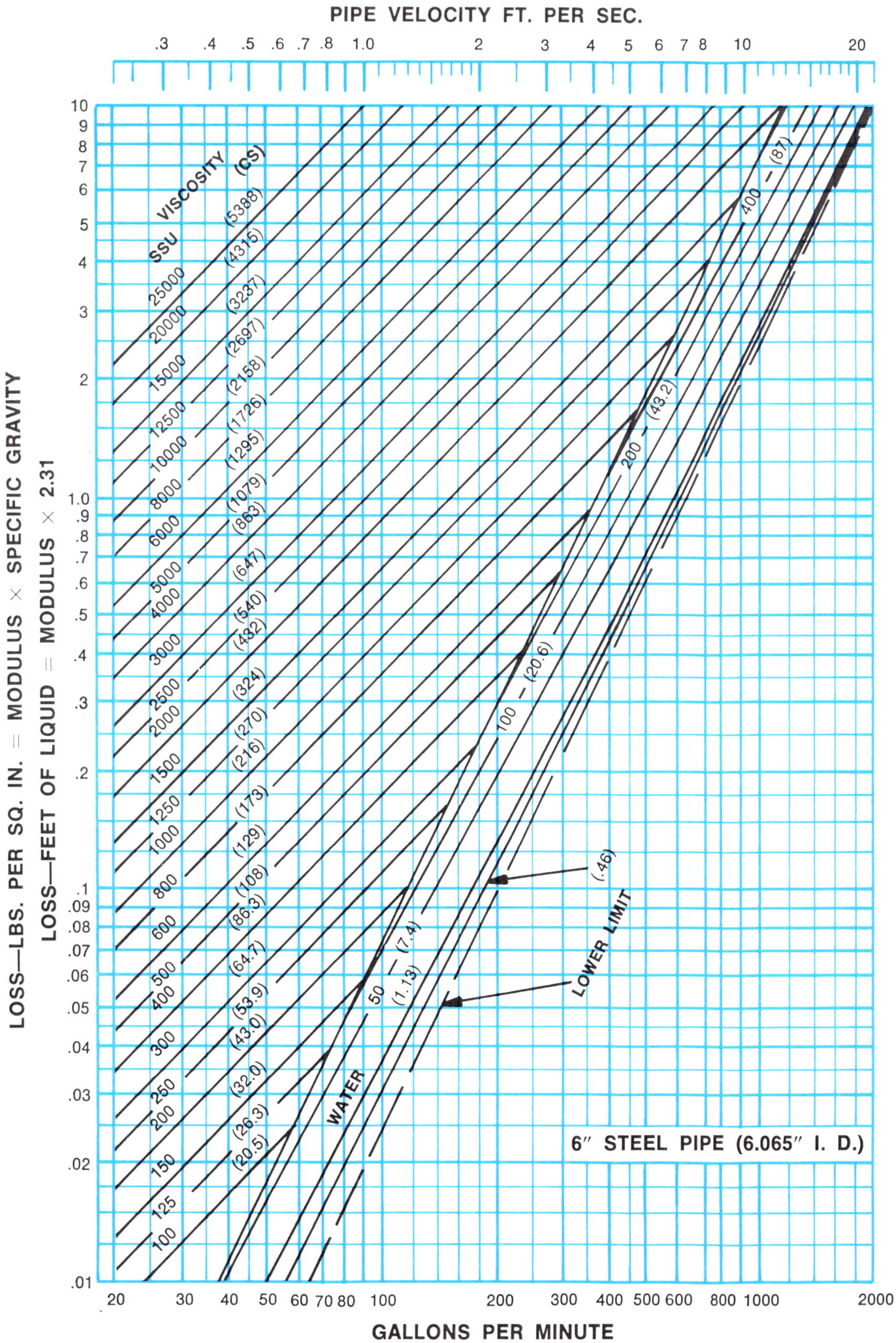

IIIC-11 FRICTION LOSS MODULUS FOR 100 FEET OF PIPE

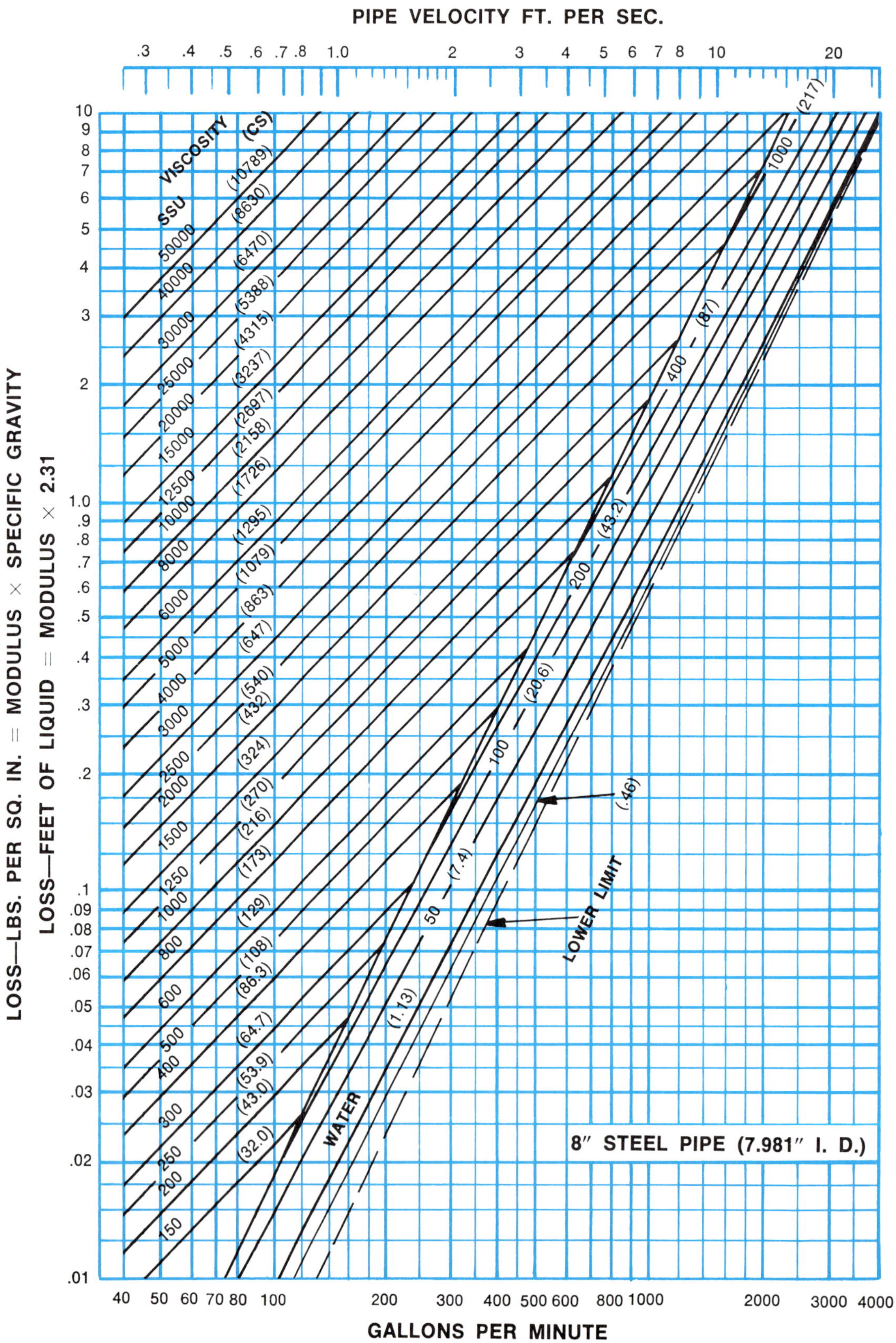

IIIC-12 FRICTION LOSS MODULUS FOR 100 FEET OF PIPE

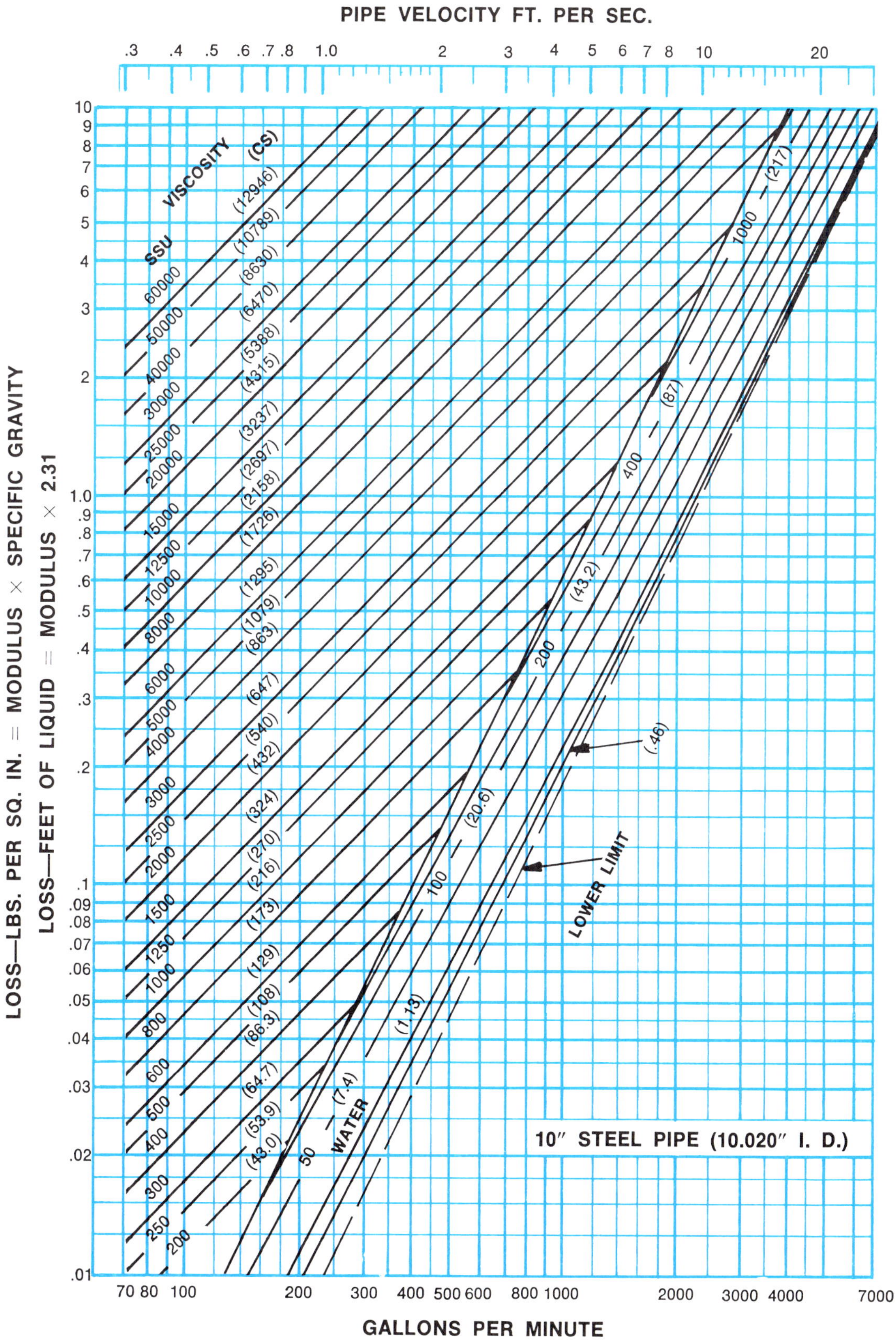

IIIC-13 FRICTION LOSS MODULUS FOR 100 FEET OF PIPE

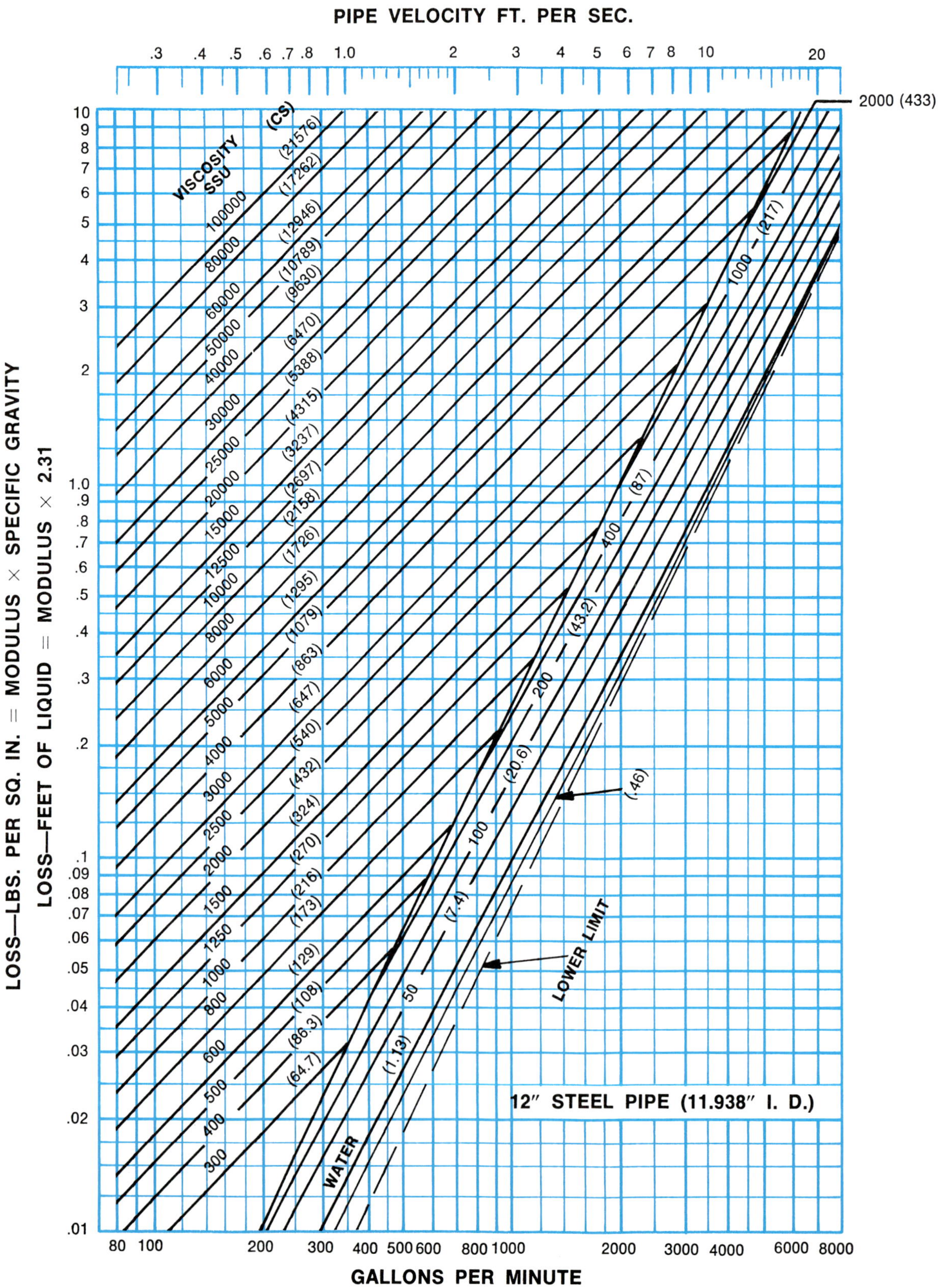

IIIC-15 TABLE 34—FRICTION LOSS FOR VISCOUS LIQUIDS

Loss in Pounds Per Square Inch Per 100 Feet of New Schedule 40 Steel Pipe Based on Specific Gravity of 1.00

GPM	Pipe Size	VISCOSITY—SAYBOLT SECONDS UNIVERSAL																	
		100	200	300	400	500	1000	1500	2000	2500	3000	4000	5000	6000	7000	8000	9000	10,000	15,000
3	½	11.2	23.6	35.3	47.1	59	118	177	236	294	353	471	589	706	824	942	—	—	—
	¾	3.7	7.6	11.5	15.3	19.1	38.2	57	76	96	115	153	191	229	268	306	344	382	573
	1	1.4	2.9	4.4	5.8	7.3	14.5	21.8	29.1	36.3	43.6	58	73	87	101	116	131	145	218
5	¾	6.1	12.7	19.1	25.5	31.9	64	96	127	159	191	255	319	382	446	510	573	637	956
	1	2.3	4.9	7.3	9.7	12.1	24.2	36.3	48.5	61	73	97	121	145	170	194	218	242	363
	1¼	0.77	1.6	2.4	3.3	4.1	8.1	12.2	16.2	20.3	24.3	32.5	40.6	48.7	57	65	73	81	122
7	¾	8.5	17.9	26.8	35.7	44.6	89	134	178	223	268	357	446	535	624	713	803	892	—
	1	3.2	6.8	10.2	13.6	17	33.9	51	68	85	102	136	170	203	237	271	305	339	509
	1¼	1.1	2.3	3.4	4.5	5.7	11.4	17	22.7	28.4	34.1	45.4	57	68	80	91	102	114	170
10	1	4.9	9.7	14.5	19.4	24.2	48.5	73	97	121	145	194	242	291	339	388	436	485	727
	1¼	1.6	3.3	4.9	6.5	8.1	16.2	24.3	32.5	40.6	48.7	65	81	97	114	130	146	162	243
	1½	0.84	1.8	2.6	3.5	4.4	8.3	13.1	17.5	21.9	26.3	35	43.8	53	61	70	79	88	131
15	1	11	14.5	21.8	29.1	36.3	73	109	145	182	218	291	363	436	509	581	654	727	—
	1¼	2.8	4.9	7.3	9.7	12.2	24.3	36.5	48.7	61	73	97	122	146	170	195	219	243	365
	1½	1.3	2.6	3.9	5.3	6.6	13.1	19.7	26.3	32.8	39.4	53	66	79	92	105	118	131	197
20	1	18	18	29.1	38.8	48.5	97	145	194	242	291	388	485	581	678	775	872	—	—
	1¼	4.9	6.4	9.7	13	16.2	32.5	48.7	65	81	97	130	162	195	227	260	292	325	487
	1½	2.3	3.5	5.3	7	8.8	17.5	26.3	35	43.8	53	70	88	105	123	140	158	175	263
	2	0.64	1.3	1.9	2.6	3.2	6.4	9.6	12.9	16.1	19.3	25.7	32.1	38.5	45	51	58	64	96
25	1½	3.5	4.4	6.6	8.8	11	21.9	32.8	43.8	55	66	88	110	131	153	176	197	219	328
	2	1	1.6	2.4	3.2	4	8	12.1	16.1	20.1	24.1	32.1	40.2	48.2	56	64	72	80	121
	2½	0.4	0.79	1.2	1.6	2	4	5.9	7.9	9.9	11.8	15.8	19.7	23.7	27.6	31.6	35.5	39.5	59
30	1½	5	5.3	7.9	10.5	13.1	26.3	39.4	53	66	79	105	131	158	184	210	237	263	394
	2	1.4	1.9	2.9	3.9	4.8	9.6	14.5	19.3	24.1	28.9	38.5	48.2	58	67	77	87	96	145
	2½	0.6	0.95	1.4	1.9	2.4	4.7	7.1	9.5	11.8	14.2	19	23.7	28.4	33.2	37.9	42.6	47.4	71
40	1½	8.5	9	10.5	14	17.5	35	53	70	88	105	140	175	210	245	280	315	350	526
	2	2.5	2.5	3.9	5.1	6.4	12.9	19.3	25.7	32.1	38.5	51	64	77	90	103	116	129	193
	2½	1.1	1.3	1.9	2.5	3.2	6.3	9.5	12.6	15.8	19	25.3	31.6	37.9	44.2	51	57	63	95
50	1½	12.5	14	14.0	17.5	21.9	43.8	66	88	110	131	175	219	263	307	350	394	438	657
	2	3.7	4	4.8	6.4	8	16.1	24.1	32.1	40.2	48.2	64	80	96	112	129	145	161	241
	2½	1.6	1.7	2.4	3.2	4	7.9	11.8	15.8	19.7	23.7	31.6	39.5	47.4	55	63	71	79	118
60	2	5	5.8	5.8	7.7	9.6	19.3	28.9	38.5	48.2	58	77	96	116	135	154	173	193	289
	2½	2.2	2.4	2.8	3.8	4.7	9.5	14.2	19	23.7	28.4	37.9	47.4	57	66	76	85	95	142
	3	0.8	0.8	1.2	1.6	2	4	6	8	9.9	11.9	15.9	19.9	23.9	27.9	31.8	35.8	39.8	60
70	2½	2.8	3.2	3.4	4.4	5.5	11.1	16.6	22.1	27.6	33.2	44.2	55	66	77	88	100	111	166
	3	1	1.1	1.4	1.9	2.3	4.6	7	9.3	11.6	13.9	18.6	23.2	27.8	32.5	37.1	41.7	46.4	70
	4	0.27	0.31	0.47	0.63	0.73	1.6	2.4	3.1	3.9	4.7	6.3	7.8	9.4	11	12.5	14.1	15.6	23.5
80	2½	3.6	4.2	4.2	5.1	6.3	12.6	19	25.3	31.6	37.9	51	63	76	88	101	114	126	190
	3	1.3	1.4	1.6	2.1	2.7	5.3	8	10.6	13.3	15.9	21.2	26.5	31.8	37.1	42.4	47.7	53	80
	4	0.36	0.36	0.54	0.72	0.89	1.8	2.7	3.6	4.5	5.4	7.2	8.9	10.7	12.5	14.3	16.1	17.9	26.8
100	2½	5.3	6.1	6.4	6.4	8	15.8	23.7	31.6	39.5	47.4	63	79	95	111	127	142	158	237
	3	1.9	2.2	2.2	2.7	3.3	6.6	9.9	13.3	16.6	19.9	26.5	33.1	39.8	46.4	53	60	66	99
	4	0.52	0.57	0.67	0.89	1.1	2.2	3.4	4.5	5.6	6.7	8.9	11.2	13.4	15.6	17.9	20.1	22.3	33.5

TURBULENT FLOW ⟷ LAMINAR FLOW

NOTE: For a liquid having a specific gravity other than 1.00 multiply the value from the table by the specific gravity of that liquid. For commercial installations, it is recommended that 15% be added to the above values. No allowance for aging of pipe is included.

IIIC-15 TABLE 34 (Continued)—FRICTION LOSS FOR VISCOUS LIQUIDS

Loss in Pounds Per Square Inch Per 100 Feet of New Schedule 40 Steel Pipe Based on Specific Gravity of 1.00

GPM	Pipe Size	VISCOSITY—SAYBOLT SECONDS UNIVERSAL																	
		100	200	300	400	500	1000	1500	2000	2500	3000	4000	5000	6000	7000	8000	9000	10,000	15,000
	3	2.7	3.1	3.2	3.2	4	8	11.9	15.9	19.9	23.9	31.8	39.8	47.7	56	64	72	80	119
120	4	0.73	0.81	0.81	1.1	1.3	2.7	4	5.4	6.7	8	10.7	13.4	16.1	18.8	21.4	24.1	26.8	40.2
	6	.098	0.11	0.16	0.21	0.26	0.52	0.78	1.0	1.3	1.6	2.1	2.6	3.1	3.6	4.2	4.7	5.2	7.8
	3	3.4	4	4.3	4.3	4.6	9.3	13.9	18.6	23.2	27.8	37.1	46.4	56	65	74	84	93	139
140	4	0.95	1.1	1.1	1.3	1.6	3.1	4.7	6.3	7.8	9.4	12.5	15.6	18.8	21.9	25	28.2	31.3	46.9
	6	0.13	0.15	0.18	0.24	0.30	0.61	0.91	1.2	1.5	1.8	2.4	3.0	3.6	4.2	4.9	5.5	6.1	9.1
	3	4.4	5	5.7	5.7	5.7	10.6	15.9	21.2	26.5	31.8	42.4	53	64	74	85	95	106	159
160	4	1.2	1.4	1.4	1.4	1.8	3.6	5.4	7.2	8.9	10.7	14.3	17.9	21.5	25	28.6	32.2	35.7	54
	6	0.17	0.18	0.21	0.28	0.35	0.69	1.0	1.4	1.7	2.1	2.8	3.5	4.2	4.9	5.5	6.2	6.9	10.4
	3	5.3	6.3	7	7	7	11.9	17.9	23.9	29.8	35.8	47.7	60	72	84	95	107	119	179
180	4	1.5	1.8	1.8	1.8	2	4	6	8	10.1	12.1	16.1	20.1	24.1	28.1	32.2	36.2	40.2	60
	6	0.2	0.24	0.24	0.31	0.39	0.78	1.2	1.6	2	2.3	3.1	3.9	4.7	5.5	6.2	7	7.8	11.7
	3	6.5	7.7	8.8	8.8	8.8	13.3	19.9	26.5	33.1	39.8	53	66	80	93	106	119	133	199
200	4	1.8	2.2	2.2	2.2	2.2	4.5	6.7	8.9	11.2	13.4	17.9	22.3	26.8	31.3	35.7	40.2	44.7	67
	6	0.25	0.3	0.3	0.35	0.43	0.87	1.3	1.7	2.2	2.6	3.5	4.3	5.2	6.1	6.9	7.8	8.7	13
	4	2.6	3.2	3.5	3.5	3.5	5.6	8.4	11.2	14	16.8	22.3	27.9	33.5	39.1	44.7	50	56	84
250	6	0.36	0.43	0.45	0.45	0.54	1.1	1.6	2.2	2.7	3.3	4.3	5.4	6.5	7.6	8.7	9.8	10.8	16.3
	8	.095	0.12	0.12	0.15	0.18	0.36	0.54	0.72	0.9	1.1	1.5	1.8	2.2	2.5	2.9	3.3	3.6	5.4
	4	3.7	4.3	5	5	5	6.7	10.1	13.4	16.8	20.1	26.8	33.5	40.2	47	54	60	67	101
300	6	0.5	0.6	0.65	0.65	0.65	1.3	2	2.6	3.3	3.9	5.2	6.5	7.8	9.1	10.4	11.7	13	19.5
	8	0.13	0.17	0.17	0.18	0.22	0.43	0.65	0.87	1.1	1.3	1.7	2.2	2.6	3	3.5	3.9	4.3	6.5
	6	0.82	1	1.1	1.2	1.2	1.7	2.6	3.5	4.3	5.2	6.9	8.7	10.4	12.1	13.9	15.6	17.3	26
400	8	0.23	0.27	0.29	0.29	0.29	0.58	0.87	1.2	1.5	1.7	2.3	2.9	3.5	4.1	4.6	5.2	5.8	8.7
	10	0.08	0.09	0.1	0.1	0.12	0.23	0.35	0.47	0.58	0.7	0.93	1.2	1.4	1.6	1.9	2.1	2.3	3.5
	6	1.2	1.5	1.6	1.8	1.8	2.2	3.2	4.3	5.4	6.5	8.7	10.8	13	15.2	17.3	19.5	21.7	32.5
500	8	0.33	0.39	0.44	0.47	0.47	0.72	1.1	1.5	1.8	2.2	2.9	3.6	4.3	5.1	5.8	6.5	7.2	10.8
	10	0.11	0.14	0.15	0.15	0.15	0.29	0.44	0.58	0.73	0.87	1.2	1.5	1.8	2	2.3	2.6	2.9	4.4
	6	1.8	2.2	2.3	2.4	2.6	2.7	3.9	5.2	6.5	7.8	10.4	13	16	18.2	20.8	23.4	26	39
600	8	0.47	0.57	0.62	0.67	0.67	0.87	1.3	1.7	2.2	2.6	3.5	4.3	5.2	6.1	6.9	7.8	8.7	13
	10	0.16	0.18	0.2	0.22	0.22	0.35	0.52	0.7	0.87	1.1	1.4	1.8	2.1	2.4	2.8	3.3	3.5	5.2
	6	2.3	2.7	3	3.2	3.5	3.6	4.6	6.1	7.6	9.1	12.1	15.2	18.4	21.2	24.3	27.3	30.3	45.5
700	8	0.6	0.74	0.82	0.89	0.93	1	1.5	2	2.5	3	4.1	5.1	6.1	7.1	8.1	9.1	10.1	15.2
	10	0.2	0.25	0.27	0.3	0.3	0.41	0.61	0.82	1	1.2	1.6	2	2.4	2.9	3.3	3.7	4.1	6.1
	6	2.8	3.5	3.7	4	4.2	4.8	5.2	6.9	8.7	10.4	13.9	17.3	20.8	24.3	27.7	31.2	34.7	52
800	8	0.78	0.94	1	1.1	1.2	1.2	1.7	2.3	2.9	3.5	4.6	5.8	6.9	8.1	9.3	10.4	11.6	17.3
	10	0.26	0.3	0.34	0.38	0.4	0.47	0.7	0.92	1.2	1.4	1.9	2.3	2.8	3.3	3.7	4.2	4.7	7
	6	3.5	4.3	4.6	5.0	5.2	6	6	7.8	9.8	11.7	15.6	19.5	23.4	27.3	31.2	35.1	39	58.5
900	8	0.95	1.1	1.3	1.4	1.5	1.5	2	2.6	3.3	3.9	5.2	6.5	7.8	9.1	10.4	11.7	13	19.5
	10	0.32	0.37	0.43	0.46	0.5	0.52	0.79	1.1	1.3	1.6	2.1	2.6	3.1	3.7	4.2	4.7	5.2	7.9
	8	1.1	1.4	1.5	1.6	1.8	1.9	2.2	2.9	3.6	4.3	5.8	7.2	8.7	10.1	11.6	13	14.5	21.7
1000	10	0.38	0.45	0.5	0.55	0.6	0.6	0.87	1.2	1.5	1.8	2.3	2.9	3.5	4.1	4.7	5.2	5.8	8.7
	12	0.17	0.2	0.22	0.24	0.25	0.29	0.43	0.58	0.72	0.87	1.2	1.5	1.7	2	2.3	2.6	2.9	4.3

TURBULENT FLOW ← → ← LAMINAR FLOW →

NOTE: For a liquid having a specific gravity other than 1.00 multiply the value from the table by the specific gravity of that liquid. For commercial installations, it is recommended that 15% be added to the above values. No allowance for aging of pipe is included.

IIIC-15 TABLE 34 (Continued)—FRICTION LOSS FOR VISCOUS LIQUIDS

Loss in Pounds Per Square Inch Per 100 Feet of New Schedule 40 Steel Pipe Based on Specific Gravity of 1.00

GPM	Pipe Size	VISCOSITY—SAYBOLT SECONDS UNIVERSAL														
		20,000	25,000	30,000	40,000	50,000	60,000	70,000	80,000	90,000	100,000	125,000	150,000	175,000	200,000	500,000
	2	19.3	24.1	28.9	38.5	48.2	58	67	77	87	96	120	145	169	193	482
3	2½	9.5	11.8	14.2	19	23.7	28.4	33.2	37.9	42.6	47.4	59	71	83	95	237
	3	4	5	6	8	9.9	11.9	13.9	15.9	17.9	19.9	24.9	29.8	34.8	39.8	99
	2	32	40	48.2	64	80	96	112	129	145	161	201	241	281	321	803
5	2½	15.8	19.7	23.7	31.6	39.5	47.4	55	63	71	79	99	118	138	158	395
	3	6.6	8.3	9.9	13.3	16.6	9.9	23.2	26.5	29.8	33	41.4	49.7	58	66	166
	2	45	56	67	90	112	135	157	180	202	225	281	337	393	450	—
7	2½	22.1	27.6	33.2	44.2	55	66	77	88	100	111	138	166	194	221	553
	3	9.3	11.6	13.9	18.6	23.2	27.8	32.5	37.1	41.7	46.4	58	70	81	93	232
	2½	31.6	39.5	47.4	63	79	95	111	126	142	158	197	237	276	316	790
10	3	13.3	16.6	19.9	26.5	33.1	39.8	46.4	53	60	66	83	99	116	133	331
	4	4.5	5.6	6.7	8.9	11.2	13.4	15.6	17.9	20.1	22.3	27.9	33.5	39.1	44.7	112
	2½	47.4	59	71	95	118	142	166	190	213	237	296	355	415	474	—
15	3	19.9	24.9	29.8	39.8	49.7	60	70	80	89	99	124	149	174	199	497
	4	6.7	8.4	10.1	13.4	16.8	20.1	23.5	26.8	30.2	33.5	41.9	50	59	67	168
	3	26.5	33.1	39.8	53	66	80	93	106	119	133	166	199	232	265	663
20	4	8.9	11.2	13.4	17.9	22.3	26.8	31.3	35.7	40.2	44.7	56	67	78	89	223
	6	1.7	2.2	2.6	3.5	4.3	5.2	6.1	6.9	7.8	8.7	10.8	13	15.2	17.3	43.3
	3	33.1	41.4	49.7	66	83	99	116	133	149	166	207	249	290	331	828
25	4	11.2	14	16.8	22.3	27.9	33.5	39.1	44.7	50	56	70	84	98	112	279
	6	2.2	2.7	3.3	4.3	5.4	6.5	7.6	8.7	9.8	10.8	13.5	16.3	19	21.7	54
	3	39.8	49.7	60	80	99	119	139	159	179	199	249	298	348	398	—
30	4	13.4	16.8	20.1	26.8	33.5	40.2	46.9	54	60	67	84	101	117	134	335
	6	2.6	3.3	3.9	5.2	6.5	7.8	9.1	10.4	11.7	13	16.3	19.5	22.7	26	65
	3	53	66	80	106	133	160	186	212	239	265	331	398	464	532	—
40	4	17.9	22.3	26.8	35.7	44.7	54	63	72	80	89	112	134	156	179	447
	6	3.5	4.3	5.2	6.9	8.7	10.4	12.1	13.9	15.6	17.3	21.7	26	30.3	34.7	87
	4	22.3	27.9	33.5	44.7	56	67	78	89	101	112	140	168	196	223	559
50	6	4.3	5.4	6.5	8.7	10.8	13	15.2	17.3	19.5	21.7	27.1	32.5	37.9	43.3	108
	8	1.5	1.8	2.7	2.9	3.6	4.3	5.1	5.8	6.5	7.2	9	10.8	12.6	14.5	36.1
	4	26.8	33.5	40.2	54	67	80	94	107	121	134	168	201	235	268	670
60	6	5.2	6.5	7.8	10.4	13	16	18.2	20.8	23.4	26	32.5	39	45.5	52	130
	8	1.7	2.2	2.6	3.5	4.3	5.2	6.1	6.9	7.8	8.7	10.8	13	15.2	17.3	43.4
	4	31.3	39.1	46.9	63	78	94	110	125	141	156	196	235	274	313	782
70	6	6.1	7.6	9.1	12.1	15.2	18.4	21.2	24.3	27.3	30.3	37.9	45.5	53	61	152
	8	2	2.5	3	4.1	5.1	6.1	7.1	8.1	9.1	10.1	12.6	15.2	17.7	20.2	51
	6	6.9	8.7	10.4	13.9	17.3	20.8	24.3	27.7	31.2	34.7	43.3	52	61	69	173
80	8	2.3	2.9	3.5	4.6	5.8	6.9	8.1	9.3	10.4	11.6	14.5	17.3	20.2	23.1	58
	10	0.93	1.2	1.4	1.9	2.3	2.8	3.3	3.7	4.2	4.7	5.8	7	8.2	9.3	23.3
	6	7.8	9.8	11.7	15.6	19.5	23.4	27.3	31.2	35.1	39	48.7	59	68	78	195
90	8	2.6	3.3	3.9	5.2	6.5	7.8	9.1	10.4	11.7	13	16.3	19.5	22.8	26	65
	10	1.1	1.3	1.6	2.1	2.6	3.1	3.7	4.2	4.7	5.2	6.6	7.9	9.2	10.5	26.2
	6	8.7	10.8	13	17.3	21.7	26	30.3	34.7	39	43.3	54	65	76	87	217
100	8	2.9	3.6	4.3	5.8	7.2	8.7	10.1	11.6	13	14.5	18.1	21.7	25.3	28.9	72
	10	1.2	1.5	1.8	2.3	2.9	3.5	4.2	4.7	5.2	5.8	7.3	8.7	10.2	11.6	29.1

← **LAMINAR FLOW** →

NOTE: For a liquid having a specific gravity other than 1.00 multiply the value from the table by the specific gravity of that liquid. For commercial installations, it is recommended that 15% be added to the above values. No allowance for aging of pipe is included.

IIIC-15 TABLE 34 (Concluded)—FRICTION LOSS FOR VISCOUS LIQUIDS

Loss in Pounds Per Square Inch Per 100 Feet of New Schedule 40 Steel Pipe Based on Specific Gravity of 1.00

GPM	Pipe Size	VISCOSITY—SAYBOLT SECONDS UNIVERSAL														
		20,000	25,000	30,000	40,000	50,000	60,000	70,000	80,000	90,000	100,000	125,000	150,000	175,000	200,000	500,000
	6	10.4	13	15.6	20.8	26	31.2	36.4	41.6	46.8	52	65	78	91	104	260
120	8	3.5	4.3	5.2	6.9	8.7	10.4	12.1	13.9	15.6	17.3	21.7	26	30.4	34.7	87
	10	1.4	1.8	2.1	2.8	3.5	4.2	4.9	5.6	6.3	7	8.7	10.5	12.2	14	34.9
	6	12.1	15.2	18.2	24.3	30.3	36.4	42.5	48.5	55	61	76	91	106	121	303
140	8	4	5.1	6.1	8.1	10.1	12.1	14.2	16.2	18.2	20.2	25.3	30.4	35.4	40.5	101
	10	1.7	2	2.4	3.3	4.1	4.9	5.7	6.5	7.3	8.1	10.2	12.2	14.3	16.3	40.7
	6	13.9	17.3	20.8	27.7	34.7	41.6	48.5	56	62	69	87	104	121	139	347
160	8	4.6	5.8	6.9	9.3	11.6	13.8	16.2	18.5	20.8	23.1	28.9	34.7	40.5	46.2	116
	10	1.9	2.3	2.8	3.7	4.7	5.6	6.5	7.5	8.4	9.3	11.6	14	16.3	18.6	46.6
	6	15.6	19.5	23.4	31.2	39	46.8	55	62	70	78	98	117	137	156	390
180	8	5.2	6.5	7.8	10.4	13	15.6	18.2	20.8	23.4	26	32.5	39	45.5	52	130
	10	2.1	2.6	3.1	4.2	5.2	6.3	7.3	8.4	9.4	10.5	13.1	15.7	18.3	21	52
	8	5.8	7.2	8.7	11.6	14.5	17.3	20.2	23.1	26	28.9	36.1	43.4	51	58	145
200	10	2.3	2.9	3.5	4.7	5.8	7	8.2	9.3	10.5	11.6	14.6	17.5	20.4	23.3	58
	12	1.2	1.5	1.7	2.3	2.9	3.5	4.1	4.6	5.2	5.8	7.2	8.7	10.1	11.6	28.9
	8	7.2	9	10.8	14.5	18.1	21.7	25.3	28.9	32.5	36.1	45.2	54	63	72	181
250	10	2.9	3.6	4.4	5.8	7.3	8.7	10.2	11.6	13.1	14.6	18.2	21.8	25.5	29.1	73
	12	1.5	1.8	2.2	2.9	3.6	4.3	5.1	5.8	6.5	7.2	9	10.9	12.7	14.5	36.2
	8	8.7	10.8	13	17.3	21.7	26	30.4	34.7	39	43.4	54	65	76	87	217
300	10	3.5	4.4	5.2	7	8.7	10.5	12.2	14	15.7	17.5	21.8	26.2	30.6	34.9	87
	12	1.7	2.2	2.6	3.5	4.3	5.2	6.1	7	7.8	8.7	10.9	13	15.2	17.4	43.4
	8	11.6	14.5	17.3	23	28.9	34.7	40.5	46.2	52	58	72	87	101	116	289
400	10	4.7	5.8	7	9.3	11.6	14	16.3	18.6	21	23.3	29.6	34.9	40.7	46.6	116
	12	2.3	2.9	3.5	4.6	5.8	7	8.1	9.3	10.4	11.6	14.5	17.4	20.3	23.2	58
	8	14.5	18.1	21.7	28.9	36.1	43.4	51	58	65	72	90	108	126	145	361
500	10	5.8	7.3	8.7	11.6	14.6	17.5	20.4	23.3	26.2	29.1	36.4	43.7	51	58	146
	12	2.9	3.6	4.3	5.8	7.2	8.7	10.1	11.6	13	14.5	18.1	21.7	25.3	28.9	72
	8	17.3	21.7	26	34.7	43.4	52	61	69	78	87	108	130	152	173	434
600	10	7	8.7	10.5	14	17.5	21	24.4	27.9	31.4	34.9	43.7	52	61	70	175
	12	3.5	4.3	5.2	7	8.7	10.4	12.2	13.9	15.6	17.4	21.7	26.1	30.4	34.7	87
	8	20.2	25.3	30.3	40.5	51	61	71	81	91	101	126	152	177	202	506
700	10	8.2	10.2	12.2	16.3	20.4	24.4	28.5	32.6	36.7	40.7	51	61	71	82	204
	12	4.1	5.1	6.1	8.1	10.1	12.2	14.2	16.2	18.2	20.3	25.3	30.4	35.5	40.5	101
	8	23.1	28.9	34.7	46.2	58	69	81	93	104	116	145	173	202	231	578
800	10	9.3	11.6	14	18.6	23.3	27.9	32.6	37.3	41.9	46.6	58	70	82	93	233
	12	4.6	5.8	7	9.3	11.6	13.9	16.2	18.5	20.8	23.1	28.9	34.7	40.5	46.3	116
	8	26	32.5	39	52	65	78	91	104	117	130	163	195	228	260	650
900	10	10.5	13.1	15.7	21	26.2	31.4	36.7	41.9	47.1	52	66	79	92	105	262
	12	5.2	6.5	7.8	10.4	13	15.6	18.2	20.8	23.4	26.1	32.6	39.1	45.6	52	130
	8	28.9	36.1	43.4	58	72	87	101	116	130	145	181	217	253	289	723
1000	10	11.6	14.6	17.5	23.3	29.1	34.9	40.7	46.6	52	58	73	87	102	116	291
	12	5.8	7.2	8.7	11.6	14.5	17.4	20.3	23.2	26.1	28.9	36.2	43.4	51	58	145

← LAMINAR FLOW →

NOTE: For a liquid having a specific gravity other than 1.00 multiply the value from the table by the specific gravity of that liquid. For commercial installations, it is recommended that 15% be added to the above values. No allowance for aging of pipe is included.

Fluid Flow

Discharge From Nozzles

The actual discharge from nozzles will vary from 90% to 95% of the theoretical discharge.

For hydraulic sluicing the following pressures at base of nozzles are recommended:

Light fine material and loose sandy soil
50 to 75 PSIG

Stiff loam overburden ....100 to 125 PSIG

Hard clay ....150 to 200 PSIG

The following amounts of water are recommended to move 1 cubic foot.

Heavy sand .... 20 cubic feet of water

Average quarry overburden, 8 to 10 cu ft of water

For sprinkling golf courses it is reasonably safe to figure 10 gallons per minute for each Green with 35 PSIA at hose valve outlet and 25 gallons per minute for each Fairway with 35 PSIA at hose valve outlet. The standard conditions are 100 feet of ¾ inch hose for Greens sprinklers and 100 feet of 1 inch hose for Fairways sprinklers. Use gate valves with a hose nipple at all outlets instead of the customary hose faucets on account of the friction loss in the hose faucets.

IIID-1 —THEORETICAL DISCHARGE OF NOZZLES IN U.S. GALLONS PER MINUTE

TABLE 35

Head		Veloc'y of Disch. Feet per Sec.	Diameter of Nozzle in Inches												
PSIG	Feet		1/16	1/8	3/16	1/4	3/8	1/2	5/8	3/4	7/8	1	1 1/8	1 1/4	1 3/8
10	23.1	38.6	0.37	1.48	3.32	5.91	13.3	23.6	36.9	53	72	94	120	148	179
15	34.6	47.25	0.45	1.81	4.06	7.24	16.3	28.9	45.2	65	88	116	147	181	219
20	46.2	54.55	0.52	2.09	4.69	8.35	19.8	33.4	52.2	75	102	134	169	209	253
25	57.7	61.0	0.58	2.34	5.25	9.34	21.0	37.3	58.3	84	114	149	189	234	283
30	69.3	66.85	0.64	2.56	5.75	10.2	23.0	40.9	63.9	92	125	164	207	256	309
35	80.8	72.2	0.69	2.77	6.21	11.1	24.8	44.2	69.0	99	135	177	224	277	334
40	92.4	77.2	0.74	2.96	6.64	11.8	26.6	47.3	73.8	106	145	189	239	296	357
45	103.9	81.8	0.78	3.13	7.03	12.5	28.2	50.1	78.2	113	153	200	253	313	379
50	115.5	86.25	0.83	3.30	7.41	13.2	29.7	52.8	82.5	119	162	211	267	330	399
55	127.0	90.5	0.87	3.46	7.77	13.8	31.1	55.3	86.4	125	169	221	280	346	418
60	138.6	94.5	0.90	3.62	8.12	14.5	32.5	57.8	90.4	130	177	231	293	362	438
65	150.1	98.3	0.94	3.77	8.45	15.1	33.8	60.2	94.0	136	184	241	305	376	455
70	161.7	102.1	0.98	3.91	8.78	15.7	35.2	62.5	97.7	141	191	250	317	391	473
75	173.2	105.7	1.01	4.05	9.08	16.2	36.4	64.7	101.0	146	198	259	327	404	489
80	184.8	109.1	1.05	4.18	9.39	16.7	37.6	66.8	104.0	150	205	267	338	418	505
85	196.3	112.5	1.08	4.31	9.67	17.3	38.8	68.9	108.0	155	211	276	349	431	521
90	207.9	115.8	1.11	4.43	9.95	17.7	39.9	70.8	111.0	160	217	284	359	443	536
95	219.4	119.0	1.14	4.56	10.2	18.2	41.0	72.8	114.0	164	223	292	369	456	551
100	230.9	122.0	1.17	4.67	10.5	18.7	42.1	74.7	117.0	168	229	299	378	467	565
105	242.4	125.0	1.20	4.79	10.8	19.2	43.1	76.5	120.0	172	234	306	388	479	579
110	254.0	128.0	1.23	4.90	11.0	19.6	44.1	78.4	122.0	176	240	314	397	490	593
115	265.5	130.9	1.25	5.01	11.2	20.0	45.1	80.1	125.0	180	245	320	406	501	606
120	277.1	133.7	1.28	5.12	11.5	20.5	46.0	81.8	128.0	184	251	327	414	512	619
125	288.6	136.4	1.31	5.22	11.7	20.9	47.0	83.5	130.0	188	256	334	423	522	632
130	300.2	139.1	1.33	5.33	12.0	21.3	48.0	85.2	133.0	192	261	341	432	533	645
135	311.7	141.8	1.36	5.43	12.2	21.7	48.9	86.7	136.0	195	266	347	439	543	656
140	323.3	144.3	1.38	5.53	12.4	22.1	49.8	88.4	138.0	199	271	354	448	553	668
145	334.8	146.9	1.41	5.62	12.6	22.5	50.6	89.9	140.0	202	275	360	455	562	680
150	346.4	149.5	1.43	5.72	12.9	22.9	51.6	91.5	143.0	206	280	366	463	572	692
175	404.1	161.4	1.55	6.18	13.9	24.7	55.6	98.8	154.0	222	302	395	500	618	747
200	461.9	172.6	1.65	6.61	14.8	26.4	59.5	106.0	165.0	238	323	423	535	660	799
250	577.4	193.0	1.85	7.39	16.6	29.6	66.5	118.0	185.0	266	362	473	598	739	894
300	692.8	211.2	2.02	8.08	18.2	32.4	72.8	129.0	202.0	291	396	517	655	808	977

Note:—The actual quantities will vary from these figures, the amount of variation depending upon the shape of nozzle and size of pipe at the point where the pressure is determined. With smooth taper nozzles the actual discharge is about 94% of the figures given in the tables.

IIID-1 TABLE 35—THEORETICAL DISCHARGE OF NOZZLES IN U.S. GALLONS PER MINUTE (CONT'D)

Head		Veloc'y of Disch. Feet per Sec.	Diameter of Nozzle in Inches												
PSIG	Feet		1½	1¾	2	2¼	2½	2¾	3	3½	4	4½	5	5½	6
10	23.1	38.6	213	289	378	479	591	714	851	1158	1510	1915	2365	2855	3405
15	34.6	47.2	260	354	463	585	723	874	1041	1418	1850	2345	2890	3490	4165
20	46.2	54.5	301	409	535	676	835	1009	1203	1638	2135	2710	3340	4040	4810
25	57.7	61.0	336	458	598	756	934	1128	1345	1830	2385	3025	3730	4510	5380
30	69.3	66.8	368	501	655	828	1023	1236	1473	2005	2615	3315	4090	4940	5895
35	80.8	72.2	398	541	708	895	1106	1335	1591	2168	2825	3580	4415	5340	6370
40	92.4	77.2	425	578	756	957	1182	1428	1701	2315	3020	3830	4725	5610	6810
45	103.9	81.8	451	613	801	1015	1252	1512	1802	2455	3200	4055	5000	6050	7210
50	115.5	86.2	475	647	845	1070	1320	1595	1900	2590	3375	4275	5280	6380	7600
55	127.0	90.5	498	678	886	1121	1385	1671	1991	2710	3540	4480	5530	6690	7970
60	138.6	94.5	521	708	926	1172	1447	1748	2085	2835	3700	4685	5790	6980	8330
65	150.1	98.3	542	737	964	1220	1506	1819	2165	2950	3850	4875	6020	7270	8670
70	161.7	102.1	563	765	1001	1267	1565	1888	2250	3065	4000	5060	6250	7560	9000
75	173.2	105.7	582	792	1037	1310	1619	1955	2330	3170	4135	5240	6475	7820	9320
80	184.8	109.1	602	818	1070	1354	1672	2020	2405	3280	4270	5410	6690	8080	9630
85	196.3	112.5	620	844	1103	1395	1723	2080	2480	3375	4400	5575	6890	8320	9920
90	207.9	115.8	638	868	1136	1436	1773	2140	2550	3475	4530	5740	7090	8560	10210
95	219.4	119.0	656	892	1168	1476	1824	2200	2625	3570	4655	5900	7290	8800	10500
100	230.9	122.0	672	915	1196	1512	1870	2255	2690	3660	4775	6050	7470	9030	10770
105	242.4	125.0	689	937	1226	1550	1916	2312	2755	3750	4890	6200	7650	9260	11020
110	254.0	128.0	705	960	1255	1588	1961	2366	2820	3840	5010	6340	7840	9470	11300
115	265.5	130.9	720	980	1282	1621	2005	2420	2885	3930	5120	6490	8010	9680	11550
120	277.1	133.7	736	1002	1310	1659	2050	2470	2945	4015	5225	6630	8180	9900	11800
125	288.6	136.4	751	1022	1338	1690	2090	2520	3005	4090	5340	6760	8350	10100	12030
130	300.2	139.1	767	1043	1365	1726	2132	2575	3070	4175	5450	6900	8530	10300	12290
135	311.7	141.8	780	1063	1390	1759	2173	2620	3125	4250	5550	7030	8680	10490	12510
140	323.3	144.3	795	1082	1415	1790	2212	2670	3180	4330	5650	7160	8850	10690	12730
145	334.8	146.9	809	1100	1440	1820	2250	2715	3235	4410	5740	7280	8990	10880	12960
150	346.4	149.5	824	1120	1466	1853	2290	2760	3295	4485	5850	7410	9150	11070	13200
175	404.1	161.4	890	1210	1582	2000	2473	2985	3560	4840	6310	8000	9890	11940	14250
200	461.9	172.6	950	1294	1691	2140	2645	3190	3800	5175	6760	8550	10580	12770	15220
250	577.4	193.0	1063	1447	1891	2392	2955	3570	4250	5795	7550	9570	11820	14290	17020
300	692.8	211.2	1163	1582	2070	2615	3235	3900	4650	6330	8260	10480	12940	15620	18610

IIIE-1 FRICTION OF PAPER STOCK SUSPENSIONS IN PIPES

I. INTRODUCTION

In any stock piping system, the pump provides flow and develops hydraulic pressure (head) to overcome the differential in head between two points. This total head differential consists of pressure head, static head, velocity head and total friction head produced by friction between the pulp suspension and the pipe, bends, and fittings. The total friction head is the most difficult to determine because of the complex, non-linear nature of the friction loss curve. This curve can be affected by many factors.

The following analytical method for determining pipe friction loss is based on the recently published TAPPI Technical Information Sheet (TIS) 408-4 (Reference I), and is applicable to stock consistencies (oven-dried) from 2 to 6 percent. Normally, stock consistencies of less than 2% (oven-dried) are considered to have the same friction loss characteristic as water. This paper only applies to systems using centrifugal pumps which is normal for these consistencies. The method for determining the friction loss of pulp suspensions in pipe, as presented here, is intended to supersede the various methods previously issued.

II. BACKGROUND

Figure 1 and Figure 2 show typical friction loss curves for two different consistencies ($C_2 > C_1$) of chemical pulp and mechanical pulp, respectively.

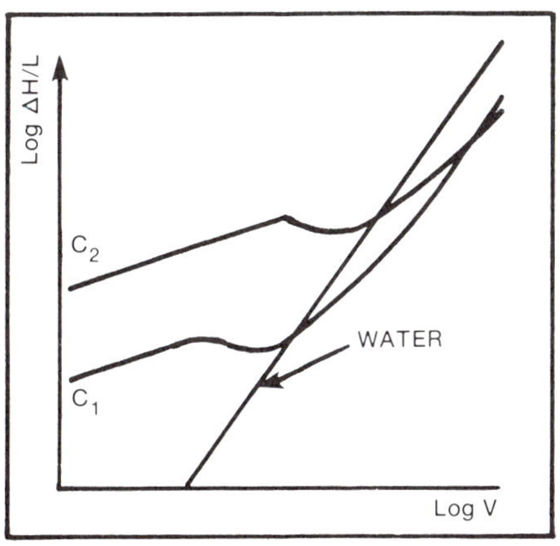

Figure 1—Friction loss curves for chemical pulp ($C_2 > C_1$).

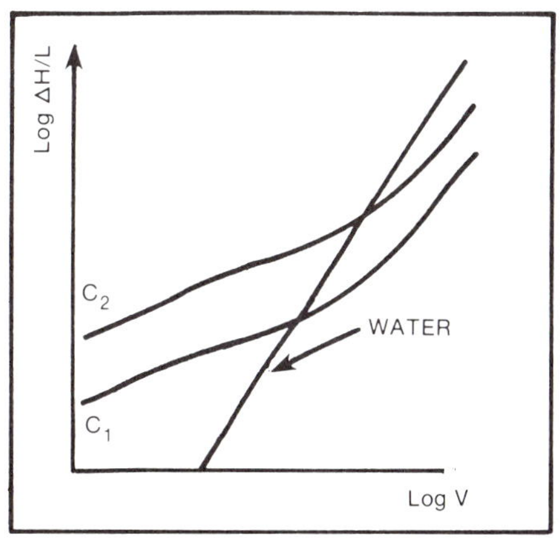

Figure 2—Friction loss curves for mechanical pulp ($C_2 > C_1$).

The friction loss curve for chemical pulp can be conveniently divided into three regions, as illustrated by the shaded areas of Figure 3.

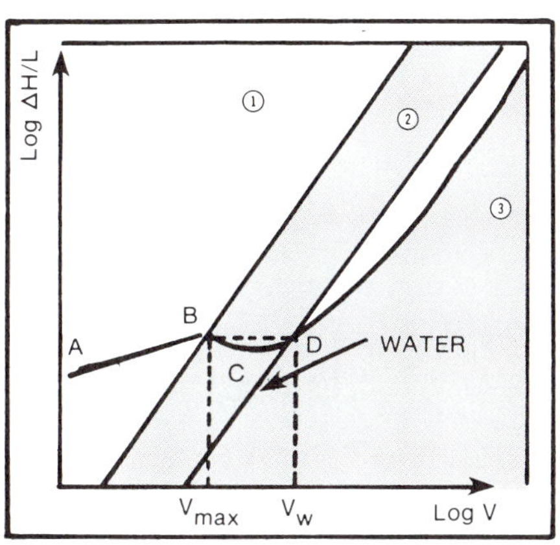

Figure 3.—Friction loss curve for chemical pulp, shaded to show individual regions.

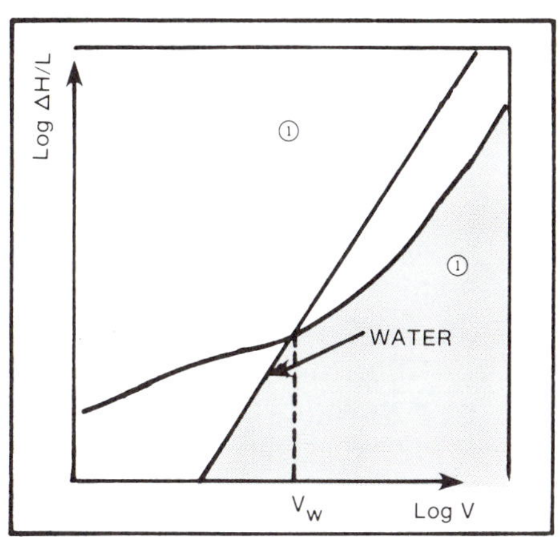

Figure 4—Friction loss curve for mechanical pulp, shaded to show individual regions.

Regions shown in Fig. 3 may be described as follows:

Region 1 (Curve AB) is a linear region where friction loss for a given pulp is a function of consistency, velocity, and pipe diameter. The velocity at the upper limit of this linear region (Point B) is designated V_{max}.

Region 2 (Curve BCD) shows an initial decrease in friction loss (to Point C) after which the friction loss again increases. The intersection of the pulp friction loss curve and the water friction loss curve (Point D) is termed the onset of drag reduction. The velocity at this point is designated V_w.

Region 3 (Curve DE) shows the friction loss curve for pulp fiber suspensions below the water curve. This is due to a phenomenon called drag reduction. Reference 2 describes the mechanisms which occur in this region.

Regions 2 and 3 are separated by the friction loss curve for water, which is a straight line with a slope approximately equal to 2.

The friction loss curve for mechanical pulp, as illustrated in Figure 4, is divided into only two regions: Regions 1 and 3. For this pulp type, the friction loss curve crosses the water curve V_w and there is no true V_{max}.

III. DESIGN PARAMETERS

To determine the pipe friction loss component for a specified design basis (usually daily mass flow rate), the following parameters must be defined:

a) Pulp Type — Chemical or mechanical pulp, long or short fibered, never dried or dried and reslurred , etc. This is required to choose the proper coefficients which define the pulp friction curve.

b) Consistency, C (oven-dried) — Often a design constraint in an existing system. NOTE: If air-dried consistency is known, multiply by 0.9 to convert to oven-dried consistency.

c) Internal pipe diameter, D — Lowering D reduces initial capital investment, but increases pump operating costs. Once the pipe diameter is selected, it fixes the velocity for a prespecified mass flow rate.

d) Bulk velocity, V — Usually based on a prespecified daily mass flow rate. Note that both V and D are interdependent for a constant mass flow rate.

e) Stock temperature, T — Required to adjust for the effect of changes in viscosity of water (the suspending medium) on pipe friction loss.

f) Freeness — Used to indicate the degree of refining or to define the pulp for comparison purposes.

g) Pipe material — Important to specify design correlations and compare design values.

IV. PIPE FRICTION ESTIMATION PROCEDURE

The bulk velocity (V) will depend on the daily mass flow rate and the pipe diameter (D) selected. The final value of V can be optimized to give the lowest capital investment and operating cost with due consideration of future demands or possible system expansion.

The bulk velocity will fall into one of the regions previously discussed. Once it has been determined in which region the design velocity will occur, the appropriate correlations for determining pipe friction loss value(s) may be selected. The following describes the procedure to be used for estimating pipe friction loss in each of the regions.

Region 1 The upper limit of Region 1 in Figure 3 (Point B) is designated V_{max}. The value of V_{max} is determined using Equation (1) and data given in Table I.

$$V_{max} = K' \, C^{\sigma} \text{ (ft/s)}, \qquad (1)$$

where

K′ = numerical coefficient (constant for a given pulp), obtained from Table I

C = consistency (oven-dried, expressed as a percentage, *not* decimally), 2-6% limit.

Fluid Flow
Friction Loss of Paper Stock

TABLE I

Data for use with Equation (1) to determine velocity limit, V_{max}.

Pulp Type (5)	Pipe Material	K	σ
Unbeaten aspen sulfite never dried (2)	Stainless Steel	0.85	1.6
Long fibered kraft never dried CSF = 725 (6) (2)	PVC Stainless Steel	0.98 0.89	1.85 1.5
Long fibered kraft never dried CSF = 650 (6) (2)	PVC	0.85	1.9
Long fibered kraft never dried CSF = 550 (6) (2)	PVC	0.75	1.65
Long fibered kraft never dried CSF = 260 (6) (2)	PVC	0.75	1.8
Bleached kraft pine dried and reslurried (6) (2)	PVC Stainless Steel	0.79 0.59	1.5 1.45
Long fibered kraft dried and reslurried (6) (2)	PVC	0.49	1.8
Kraft birch dried and reslurried (6) (2)	PVC	0.69	1.3
Stone groundwood CSF = 114 (2)	PVC	4.0	1.40
Refiner groundwood CSF = 150 (2)	PVC	4.0	1.40
Newsprint broke CSF = 75 (2)	PVC	4.0	1.40
Refiner groundwood (hardboard) (2)	PVC	4.0	1.40
Refiner groundwood (insulating board) (2)	PVC	4.0	1.40
Hardwood NSSC CSF = 620 (2)	PVC	0.59	1.8
Unbleached sulfite (1)	Copper	0.98	1.2
Bleached sulfite (1)	Copper	0.98	1.2
Kraft (1)	Copper	0.98	1.2
Bleached straw (1)	Copper	0.98	1.2
Unbleached straw (1)	Copper	0.98	1.2
Cooked groundwood (1)	Copper	0.75	1.8
Soda (1)	Steel	4.0	1.4

NOTES:
1. Estimates for pulps based on published literature.
2. Original data obtained in stainless steel and PVC pipe.
3. Stainless steel may be hydraulically smooth although some manufacturing processes may destroy the surface and hydraulic smoothness is lost. PVC is taken to be hydraulically smooth pipe.
4. For cast iron and galvanized pipe, the K^9 values will be reduced. No systematic data are available for the effects of surface roughness.
5. If pulps are not identical to those shown, some engineering judgement is required.
6. Wood is New Zealand Kraft pulp.

σ = exponent (constant for a given pulp), obtained from Table I

If the proposed design velocity (V) is less than V_{max}, the value of flow resistance ($\Delta H/L$) may be calculated using Equation (2) and data given in Table II and the appendices.

$$\Delta H/L = F\,K\,V^{\alpha}\,C^{\beta}\,D^{\gamma}\ \text{(ft/100 ft)}, \qquad (2)$$

where

F = factor to correct for temperature, pipe roughness, pulp type, freeness, or safety factor (refer to Appendix A)

K = numerical coefficient (constant for a given pulp), obtained from Table II

V = bulk velocity (ft/s)

C = consistency (oven-dried, expressed as a percentage, *not* decimally), 2-6% limit

D = pipe inside diameter (in), and

α, β, γ = exponents (constant for a given pulp), obtained from Table II

For mechancial pulps, there is no true V_{max}. The upper limit of the correlation equation (Equation (2)) is also given by Equation (1) in this case, the upper velocity is actually V_W.

Region 2 The lower limit of Region 2 in Figure 3 (Point B) is V_{max} and the upper limit (Point D) is V_W. The velocity of the stock at the onset of drag reduction is determined using Equation (3).

$$V_W = 4.00\,C^{1.40}\ \text{(ft/s)}, \qquad (3)$$

where

C = consistency (oven-dried, expressed as a percentage, *not* decimally).

If V is between V_{max} and V_W, Equation (2) may be used to determine $\Delta H/L$ at the maximum point (V_{max}). Because the system must cope with the worst flow condition, $\Delta H/L$ at the maximum point (V_{max}) can be used for all design velocities between V_{max} and V_W.

Region 3 A conservative estimate of friction loss is obtained by using the water curve. $(\Delta H/L)_W$ can be obtained from a Friction Factor vs. Reynolds Number plot (Reference 3, for example), or approximated from the following equation (based on the Blasius equation).

$$(\Delta H/L)_W = 0.58\,V^{1.75}\,D^{-1.25}\ \text{(ft/100 ft)}, \qquad (4)$$

where

V = bulk velocity (ft/s), and

D = pipe diameter (in).

Previously published methods for calculating pipe friction loss of pulp suspensions gave a very conservative estimate of head loss. The method just described gives a more accurate estimate of head loss due to friction, and has been used successfully in systems in North America and world-wide.

Pertinent equations, in addition to those herein presented, are located in Appendix A. Example problems are located in Appendix B.

V. HEAD LOSSES IN BENDS AND FITTINGS

The friction head loss of pulp suspensions in bends and fittings may be determined from the basic equation for head loss, Equation (5).

$$H = K\,V_1^2/2g\ \text{(ft)}, \qquad (5)$$

where

K = loss coefficient for a given fitting,

V_1 = inlet velocity (ft/s), and

g = acceleration due to gravity (32.2 ft/s²).

Values of K for the flow of water through various types of bends and fittings are tabulated in numerous reference sources (Reference 3, for example). The loss coefficient for valves may be obtained from the valve manufacturer.

The loss coefficient for pulp suspensions in a given bend or fitting generally exceeds the loss coefficient for water in the same bend or fitting. As an approximate rule, the loss coefficient (K) increases 20 percent for each 1 percent increase in oven-dried stock consistency. Please note that this is an approximation; actual values of K may differ, depending on the type of bend or fitting under consideration (4).

Friction Loss of Paper Stock

TABLE II

Data for use with Equation (2) to determine head loss. ΔH/L.

Pulp Type	K	α	β	γ (3), (4)
Unbeaten aspen sulfite never dried (2)	5.30	0.36	2.14	-1.04
Long fibered kraft never dried CSF = 725 (5) (2)	11.80	0.31	1.81	-1.34
Long fibered kraft never dried CSF = 650 (5) (2)	11.30	0.31	1.81	-1.34
Long fibered kraft never dried CSF = 550 (5) (2)	12.10	0.31	1.81	-1.34
Long fibered kraft never dried CSF = 260 (5) (2)	17.00	0.31	1.81	-1.34
Bleached draft pine dried and reslurried (5) (2)	8.80	0.31	1.81	-1.34
Long fibered kraft dried and reslurried (5) (2)	9.40	0.31	1.81	-1.34
Kraft birch dried and reslurried (5) (2)	5.20	0.27	1.78	-1.08
Stone groundwood CSF = 114 (2)	3.81	0.27	2.37	-0.85
Refiner groundwood CSF = 150 (2)	3.40	0.18	2.34	-1.09
Newsprint broke CSF = 75 (2)	5.19	0.36	1.91	-0.82
Refiner groundwood (hardboard) (2)	2.30	0.23	2.21	-1.29
Refiner groundwood (insulating board) (2)	1.40	0.32	2.19	-1.16
Hardwood NSSC CSF = 620 (2)	4.56	0.43	2.31	-1.20
Unbleached sulfite (1)	12.69	0.36	1.89	-1.33
Bleached sulfite (1)	11.40	0.36	1.89	-1.33
Kraft (1)	11.40	0.36	1.89	-1.33
Bleached straw (1)	11.40	0.36	1.89	-1.33
Unbleached straw (1)	5.70	0.36	1.89	-1.33
Cooked groundwood (1)	6.20	0.43	2.31	-1.20
Soda (1)	6.50	0.36	1.85	-1.04

NOTES:
1. Estimates for pulbs based on published literature.
2. Original data obtained in stainless steel and PVC pipe (7. 8, 9).
3. No safety factors are included in the above correlations.
4. The friction loss depends considerably on the condition of the inside of the pipe surface (10).
5. Wood is New Zealand Kraft pulp.

APPENDIX A

The following gives supplemental information to that provided in the main text.

1. Capacity (flow), Q —

$$Q = \frac{16.65\ (\text{T.P.D.})}{C}\ (\text{U.S. GPM}), \qquad \text{(i)}$$

where

T.P.D. = mill capacity (short tons per day),

and

C = consistency (oven-dried, expressed as a percentage, *not* decimally).

2. Bulk velocity, V —

$$V = \frac{0.321\ Q}{A}\ (\text{ft/s}), \text{ or} \qquad \text{(ii)}$$

$$V = \frac{0.4085\ Q}{D^2}\ (\text{ft/s}), \qquad \text{(iii)}$$

where

Q = capacity (U.S. GPM),

A = inside area of pipe (in^2), and

D = inside diameter of pipe (in).

3. Multiplication Factor, F (included in Equation (2)—

$$F = F_1 \bullet F_2 \bullet F_3 \bullet F_4 \bullet F_5, \qquad \text{(iv)}$$

where

F_1 = correction factor for temperature. Friction loss calculations are normally based on a reference pulp temperature of 95° F (35° C). The flow resistance may be increased or decreased by 1 percent for each 1.8° F (1° C) below or above 95° F (35° C), respectively. This may be expressed as follows:

$$F_1 = 1.526 - 0.00556\ T, \qquad \text{(v)}$$

where
T = pulp temperature (° F)

F_2 = correction factor for pipe roughness. This factor may vary due to manufacturing processes of the piping, surface roughness, age, etc. Typical values for PVC and stainless steel piping are listed below:
F_2 = 1.0 for PVC piping,
F_2 = 1.25 for stainless steel piping.

Please note that the above are typical values; experience and/or additional data may modify the above factors.

F_3 = correction factor for pulp type. Typical values are listed below:
F_3 = 1.0 for pulps that have never been dried reslurried,
F_3 = 0.8 for pulps that have been dried and reslurried.

Note: This factor has been incorporated in the numerical coefficient, K, for the pulps listed in Table II. When using Table II, F_3 should *not* be used.

F_4 = correction factor for beating. Data have shown that progressive beating causes, initially, a small decrease in friction loss, followed by a substantial increase. For a kraft pine pulp initially at 725 CSF and F_4 = 1.0, beating caused the freeness to decrease to 636 CSF and F_4 to decrease to 0.96. Progressive beating decreased the freeness to 300 CSF and increased F_4 to 1.37 (see K values in Table II). Some engineering judgement may be required.

F_5 = design safety factor. This is usually specified by company policy with consideration given to future requirements.

APPENDIX B

The following are three examples which illustrate the method for determination of pipe friction loss in each of the three regions shown in Figure 3.

Example 1.
Determine the friction loss (per 100 ft of pipe) for 1000 U.S. GPM of 4.5% oven-dried unbeaten aspen sulfite stock, never dried, in 8 inch schedule 40 stainless steel pipe (pipe inside diameter = 7.981 in). Assume the pulp temperature to be 95° F.

Solution:
a) The bulk velocity, V, is

$$V = \frac{0.4085\ Q}{D^2}, \qquad \text{(ii)}$$

and

Q = flow = 1000 U.S. GPM.

D = pipe inside diameter = 7.981 in.

$$V = \frac{0.4085\ (1000)}{7.981^2} = 6.41 \text{ ft/s.}$$

b) It must be determined in which region (1, 2, or 3) this velocity falls. Therefore, the next step is to determine the velocity at the upper limit of the linear region, V_{max}.

$$V_{max} = K'\ C^{\sigma} \qquad (1)$$

and

K' = numerical coefficient = 0.85 (from Table I).

C = consistency = 4.5%,

σ = exponent 1.6 (from Table I).

$$V_{max} = 0.85\ (4.5^{1.6}) = 9.43 \text{ ft/s.}$$

c) Since V_{max} exceeds V, the friction loss, $\Delta H/L$, falls within the linear region, Region 1. The friction loss is given by the correlation:

$$\Delta H/L = F\ K\ V^{\alpha}\ C^{\beta}\ D^{\gamma} \qquad (2)$$

and

F = correction factor = $F_1 \bullet F_2 \bullet F_3 \bullet F_4 \bullet F_5$,

F_1 = correction factor for pulp temperature. Since the pulp temperature is 95° F,

F_1 = 1.0,

F_2 = correction factor for pipe roughness. For stainless steel pipe,

F_2 = 1.25 (from Appendix A).

F_3 = correction factor for pulp type. Numerical coefficients for this pulp are contained in Table II, and have already incorporated this factor.

F_4 = correction factor for beating. No additional beating has taken place, therefore

F_4 = 1.0 (from Appendix A).

F_5 = design safety factor. This has been assumed to be unity.

F_5 = 1.0.

F = (1.0) (1.25) (1.0) (1.0) (1.0) = 1.25.

K = numerical coefficient = 5.30 (from Table II).

α, β, γ = exponents = 0.36, 2.14, and -1.04, respectively (from Table II).

V, C, D have been evaluated previously.

$$\begin{aligned}\Delta H/L &= (1.25)\ (5.30)\ (6.41^{0.36})\ (4.5^{2.14})\ (7.981^{-1.04}) \\ &= (1.25)\ (5.30)\ (1.952)\ (25.0)\ (0.1153) \\ &= 37.28 \text{ ft head loss/100 ft of pipe.}\end{aligned}$$

This is a rather substantial head loss, but may be acceptable for short piping runs. In a large system, the economics of initial piping costs versus power costs should be weighed, however, before using piping which gives a friction loss of this magnitude.

Example 2.

Determine the friction loss (per 100 ft of pipe) of 2500 U.S. GPM of 3% oven-dried bleached kraft pine, dried and reslurried in 12 inch schedule 10 stainless steel pipe (pipe inside diameter = 12.39 in). Stock temperature is 125° F.

Solution:

a) V, the bulk velocity, is

$$V = \frac{0.4085\ Q}{D^2} \qquad \text{(ii)}$$

$$= \frac{0.4085\ (2500)}{12.39^2} = 6.65 \text{ ft/s.}$$

b) The velocity at the upper limit of the linear region, V_{max}, is

$$V_{max} = K'\ C^{\sigma} \qquad (1)$$

and

K' = 0.59 (from Table I),

σ = 1.45 (from Table I).

$$V_{max} = 0.59\ (3.0^{1.45}) = 2.90 \text{ ft/s.}$$

c) Region 1 (the linear region) has been eliminated, since the bulk velocity, V, exceeds V_{max}.

The next step requires calculation of V_w.

$$V_w = 4.00\ C^{1.40} \quad (3)$$

$$= 4.00\ (3.0^{1.40}) = 18.62 \text{ ft/s.}$$

d) V exceeds V_{max}, but is less than V_w, indicating that it falls in Region 2. The friction loss in this region is calculated by substituting V_{max} into the equation for head loss, Equation (2).

$$\Delta H/L = F\ K\ (V_{max})^{\alpha}\ C^{\beta}\ D^{\gamma},$$

and

$$F = F_1 \bullet F_2 \bullet F_3 \bullet F_5; \quad \text{(iv)}$$

$$F_1 = 1.528 - 0.00556T, \quad \text{(v)}$$

and

$$T = \text{stock temperature} = 125°\text{F}$$

$$F_1 = 1.528 - 0.00556\ (125) = 0.833,$$

$$F_2 = 1.25 \text{ (from Appendix A)}$$

$$F_3 = F_4 = F_5 = 1.0,$$

$$F = 0.833\ (1.25)\ (1.0) = 1.041,$$

$$K = 8.80 \text{ (from Table II).}$$

α, β, γ = 0.31, 1.81, and -1.34, respectively (from Table II),

V_{max}, C, and D have been defined previously.

$$\Delta H/L = 1.041\ (8.80)\ (2.90^{0.31})\ (3.0^{1.81})\ (12.39^{-1.34})$$
$$= 1.041\ (8.80)\ (1.391)\ (7.304)\ (0.03430)$$
$$= 3.19 \text{ ft head loss/100 ft of pipe.}$$

Example 3.

Determine the friction loss (per 100 ft of pipe) for 2% oven-dried bleached kraft pine, dried and reslurred, through 6 inch schedule 40 stainless steel pipe (inside diameter = 6.065 in). The pulp temperature is 90° F, the flow rate 1100 U.S. GPM.

Solution:

a) The bulk velocity is

$$V = \frac{0.4085\ Q}{D^2}$$

$$= \frac{0.4085\ (1100)}{6.065^2} = 12.22 \text{ ft/s.} \quad \text{(ii)}$$

b) It must be determined in which region (1, 2 or 3) this velocity falls. To obtain an initial indication, determine V_{max}.

$$V_{max} = K'\ C^{\sigma}, \quad (1)$$

and

$$K' = 0.59 \text{ (from Table I),}$$

$$\sigma = 1.45 \text{ (from Table I),}$$

$$V_{max} = 0.59\ (2.0^{1.45}) = 1.61 \text{ ft/s.}$$

c) Since V exceeds V_{max}, Region 1 (the linear region) is eliminated. To determine whether V lies in Region 2 or 3, the velocity at the onset of drag reduction, V_w, must be calculated.

$$V_w = 4.00\ C^{1.40}, \quad (3)$$

$$= 4.00\ (2.0^{1.40}) = 10.56 \text{ ft/s.}$$

d) V exceeds V_w, indicating that it falls in Region 3. The friction loss is calculated as that of water flowing at the same velocity.

$$(\Delta H/L)_w = 0.579\ V^{1.75}\ D^{-1.25}, \quad (4)$$

$$= 0.579\ (12.22^{1.75})\ (6.065^{-1.25})$$

$$= 4.85 \text{ ft head loss/100 ft of pipe.}$$

This will be a conservative estimate, as the actual friction loss curve for pulp suspensions under these conditions will be below the water curve.

REFERENCES

(1) TAPPI Technical Information Sheet (TIS) 408-4. Technical Association of the Pulp and Paper Industry, Atlanta, Georgia (1981).
(2) K. Molier and G. G. Duffy, TAPPI 61, 1, 63 (1978).
(3) Hydraulic Institute Engineering Data Book, First Edition, Hydraulic Institute, Cleveland, Ohio (1979).
(4) K. Molier and G. Elmquist, TAPPI 63, 3, 101 (1980).
(5) W. Brecht and H. Heller, TAPPI 33, 9, 14A (1950).
(6) R. E. Durst and L. C. Jenness, TAPPI 39, 5, 277 (1956).
(7) K. Molier, G. G. Duffy and A. L. Titchener, APPITA 26, 4, 278 (1973).
(8) G. G. Duffy and A. L. Tichener, TAPPI 57, 5, 162 (1974).
(9) G. G. Duffy, K. Molier, P. F. W. Lee and S. W. A. Mine, APPITA 27, 5, 327 (1974).
(10) G. G. Duffy, TAPPI 59, 8, 124 (1976).
(11) G. G. Duffy, Company Communications Goulds Pumps, Inc. (1980-1981).

Section IV

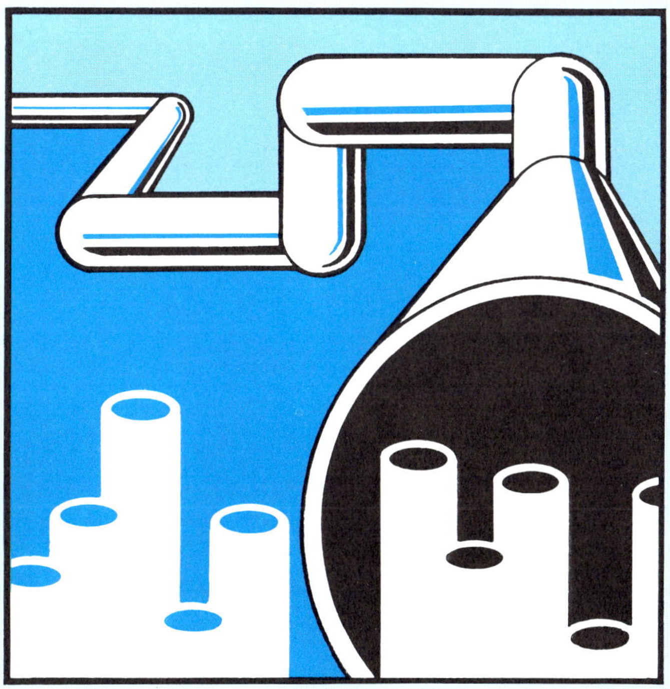

Characteristics of Piping Materials

IVA-1 PIPE DIMENSIONS; WROUGHT STEEL AND STAINLESS STEEL (ANSI B36.10MM-1985 AND ANSI B36.10M-1985)

NOTES:

1. Covers pipe manufactured to specifications noted in Table.
2. Data reprinted by permission of The American Society of Mechanical Engineers, 345 East 45th St., New York, N.Y. 10017.
3. "S" suffix to schedule number denotes wall thickness pertains only to stainless steel pipe. Stainless steel pipe schedule numbers with dimensions and schedule numbers the same as steel pipe are not suffixed in this table.
4. U.S. customary units only. For SI units see above ANSI Standard references.

Size Nominal & (Outside Diameter)	IDENTIFICATION			Wall Thickness Inches	Internal Diameter Inches	Internal Area Square Inches	Internal Diameter Feet	Internal Area Square Feet	ϵ/D ϵ=0.00015 Ft.
	API Standard	Standard X-Strong XX-Strong	Schedule Number						
1/8			10S	0.049	0.307	0.0740	0.0256	0.00051	0.00586
	5L	STD	40	0.068	0.269	0.0568	0.0224	0.00040	0.006691
(.405)	5L	XS	80	0.095	0.215	0.0363	0.0179	0.00025	0.008372
1/4			10S	0.065	0.410	0.1320	0.0342	0.00092	0.004390
	5L	STD	40	0.088	0.364	0.1041	0.0303	0.00072	0.004945
(.540)	5L	XS	80	0.119	0.302	0.0716	0.0252	0.00050	0.005960
3/8			10S	0.065	0.545	0.2333	0.0454	0.00162	0.003303
	5L	STD	40	0.091	0.493	0.1909	0.0411	0.00133	0.003651
(.675)	5L	XS	80	0.126	0.423	0.1405	0.0353	0.00098	0.004255
1/2			5S	0.065	0.710	0.3959	0.0592	0.00275	0.002535
			10S	0.083	0.674	0.3568	0.0562	0.00248	0.002671
(.840)	5L	STD	40	0.109	0.622	0.3039	0.0518	0.00211	0.002894
	5L	XS	80	0.147	0.546	0.2341	0.0455	0.00163	0.003297
			160	0.188	0.464	0.1691	0.0387	0.00117	0.003879
	5L	XXS		0.294	0.252	0.0499	0.0210	0.00035	0.007143
3/4			5S	0.065	0.920	0.6648	0.0767	0.00462	0.001957
			10S	0.083	0.884	0.6138	0.0737	0.00426	0.002036
(1.050)	5L	STD	40	0.113	0.824	0.5333	0.0687	0.00370	0.002184
	5L	XS	80	0.154	0.742	0.4324	0.0618	0.00300	0.002426
			160	0.219	0.612	0.2942	0.0510	0.00204	0.002941
	5L	XXS		0.308	0.434	0.1479	0.0362	0.00103	0.004147
1			5S	0.065	1.185	1.1029	0.0988	0.00766	0.001519
			10S	0.109	1.097	0.9452	0.0914	0.00656	0.001641
(1.315)	5L	STD	40	0.133	1.049	0.8643	0.0874	0.00600	0.001716
	5L	XS	80	0.179	0.957	0.7193	0.0798	0.00500	0.001881
			160	0.250	0.815	0.5217	0.0679	0.00362	0.002209
	5L	XXS		0.358	0.599	0.2818	0.0499	0.00196	0.003005
1 1/4			5S	0.065	1.530	1.8385	0.1275	0.01277	0.001176
			10S	0.109	1.442	1.6331	0.1202	0.01134	0.001248
(1.660)	5L	STD	40	0.140	1.380	1.4957	0.1150	0.01039	0.001304
	5L	XS	80	0.191	1.278	1.2828	0.1065	0.00891	0.001408
			160	0.250	1.160	1.0568	0.0967	0.00734	0.001552
	5L	XXS		0.382	0.896	0.6305	0.0747	0.00438	0.002009
1 1/2			5S	0.065	1.770	2.4606	0.1475	0.01709	0.001017
			10S	0.109	1.682	2.2220	0.1402	0.01543	0.001070
(1.900)	5L	STD	40	0.145	1.610	2.0358	0.1342	0.01414	0.001118
	5L	XS	80	0.200	1.500	1.7671	0.1250	0.01227	0.001200
			160	0.281	1.338	1.4061	0.1115	0.00976	0.001345
	5L	XXS		0.400	1.100	0.9503	0.0917	0.00660	0.001636

Characteristics of Piping Materials
Steel Pipe

IVA-1 PIPE DIMENSIONS; WROUGHT STEEL AND STAINLESS STEEL

Size Nominal & (Outside Diameter)	IDENTIFICATION			Wall Thickness Inches	Internal Diameter Inches	Internal Area Square Inches	Internal Diameter Feet	Internal Area Square Feet	ϵ/D ϵ=0.00015 Ft.
	API Standard	Standard X-Strong XX-Strong	Schedule Number						
2	5L		5S	0.065	2.245	3.9584	0.1871	0.02749	0.000802
			10S	0.109	2.157	3.6542	0.1798	0.02538	0.000834
(2.375)	5L	STD	40	0.154	2.067	3.3556	0.1723	0.02330	0.000871
	5L	XS	80	0.218	1.939	2.9529	0.1616	0.02051	0.000928
			160	0.344	1.687	2.2352	0.1406	0.01522	0.001067
	5L	XXS		0.436	1.503	1.7742	0.1253	0.01232	0.001198
2½	5L		5S	0.083	2.709	5.7638	0.2258	0.04003	0.000664
			10S	0.120	2.635	5.4532	0.2196	0.03787	0.000683
(2.875)	5L	STD	40	0.203	2.469	4.7878	0.2058	0.03325	0.000729
	5L	XS	80	0.276	2.323	4.2383	0.1936	0.02943	0.000775
			160	0.375	2.125	3.5466	0.1771	0.02463	0.000847
	5L	XXS		0.552	1.771	2.4634	0.1476	0.01711	0.001016
3	5L		5S	0.083	3.334	8.7301	0.2778	0.06063	0.000540
(3.500)			10S	0.120	3.260	8.3469	0.2717	0.05797	0.000552
	5L			0.125	3.250	8.2958	0.2708	0.05761	0.000554
	5L			0.156	3.188	7.9823	0.2657	0.05543	0.000565
	5L			0.188	3.124	7.6650	0.2603	0.05323	0.000576
	5L	STD	40	0.216	3.068	7.3927	0.2557	0.05134	0.000587
	5L			0.250	3.000	7.0686	0.2500	0.04909	0.000600
	5L			0.281	2.938	6.7794	0.2448	0.04708	0.000613
	5L	XS	80	0.300	2.900	6.6052	0.2417	0.04587	0.000621
			160	0.438	2.624	5.4078	0.2187	0.03755	0.000686
	5L	XXS		0.600	2.300	4.1548	0.1917	0.02885	0.000783
3½	5L		5S	0.083	3.834	11.5450	0.3195	0.08017	0.000469
(4.000)			10S	0.120	3.760	11.1036	0.3133	0.07711	0.000479
	5L			0.125	3.750	11.0447	0.3125	0.07670	0.000480
	5L			0.156	3.688	10.6825	0.3073	0.07418	0.000488
	5L			0.188	3.624	10.3149	0.3020	0.07163	0.000497
	5L	STD	40	0.226	3.548	9.8868	0.2957	0.06866	0.000507
	5L			0.250	3.500	9.6211	0.2917	0.06681	0.000514
	5L			0.281	3.438	9.2833	0.2865	0.06447	0.000524
	5L	XS	80	0.318	3.364	8.8880	0.2803	0.06172	0.000535
4	5L		5S	0.083	4.334	14.7526	0.3612	0.10245	0.000415
(4.500)			10S	0.120	4.260	14.2531	0.3550	0.09898	0.000423
	5L			0.125	4.250	14.1863	0.3542	0.09852	0.000424
	5L			0.156	4.188	13.7754	0.3490	0.09566	0.000430
	5L			0.188	4.124	13.3576	0.3437	0.09276	0.000436
	5L			0.219	4.062	12.9589	0.3385	0.08999	0.000443
	5L	STD	40	0.237	4.026	12.7303	0.3355	0.08841	0.000447
	5L			0.250	4.000	12.5664	0.3333	0.08727	0.000450
	5L			0.281	3.938	12.1798	0.3282	0.08458	0.000457
	5L			0.312	3.876	11.7993	0.3230	0.08194	0.000464
	5L	XS	80	0.337	3.826	11.4969	0.3188	0.07984	0.000470
	5L		120	0.438	3.624	10.3149	0.3020	0.07163	0.000497
	5L		160	0.531	3.438	9.2833	0.2865	0.06447	0.000524
	5L	XXS		0.674	3.152	7.8030	0.2627	0.05419	0.000571
5			5S	0.109	5.345	22.4381	0.4454	0.15582	0.000337
(5.563)			10S	0.134	5.295	22.0202	0.4413	0.15292	0.000340
	5L			0.156	5.251	21.6558	0.4376	0.15039	0.000343
	5L			0.188	5.187	21.1311	0.4323	0.14674	0.000347
	5L			0.219	5.125	20.6290	0.4271	0.14326	0.000351
	5L	STD	40	0.258	5.047	20.0058	0.4206	0.13893	0.000357
	5L			0.281	5.001	19.6428	0.4163	0.13641	0.000360
	5L			0.312	4.939	19.1588	0.4116	0.13305	0.000364
	5L			0.344	4.875	18.6655	0.4063	0.12962	0.000369
	5L	XS	80	0.375	4.813	18.1937	0.4011	0.12635	0.000374
	5L		120	0.500	4.563	16.3528	0.3803	0.11356	0.000394
	5L		160	0.625	4.313	14.6100	0.3594	0.10146	0.000417
	5L	XXS		0.750	4.063	12.9653	0.3386	0.09004	0.000443

Characteristics of Piping Materials
Steel Pipe

IVA-1 PIPE DIMENSIONS; WROUGHT STEEL AND STAINLESS STEEL

Size Nominal & (Outside Diameter)	IDENTIFICATION			Wall Thickness Inches	Internal Diameter Inches	Internal Area Square Inches	Internal Diameter Feet	Internal Area Square Feet	ϵ/D ϵ=0.00015 Ft.
	API Standard	Standard X-Strong XX-Strong	Schedule Number						
6 (6.625)	5L		5S	0.109	6.407	32.2403	0.5339	0.22389	0.000281
			10S	0.134	6.357	31.7391	0.5298	0.22041	0.000283
	5L			0.188	6.249	30.6698	0.5208	0.21299	0.000288
	5L			0.219	6.187	30.0642	0.5156	0.20878	0.000291
	5L			0.250	6.125	29.4647	0.5104	0.20462	0.000294
	5L	STD	40	0.280	6.065	28.8903	0.5054	0.20063	0.000297
	5L			0.312	6.001	28.2838	0.5001	0.19642	0.000300
	5L			0.344	5.937	27.6837	0.4948	0.19225	0.000303
	5L			0.375	5.875	27.1085	0.4896	0.18825	0.000306
	5L	XS	80	0.432	5.761	26.0667	0.4801	0.18102	0.000312
	5L		120	0.562	5.501	23.7669	0.4584	0.16505	0.000327
	5L		160	0.719	5.187	21.1311	0.4323	0.14674	0.000347
	5L	XXS		0.864	4.897	18.8343	0.4081	0.13079	0.000368
8 (8.625)			5S	0.109	8.407	55.5101	0.7006	0.38549	0.000214
			10S	0.148	8.329	54.4848	0.6941	0.37837	0.000216
	5L			0.188	8.249	53.4432	0.6874	0.37113	0.000218
				0.203	8.219	53.0552	0.6849	0.36844	0.000219
	5L			0.219	8.187	52.6429	0.6823	0.36558	0.000220
	5L		20	0.250	8.125	51.8486	0.6771	0.36006	0.000222
	5L		30	0.277	8.071	51.1617	0.6726	0.35529	0.000223
	5L			0.312	8.001	50.2781	0.6668	0.34915	0.000225
	5L	STD	40	0.322	7.981	50.0270	0.6651	0.34741	0.000226
	5L			0.344	7.937	49.4769	0.6614	0.34359	0.000227
	5L			0.375	7.875	48.7070	0.6563	0.33824	0.000229
	5L		60	0.406	7.813	47.9430	0.6511	0.33294	0.000230
	5L			0.438	7.749	47.1608	0.6458	0.32751	0.000232
	5L	XS	80	0.500	7.625	45.6635	0.6354	0.31711	0.000236
8 (8.625)	5L		100	0.594	7.437	43.4396	0.6198	0.30166	0.000242
	5L		120	0.719	7.187	40.5681	0.5989	0.28172	0.000250
	5L		140	0.812	7.001	38.4955	0.5834	0.26733	0.000257
	5L	XXS		0.875	6.875	37.1223	0.5729	0.25779	0.000262
			160	0.906	6.813	36.4558	0.5678	0.25317	0.000264
10 (10.750)			5S	0.134	10.482	86.2935	0.8735	0.59926	0.000172
			10S	0.165	10.420	85.2757	0.8683	0.59219	0.000173
	5L			0.188	10.374	84.5245	0.8645	0.58698	0.000174
	5L			0.203	10.344	84.0363	0.8620	0.58359	0.000174
	5L			0.219	10.312	83.5172	0.8593	0.57998	0.000175
	5L		20	0.250	10.250	82.5159	0.8542	0.57303	0.000176
	5L			0.279	10.192	81.5847	0.8493	0.56656	0.000177
	5L		30	0.307	10.136	80.6906	0.8447	0.56035	0.000178
	5L			0.344	10.062	79.5167	0.8385	0.55220	0.000179
	5L	STD	40	0.365	10.020	78.8543	0.8350	0.54760	0.000180
	5L			0.438	9.874	76.5731	0.8228	0.53176	0.000182
	5L	XS	60	0.500	9.750	74.6619	0.8125	0.51849	0.000185
			80S*	0.500	9.750	74.6619	0.8125	0.51849	0.000185
			80	0.594	9.562	71.8104	0.7968	0.49868	0.000188
	5L		100	0.719	9.312	68.1045	0.7760	0.47295	0.000193
			120	0.844	9.062	64.4968	0.7552	0.44789	0.000199
	5L	XXS	140	1.000	8.750	60.1320	0.7292	0.41758	0.000206
			160	1.125	8.500	56.7450	0.7083	0.39406	0.000212

*Does Not Conform To USA B36.10-1985

Characteristics of Piping Materials

Steel Pipe

IVA-1 PIPE DIMENSIONS; WROUGHT STEEL AND STAINLESS STEEL

Size Nominal & (Outside Diameter)	IDENTIFICATION			Wall Thickness Inches	Internal Diameter Inches	Internal Area Square Inches	Internal Diameter Feet	Internal Area Square Feet	ε/D ε=0.00015 Ft.
	API Standard	Standard X-Strong XX-Strong	Schedule Number						
12 (12.750)			5S	0.156	12.438	121.5041	1.0365	0.84378	0.000145
			10S	0.180	12.390	120.5681	1.0325	0.83728	0.000145
				0.203	12.344	119.6745	1.0287	0.83107	0.000146
	5L			0.219	12.312	119.0549	1.0260	0.82677	0.000146
	5L		20	0.250	12.250	117.8588	1.0208	0.81846	0.000147
	5L			0.281	12.188	116.6688	1.0157	0.81020	0.000148
	5L			0.312	12.126	115.4848	1.0105	0.80198	0.000148
	5L		30	0.330	12.090	114.8002	1.0075	0.79722	0.000149
	5L			0.344	12.062	114.2690	1.0052	0.79354	0.000149
	5L	STD		0.375	12.000	113.0973	1.0000	0.78540	0.000150
			40S*	0.375	12.000	113.0973	1.0000	0.78540	0.000150
	5L		40	0.406	11.938	111.9317	0.9948	0.77730	0.000151
	5L			0.438	11.874	110.7348	0.9895	0.76899	0.000152
	5L	XS		0.500	11.750	108.4340	0.9792	0.75301	0.000153
			80S*	0.500	11.750	108.4340	0.9792	0.75301	0.000153
	5L		60	0.562	11.626	106.1575	0.9688	0.73721	0.000155
	5L		80	0.688	11.374	101.6053	0.9478	0.70559	0.000158
			100	0.844	11.062	96.1075	0.9218	0.66741	0.000163
	5L	XXS	120	1.000	10.750	90.7626	0.8958	0.63030	0.000167
	5L		140	1.125	10.500	86.5901	0.8750	0.60132	0.000171
	5L		160	1.312	10.126	80.5315	0.8438	0.55925	0.000178
14 (14.000)			5S	0.156	13.688	147.1533	1.1407	1.0219	0.000132
	5L		10S*	0.188	13.624	145.7804	1.1353	1.0124	0.000132
			10	0.210	13.580	144.8403	1.1317	1.0058	0.000133
				0.219	13.562	144.4566	1.1302	1.0032	0.000133
	5L			0.250	13.500	143.1388	1.1250	0.9940	0.000133
	5L			0.281	13.438	141.8271	1.1198	0.9849	0.000134
	5L		20	0.312	13.376	140.5214	1.1147	0.9758	0.000135
	5L			0.344	13.312	139.1799	1.1093	0.9665	0.000135
	5L	STD	30	0.375	13.250	137.8865	1.1042	0.9575	0.000136
	5L			0.438	13.124	135.2765	1.0937	0.9394	0.000137
				0.469	13.062	134.00	1.0885	0.9306	0.000138
	5L	XS		0.500	13.000	132.73	1.0833	0.9218	0.000138
			60	0.594	12.812	128.92	1.0677	0.8953	0.000140
	5L		80	0.750	12.500	122.72	1.0417	0.8522	0.000144
	5L		100	0.938	12.124	115.45	1.0103	0.8017	0.000148
			120	1.094	11.812	109.58	0.9843	0.7610	0.000152
	5L		140	1.250	11.500	103.87	0.9583	0.7213	0.000157
			160	1.406	11.188	98.31	0.9323	0.6827	0.000161
				2.000	10.000	78.54	0.8333	0.5454	0.000180
				2.125	9.750	74.66	0.8125	0.5185	0.000185
				2.200	9.600	72.38	0.8000	0.5027	0.000188
				2.500	9.000	63.62	0.7500	0.4418	0.000200
16 (16.000)			5S	0.165	15.670	192.85	1.3058	1.3393	0.000115
	5L		10S*	0.188	15.624	191.72	1.3020	1.3314	0.000115
	5L			0.219	15.562	190.20	1.2968	1.3209	0.000116
	5L		10	0.250	15.500	188.69	1.2917	1.3104	0.000116
	5L			0.281	15.438	187.19	1.2865	1.2999	0.000117
	5L		20	0.312	15.376	185.68	1.2813	1.2895	0.000117
	5L			0.344	15.312	184.14	1.2760	1.2788	0.000118
	5L	STD	30	0.375	15.250	182.65	1.2708	1.2684	0.000118
	5L			0.438	15.124	179.65	1.2603	1.2476	0.000119
	5L			0.469	15.062	178.18	1.2552	1.2374	0.000120
	5L	XS	40	0.500	15.000	176.71	1.2500	1.2272	0.000120
			60	0.656	14.688	169.44	1.2240	1.1767	0.000123
			80	0.844	14.312	160.88	1.1927	1.1172	0.000126
			100	1.031	13.938	152.58	1.1615	1.0596	0.000129
			120	1.219	13.562	144.46	1.1302	1.0032	0.000133
			140	1.438	13.124	135.28	1.0937	0.9394	0.000137
			160	1.594	12.812	128.92	1.0677	0.8953	0.000140

*Does Not Conform To USA B36.10M-1985

Characteristics of Piping Materials

Steel Pipe

IVA-1 PIPE DIMENSIONS; WROUGHT STEEL AND STAINLESS STEEL

Size Nominal & (Outside Diameter)	IDENTIFICATION			Wall Thickness Inches	Internal Diameter Inches	Internal Area Square Inches	Internal Diameter Feet	Internal Area Square Feet	ε/D ε=0.00015 Ft.
	API Standard	Standard X-Strong XX-Strong	Schedule Number						
18 (18.000)			5S	0.165	17.670	245.22	1.4725	1.7029	0.000102
	5L		10S*	0.188	17.624	243.95	1.4687	1.6941	0.000102
	5L		10	0.250	17.500	240.53	1.4583	1.6703	0.000103
	5L			0.281	17.438	238.83	1.4532	1.6585	0.000103
	5L		20	0.312	17.376	237.13	1.4480	1.6467	0.000104
	5L			0.344	17.312	235.39	1.4427	1.6346	0.000104
	5L	STD		0.375	17.250	233.71	1.4375	1.6230	0.000104
	5L			0.406	17.188	232.03	1.4323	1.6113	0.000105
	5L		30	0.438	17.124	230.30	1.4270	1.5993	0.000105
	5L			0.469	17.062	228.64	1.4218	1.5878	0.000105
	5L	XS		0.500	17.000	226.98	1.4167	1.5763	0.000106
	5L		40	0.562	16.876	223.68	1.4063	1.5533	0.000107
	5L		60	0.750	16.500	213.82	1.3750	1.4849	0.000109
	5L		80	0.938	16.124	204.19	1.3437	1.4180	0.000112
			100	1.156	15.688	193.30	1.3073	1.3423	0.000115
			120	1.375	15.250	182.65	1.2708	1.2684	0.000118
			140	1.562	14.876	173.80	1.2397	1.2070	0.000121
			160	1.781	14.438	163.72	1.2032	1.1370	0.000125
20 (20.000)			5S	0.188	19.624	302.46	1.6353	2.1004	0.0000917
			10S*	0.218	19.564	300.61	1.6303	2.0876	0.0000920
	5L		10	0.250	19.500	293.65	1.6250	2.0739	0.0000923
	5L			0.281	19.438	296.75	1.6198	2.0608	0.0000926
	5L			0.312	19.376	294.86	1.6147	2.0476	0.0000929
	5L			0.344	19.312	292.92	1.6093	2.0341	0.0000932
	5L	STD	20	0.375	19.250	291.04	1.6042	2.0211	0.0000935
	5L			0.406	19.188	289.17	1.5990	2.0081	0.0000938
	5L			0.438	19.124	287.24	1.5937	1.9947	0.0000941
	5L			0.469	19.062	285.38	1.5885	1.9818	0.0000944
	5L	XS	30	0.500	19.000	283.53	1.5833	1.9689	0.0000947
			40	0.594	18.812	277.95	1.5677	1.9302	0.0000957
	5L		60	0.812	18.375	265.21	1.5313	1.8417	0.0000980
			80	1.031	17.938	252.72	1.4948	1.7550	0.0001009
			100	1.281	17.438	238.83	1.4532	1.6585	0.0001032
			120	1.500	17.000	226.98	1.4167	1.5763	0.0001059
			140	1.750	16.500	213.82	1.3750	1.4849	0.0001091
			160	1.969	16.062	202.62	1.3385	1.4071	0.0001121
22 (22.000)			5S	0.188	21.624	367.25	1.8020	2.5503	0.0000832
			10S*	0.218	21.564	365.21	1.7970	2.5362	0.0000835
	5L		10	0.250	21.500	363.05	1.7917	2.5212	0.0000837
	5L			0.281	21.438	360.96	1.7865	2.5067	0.0000840
	5L			0.312	21.376	358.87	1.7813	2.4922	0.0000842
	5L			0.344	21.312	356.73	1.7760	2.4773	0.0000845
	5L	STD	20	0.375	21.250	354.66	1.7708	2.4629	0.0000847
	5L			0.406	21.188	352.59	1.7657	2.4485	0.0000850
	5L			0.438	21.124	350.46	1.7603	2.4338	0.0000852
	5L			0.469	21.062	348.41	1.7552	2.4195	0.0000855
	5L	XS	30	0.500	21.000	346.36	1.7500	2.4053	0.0000857
	5L			0.625	20.750	338.16	1.7292	2.3484	0.0000867
	5L		60	0.875	20.250	322.06	1.6875	2.2365	0.0000889
	5L		80	1.125	19.750	306.35	1.6458	2.1275	0.0000911
	5L		100	1.375	19.250	291.04	1.6042	2.0211	0.0000935
			120	1.625	18.750	276.12	1.5625	1.9175	0.0000960
			140	1.875	18.250	261.59	1.5208	1.8166	0.0000986
			160	2.125	17.750	247.45	1.4792	1.7184	0.0001014

*Does Not Conform To USA B36.10M-1985

Characteristics of Piping Materials

Steel Pipe

IVA-1 PIPE DIMENSIONS; WROUGHT STEEL AND STAINLESS STEEL

Size Nominal & (Outside Diameter)	IDENTIFICATION			Wall Thickness Inches	Internal Diameter Inches	Internal Area Square Inches	Internal Diameter Feet	Internal Area Square Feet	ε/D ε=0.00015 Ft.
	API Standard	Standard X-Strong XX-Strong	Schedule Number						
24 (24.000)			5S	0.218	23.564	436.10	1.9637	3.0285	0.0000764
	5L 5LX		10	0.250	23.500	433.74	1.9583	3.0121	0.0000766
	5L 5LX			0.281	23.438	431.45	1.9532	2.9962	0.0000768
	5L 5LX			0.312	23.376	429.17	1.9480	2.9804	0.0000770
	5L 5LX			0.344	23.312	426.82	1.9427	2.9641	0.0000772
	5L 5LX	STD	20	0.375	23.250	424.56	1.9375	2.9483	0.0000774
	5LX			0.406	23.188	422.30	1.9323	2.9326	0.0000776
	5L 5LX			0.438	23.124	419.97	1.9270	2.9164	0.0000778
	5LX			0.469	23.062	417.72	1.9218	2.9008	0.0000781
	5L 5LX	XS		0.500	23.000	415.48	1.9167	2.8852	0.0000783
	5L 5LX		30	0.562	22.876	411.01	1.9063	2.8542	0.0000787
	5L 5LX		40	0.688	22.624	402.00	1.8853	2.7917	0.0000796
			60	0.969	22.062	382.28	1.8385	2.6547	0.0000816
			80	1.219	21.562	365.15	1.7968	2.5357	0.0000835
			100	1.531	20.938	344.32	1.7448	2.3911	0.0000860
			120	1.812	20.376	326.08	1.6980	2.2645	0.0000883
			140	2.062	19.876	310.28	1.6563	2.1547	0.0000906
			160	2.344	19.312	292.92	1.6093	2.0341	0.0000932
26 (26.000	5L 5LX			0.250	25.500	510.71	2.1250	3.5466	0.0000706
	5L 5LX			0.281	25.438	508.22	2.1198	3.5293	0.0000708
	5L 5LX		10	0.312	25.376	505.75	2.1147	3.5122	0.0000709
	5L 5LX			0.344	25.312	503.20	2.1093	3.4945	0.0000711
	5L 5LX	STD		0.375	25.250	500.74	2.1042	3.4774	0.0000713
	5LX			0.406	25.188	498.28	2.0990	3.4603	0.0000715
	5L 5LX			0.438	25.124	495.76	2.0937	3.4427	0.0000716
	5LX			0.469	25.062	493.31	2.0885	3.4258	0.0000718
	5L 5LX	XS	20	0.500	25.000	490.87	2.0833	3.4088	0.0000720
	5L 5LX			0.562	24.876	486.02	2.0730	3.3751	0.0000724
28 (28.000)	5L 5LX			0.250	27.500	593.96	2.2917	4.1247	0.0000655
	5L 5LX			0.281	27.438	591.28	2.2865	4.1061	0.0000656
	5L 5LX		10	0.312	27.376	588.61	2.2813	4.0876	0.0000658
	5LX			0.344	27.312	585.86	2.2760	4.0685	0.0000659
	5L 5LX	STD		0.375	27.250	583.21	2.2708	4.0501	0.0000661
	5LX			0.406	27.188	580.56	2.2657	4.0316	0.0000662
	5L 5LX			0.438	27.124	577.83	2.2603	4.0127	0.0000664
	5LX			0.469	27.062	575.19	2.2552	3.9944	0.0000665
	5L 5LX	XS	20	0.500	27.000	572.56	2.2500	3.9761	0.0000667
	5L 5LX		30	0.625	26.750	562.00	2.2292	3.9028	0.0000673
30 (30.000)	5L 5LX		5S	0.250	29.500	683.49	2.4583	4.7465	0.0000610
	5L 5LX			0.281	29.438	680.62	2.4532	4.7265	0.0000611
	5L 5LX		10	0.312	29.376	677.76	2.4480	4.7067	0.0000613
	5LX			0.344	29.312	674.81	2.4427	4.6862	0.0000614
	5L 5LX	STD		0.375	29.250	671.96	2.4375	4.6664	0.0000615
	5LX			0.406	29.188	669.11	2.4323	4.6466	0.0000617
	5L 5LX			0.438	29.124	666.18	2.4270	4.6263	0.0000618
	5LX			0.469	29.062	663.35	2.4218	4.6066	0.0000619
	5L 5LX	XS	20	0.500	29.000	660.52	2.4167	4.5869	0.0000621
	5L 5LX		30	0.625	28.750	649.18	2.3958	4.5082	0.0000626
32 (32.000)	5L 5LX			0.250	31.500	779.31	2.6250	5.4119	0.0000571
	5L 5LX			0.281	31.438	7.76.25	2.6198	5.3906	0.0000573
	5L 5LX		10	0.312	31.376	773.19	2.6147	5.3694	0.0000574
	5LX			0.344	31.312	770.04	2.6093	5.3475	0.0000575
	5L 5LX	STD		0.375	31.250	766.99	2.6042	5.3263	0.0000576
	5LX			0.406	31.188	763.95	2.5990	5.3052	0.0000577
	5L 5LX			0.438	31.124	760.82	2.5937	5.2835	0.0000578
	5LX			0.469	31.062	757.79	2.5885	5.2624	0.0000579
	5L 5LX	XS	20	0.500	31.000	754.77	2.5833	5.2414	0.0000581
	5L 5LX		30	0.625	30.750	742.64	2.5625	5.1572	0.0000585
	5L 5LX		40	0.688	30.624	736.57	2.5520	5.1151	0.0000588

IVA-1 PIPE DIMENSIONS; WROUGHT STEEL AND STAINLESS STEEL

Size Nominal & (Outside Diameter)	IDENTIFICATION			Wall Thickness Inches	Internal Diameter Inches	Internal Area Square Inches	Internal Diameter Feet	Internal Area Square Feet	ϵ/D $\epsilon=0.00015$ Ft.
	API Standard	Standard X-Strong XX-Strong	Schedule Number						
34 (34.000)	5L 5LX			0.250	33.500	881.41	2.7917	6.1209	0.0000537
	5L 5LX			0.281	33.438	878.15	2.7865	6.0983	0.0000538
	5L 5LX		10	0.312	33.376	874.90	2.7813	6.0757	0.0000539
	5LX			0.344	33.312	871.55	2.7760	6.0524	0.0000540
	5L 5LX	STD		0.375	33.250	868.31	2.7708	6.0299	0.0000541
	5LX			0.406	33.188	865.07	2.7657	6.0074	0.0000542
	5L 5LX			0.438	33.124	861.74	2.7603	5.9843	0.0000543
	5LX			0.469	33.062	858.52	2.7552	5.9619	0.0000544
	5L 5LX	XS	20	0.500	33.000	855.30	2.7500	5.9396	0.0000545
	5L 5LX		30	0.625	32.750	842.39	2.7292	5.8499	0.0000550
	5L 5LX		40	0.688	32.624	835.92	2.7187	5.8050	0.0000552
36 (36.000)	5L 5LX			0.250	35.500	989.80	2.9583	6.8736	0.0000507
	5L 5LX			0.281	35.438	986.34	2.9532	6.8496	0.0000508
	5L 5LX		10	0.312	35.376	982.90	2.9480	6.8257	0.0000509
	5LX			0.344	35.312	979.34	2.9427	6.8010	0.0000510
	5L 5LX	STD		0.375	35.250	975.91	2.9375	6.7771	0.0000511
	5LX			0.406	35.188	972.48	2.9323	6.7533	0.0000512
	5L 5LX			0.438	35.124	968.94	2.9270	6.7288	0.0000512
	5LX			0.469	35.062	965.52	2.9218	6.7050	0.0000513
	5L 5LX	XS	20	0.500	35.000	962.11	2.9167	6.6813	0.0000514
	5L 5LX			0.562	34.876	955.31	2.9063	6.6341	0.0000516
	5L 5LX		30	0.625	34.750	948.42	2.8958	6.5862	0.0000518
	5L 5LX		40	0.750	34.500	934.82	2.8750	6.4918	0.0000522

IVA-2 SPECIFICATIONS FOR WROUGHT STEEL AND STAINLESS STEEL PIPE

ANSI Designation	ASTM or API Designation	Title
B36.1	ASTM A53	Welded and Seamless Steel Pipe
B36.2	ASTM A72	Welded Wrought Iron Pipe
B36.3	ASTM A106	Seamless Carbon-Steel Pipe for High-Temperature Service
B36.20	ASTM A120	Black and Hot-Dipped Zinc-Coated (Galvanized) Welded and Seamless Steel Pipe for Ordinary Uses
B36.4	ASTM A134	Electric-Fusion (Arc)-Welded Steel Plate Pipe, Sizes 16 in. and Over
B36.5	ASTM A135	Electric-Resistance-Welded Steel Pipe
B36.9	ASTM A139	Electric-Fusion (Arc)-Welded Steel Pipe, Sizes 4 in. and Over
B36.11	ASTM A155	Electric-Fusion-Welded Steel Pipe for High-Temperature Service
B36.16	ASTM A211	Spiral-Welded Steel or Iron Pipe
B36.26	ASTM A312	Seamless and Welded Austenitic Stainless Steel Pipe
B36.40	ASTM A333	Seamless and Welded Steel Pipe for Low-Temperature Service
B36.42	ASTM A335	Seamless Ferritic Alloy Steel Pipe for High-Temperature Service
B36.47	ASTM A358	Electric-Fusion-Welded Austenitic Chromium-Nickel Alloy Steel Pipe for High-Temperature Service
B36.48	ASTM A369	Ferritic Alloy Steel Forged and Bored Pipe for High-Temperature Service
B36.43	ASTM A376	Seamless Austenitic Steel Pipe for High-Temperature Central-Station Service
B36.49	ASTM A381	Metal-Arc Welded Steel Pipe for High-Pressure Transmission Service
	API 5L	Line Pipe
	API 5LX	High-Test Line Pipe

Reprinted with permission from the American Society of Mechanical Engineers, 345 East 45th Street, N.Y., N.Y.; ANSI B3610-59.

IVA-3 API CASING LIST (See A.P.I. Standard 5A)

Size: Outside Diameter, in.	D, mm	Nominal Weight, Threads and Coupling, lb. per ft.	Grade	Wall Thickness t, in.	t, mm	Short	Long	Buttress	Extreme Line
4½	114,3	9.50	H, J, K	0.205	5,21	X	—	—	—
4½	114,3	10.50	J, K	0.224	5,69	X	—	X	—
4½	114,3	11.60	J, K	0.250	6,35	X	X	X	—
4½	114,3	11.60	N	0.250	6,35	—	X	X	—
4½	114,3	13.50	N	0.290	7,37	—	X	X	—
5	127,0	11.50	J, K	0.220	5,59	X	—	—	—
5	127,0	13.00	J, K	0.253	6,43	X	X	X	—
5	127,0	15.00	J, K	0.296	7,52	X	X	X	X
5	127,0	15.00	N	0.296	7,52	—	X	X	X
5	127,0	18.00	N	0.362	9,19	—	X	X	X
5½	139,7	14.00	H, J, K	0.244	6,20	X	—	—	—
5½	139,7	15.50	J, K	0.275	6,98	X	X	X	X
5½	139,7	17.00	J, K	0.304	7,72	X	X	X	X
5½	139,7	17.00	N	0.304	7,72	—	X	X	X
5½	139,7	20.00	N	0.361	9,17	—	X	X	X
5½	139,7	23.00	N	0.415	10,54	—	X	X	X
6⅝	168,3	20.00	H	0.288	7,32	X	—	—	—
6⅝	168,3	20.00	J, K	0.288	7,32	X	X	X	—
6⅝	168,3	24.00	J, K	0.352	8,94	X	X	X	X
6⅝	168,3	24.00	N	0.352	8,94	—	X	X	X
6⅝	168,3	28.00	N	0.417	10,59	—	X	X	X
6⅝	168,3	32.00	N	0.475	12,06	—	X	X	X
7	177,8	17.00	H	0.231	5,87	X	—	—	—
7	177,8	20.00	H, J, K	0.272	6,91	X	—	—	—
7	177,8	23.00	J, K	0.317	8,05	X	X	X	X
7	177,8	23.00	N	0.317	8,05	—	X	X	X
7	177,8	26.00	J, K	0.362	9,19	X	X	X	X
7	177,8	26.00	N	0.362	9,19	—	X	X	X
7	177,8	29.00	N	0.408	10,36	—	X	X	X
7	177,8	32.00	N	0.453	11,51	—	X	X	X
7	177,8	35.00	N	0.498	12,65	—	X	X	X
7	177,8	38.00	N	0.540	13,72	—	X	X	X
7⅝	193,7	24.00	H	0.300	7,62	X	—	—	—
7⅝	193,7	26.40	J, K	0.328	8,33	X	X	X	X
7⅝	193,7	26.40	N	0.328	8,33	—	X	X	X
7⅝	193,7	29.70	N	0.375	9,52	—	X	X	X
7⅝	193,7	33.70	N	0.430	10,92	—	X	X	X
7⅝	193,7	39.00	N	0.500	12,70	—	X	X	X
8⅝	219,1	24.00	J, K	0.264	6,71	X	—	—	—
8⅝	219,1	28.00	H	0.304	7,72	X	—	—	—
8⅝	219,1	32.00	H	0.352	8,94	X	—	—	—
8⅝	219,1	32.00	J, K	0.352	8,94	X	X	X	X
8⅝	219,1	36.00	J, K	0.400	10,16	X	X	X	X
8⅝	219,1	36.00	N	0.400	10,16	—	X	X	X
8⅝	219,1	40.00	N	0.450	11,43	—	X	X	X
8⅝	219,1	44.00	N	0.500	12,70	—	X	X	X
8⅝	219,1	49.00	N	0.557	14,15	—	X	X	X
9⅝	244,5	32.30	H	0.312	7,92	X	—	—	—
9⅝	244,5	36.00	H	0.352	8,94	X	—	—	—
9⅝	244,5	36.00	J, K	0.352	8,94	X	X	X	—
9⅝	244,5	40.00	J, K	0.395	10,03	X	X	X	X
9⅝	244,5	40.00	N	0.395	10,03	—	X	X	X
9⅝	244,5	43.50	N	0.435	11,05	—	X	X	X
9⅝	244,5	47.00	N	0.472	11,99	—	X	X	X
9⅝	244,5	53.50	N	0.545	13,84	—	X	X	X
10¾	273,0	32.75	H	0.279	7,09	X	—	—	—
10¾	273,0	40.50	H	0.350	8,89	X	—	—	—
10¾	273,0	40.50	J, K	0.350	8,89	X	—	X	—
10¾	273,0	45.50	J, K	0.400	10,16	X	—	X	X
10¾	273,0	51.00	J, K, N	0.450	11,43	X	—	X	X
10¾	273,0	55.50	N	0.495	12,57	X	—	X	X
11¾	298,4	42.00	H	0.333	8,46	X	—	—	—
11¾	298,4	47.00	J, K	0.375	9,52	X	—	X	—
11¾	298,4	54.00	J, K	0.435	11,05	X	—	X	—
11¾	298,4	60.00	J, K, N	0.489	12,42	X	—	X	—
13⅜	339,7	48.00	H	0.330	8,38	X	—	—	—
13⅜	339,7	54.50	J, K	0.380	9,65	X	—	X	—
13⅜	339,7	61.00	J, K	0.430	10,92	X	—	X	—
13⅜	339,7	68.00	J, K	0.480	12,19	X	—	X	—
13⅜	339,7	72.00	N	0.514	13,06	X	—	X	—
16	406,4	65.00	H	0.375	9,52	X	—	—	—
16	406,4	75.00	J, K	0.438	11,13	X	—	X	—
16	406,4	84.00	J, K	0.495	12,57	X	—	X	—
18⅝	473,1	87.50	H, J, K	0.435	11,05	X	—	—	—
20	508,0	94.00	H, J, K	0.438	11,13	X	X	—	—
20	508,0	106.50	J, K	0.500	12,70	X	X	—	—
20	508,0	133.00	J, K	0.635	16,13	X	X	—	—

Characteristics of Piping Materials
Steel Pipe

IVA-4 API PLAIN-END LINER LIST

Size: Outside Diameter. in.	D mm	Plain-End Weight lb/ft	kg/m	Grade	Wall Thickness in.	t mm
3½	88,9	9.91	14,76	J	0.289	7,34
4	101,6	11.34	16,89	J	0.286	7,26
4½	114,3	13.04	19,42	J	0.290	7,37
5	127,0	17.93	26,71	J	0.362	9,19
5½	139,7	19.81	29,51	J	0.361	9,17
6⅝	168,3	27.65	41,18	J	0.417	10,59

Reprinted with permission from the American Petroleum Institute.

IVA-5 API TUBING LIST

Size: Outside Diameter in.	D mm	Nominal Weight, Threads and Coupling lb. per ft.	Grade	Wall Thickness in.	t mm	Types of Ends[1]
††1.050	*26,7*	*1.14*	*H, J, N*	*0.113*	*2,87*	*Non-Upset*
1.050	26,7	1.20	H, J, N	0.113	2,87	Ext. Upset
1.315	33,4	1.70	H, J, N	0.133	3,38	Non-Upset
1.315	33,4	1.72	H, J, N	0.133	3,38	Integral Joint
1.315	33,4	1.80	H, J, N	0.133	3,38	Ext. Upset
1.660	42,2	2.10	H, J	0.125	3,18	Integral Joint
1.660	42,2	2.30	H, J, N	0.140	3,56	Non-Upset
1.660	42,2	2.33	H, J, N	0.140	3,56	Integral Joint
1.660	42,2	2.40	H, J, N	0.140	3,56	Ext. Upset
1.900	48,3	2.40	H, J	0.125	3,18	Integral Joint
1.900	48,3	2.75	H, J, N	0.145	3,68	Non-Upset
1.900	48,3	2.76	H, J, N	0.145	3,68	Integral Joint
1.900	48,3	2.90	H, J, N	0.145	3,68	Ext. Upset
2.063	52,4	3.25	H, J, N	0.156	3,96	Integral Joint
2⅜	60,3	4.00	H, J, N	0.167	4,24	Non-Upset
2⅜	60,3	4.60	H, J, N	0.190	4,83	Non-Upset
2⅜	60,3	4.70	H, J, N	0.190	4,83	Ext. Upset
2⅜	60,3	5.80	N	0.254	6,45	Non-Upset
2⅜	60,3	5.95	N	0.254	6,45	Ext. Upset
2⅞	73,0	6.40	H, J, N	0.217	5,51	Non-Upset
2⅞	73,0	6.50	H, J, N	0.217	5,51	Ext. Upset
2⅞	73,0	8.60	N	0.308	7,82	Non-Upset
2⅞	73,0	8.70	N	0.308	7,82	Ext. Upset
3½	88,9	7.70	H, J, N	0.216	5,49	Non-Upset
3½	88,9	9.20	H, J, N	0.254	6,45	Non-Upset
3½	88,9	9.30	H, J, N	0.254	6,45	Ext. Upset
3½	88,9	10.20	H, J, N	0.289	7,34	Non-Upset
3½	88,9	12.70	N	0.375	9,52	Non-Upset
3½	88,9	12.95	N	0.375	9,52	Ext. Upset
4	101,6	9.50	H, J, N	0.226	5,74	Non-Upset
4	101,6	11.00	H, J, N	0.262	6,65	Ext. Upset
4½	114,3	12.60	H, J, N	0.271	6,88	Non-Upset
4½	114,3	12.75	H, J, N	0.271	6.88	Ext. Upset

[1] Non-upset tubing is available with regular couplings or special bevel couplings. External-upset tubing is available with regular, special bevel or special clearance couplings.

†† For information purposes only.

Reprinted with permission from the American Petroleum Institute.

IVA-6 API DRILL PIPE LIST*†

Size: Outside Diameter, D		Nominal Weight, lb/ft	Calculated Plain-end Weight, w_{pe}		Grade	Wall Thickness, t		Upset Ends, for Weld-on Tool Joints
in.	mm		lb/ft	kg/m		in.	mm	
2⅜	*60,3*	*4.85*	*4.43*	*6,60*	*E*	*0.190*	*4,83*	*Ext. Upset*
2⅜	60,3	6.65	6.26	9,32	D, E	0.280	7,11	Ext. Upset
2⅞	*73,0*	*6.85*	*6.16*	*9,18*	*E*	*0.217*	*5,51*	*Int. Upset or Ext. Upset*
2⅞	73,0	10.40	9.72	14,48	D, E	0.362	9,19	Int. Upset or Ext. Upset
3½	*88,9*	*9.50*	*8.81*	*13,12*	*E*	*0.254*	*6,45*	*Int. Upset or Ext. Upset*
3½	88,9	13.30	12.31	18,34	D, E	0.368	9,35	Int. Upset or Ext. Upset
3½	88,9	15.50	14.63	21,79	D, E	0.449	11,40	Int. Upset or Ext. Upset
4	*101,6*	*11.85*	*10.46*	*15,58*	*E*	*0.262*	*6,65*	*Int. Upset or Ext. Upset*
4	101,6	14.00	12.93	19,26	D, E	0.330	8,38	Int. Upset or Ext. Upset
4½	*114,3*	*13.75*	*12.24*	*18,23*	*E*	*0.271*	*6,88*	*Int. Upset or Ext. Upset*
4½	114,3	16.60	14.98	22,31	D, E	0.337	8,56	Int. Upset or Ext. Upset
4½	114,3	20.00	18.69	27,84	D, E	0.430	10,92	Ext. Upset or Int.-Ext. Upset
5	*127,0*	*16.25*	*14.87*	*22,15*	*E*	*0.296*	*7,52*	*Int. Upset*
5	127,0	19.50	17.93	26,71	D, E	0.362	9,19	Int.-Ext. Upset
5½	139,7	21.90	19.81	29,51	D, E	0.361	9,17	Int.-Ext. Upset
5½	139,7	24.70	22.54	33,57	D, E	0.415	10,54	Int.-Ext. Upset
6⅝	168,3	25.20	22.19	33,05	D, E	0.330	8,38	See footnote

* Drill pipe in sizes and weights shown italicized are tentative and marking within the API monogram is not permitted.
† Drill pipe with special end finish is available under this specification. See API Std 5A.
Upset requirements for 6⅝ in.—25.20 lb. drill pipe not established.

IVB-1 CAST IRON PIPE DIMENSIONS

Data obtained from **Handbook of Cast Iron Pipe,** Cast Iron Pipe Research Association, Oak Brook, Illinois 60521.

Pipe Size Inches	Thickness Classes	Outside Diameter Inches	Wall Thickness Inches	Internal Diameter Inches	Internal Area Square Inches	Internal Diameter Feet	Internal Area Square Feet	ϵ/D ϵ=0.00085 Ft.
3	22	3.96	0.32	3.32	8.66	0.2767	0.0601	0.0030723
	23		0.35	3.26	8.35	0.2717	0.0580	0.0031288
	24		0.38	3.20	8.04	0.2667	0.0559	0.0031875
	25		0.41	3.14	7.74	0.2617	0.0538	0.0032484
	26		0.44	3.08	7.45	0.2567	0.0517	0.0033117
	27		0.48	3.00	7.07	0.2500	0.0491	0.0034000
	28		0.52	2.92	6.70	0.2433	0.0465	0.0034932
	29		0.56	2.84	6.33	0.2367	0.0440	0.0035915
	30		0.60	2.76	5.98	0.2300	0.0415	0.0036957
4	22	4.80	0.35	4.10	13.20	0.3417	0.0917	0.0024878
	23		0.38	4.04	12.82	0.3367	0.0890	0.0025248
	24		0.41	3.98	12.44	0.3317	0.0864	0.0025628
	25		0.44	3.92	12.07	0.3267	0.0838	0.0026020
	26		0.48	3.84	11.58	0.3200	0.0804	0.0026563
	27		0.52	3.76	11.10	0.3133	0.0771	0.0027128
	28		0.56	3.68	10.64	0.3067	0.0739	0.0027717
	29		0.60	3.60	10.18	0.3000	0.0707	0.0028333
	30		0.65	3.50	9.62	0.2917	0.0668	0.0029143
6	21	6.90	.35	6.20	30.19	.5167	.2097	0.0016452
	22		.38	6.14	29.61	.5117	.2056	0.0016612
	23		.41	6.08	29.03	.5067	.2016	0.0016776
	24		.44	6.02	28.46	.5017	.1977	0.0016944
	25		.48	5.94	27.71	.4950	.1924	0.0017172
	26		.52	5.86	26.97	.4883	.1873	0.0017406
	27		.56	5.78	26.24	.4817	.1822	0.0017647
	28		.60	5.70	25.52	.4750	.1772	0.0017895
	29		.65	5.60	24.63	.4667	.1710	0.0018214
	30		.70	5.50	23.76	.4583	.1650	0.0018545
8	20	9.05	.35	8.35	54.76	.6958	.3803	0.0012216
	21		.38	8.29	53.96	.6908	.3748	0.0012304
	22		.41	8.23	53.20	.6858	.3694	0.0012394
	23		.44	8.17	52.42	.6808	.3641	0.0012485
	24		.48	8.09	51.40	.6742	.3570	0.0012608
	25		.52	8.01	50.39	.6675	.3499	0.0012734
	26		.56	7.93	49.39	.6608	.3430	0.0012863
	27		.60	7.85	48.40	.6542	.3361	0.0012994
	28		.65	7.75	47.17	.6458	.3276	0.0013161
	29		.70	7.65	45.96	.6375	.3192	0.0013333
	30		.76	7.57	44.53	.6275	.3093	0.0013546
10	20	11.10	.38	10.34	83.97	.8617	.5831	0.0009865
	21		.41	10.28	83.00	.8567	.5764	0.0009922
	22		.44	10.22	82.03	.8517	.5697	0.0009980
	23		.48	10.14	80.75	.8450	.5608	0.0010059
	24		.52	10.06	79.49	.8383	.5520	0.0010139
	25		.56	9.98	78.23	.8317	.5432	0.0010220
	26		.60	9.90	76.98	.8250	.5346	0.0010303
	27		.65	9.80	75.43	.8167	.5238	0.0010408
	28		.70	9.70	73.90	.8083	.5132	0.0010515
	29		.76	9.58	72.08	.7983	.5006	0.0010647
	30		.82	9.46	70.29	.7883	.4881	0.0010782
12	20	13.20	.41	12.38	120.37	1.0317	.8359	0.0008239
	21		.44	12.32	119.21	1.0267	.8278	0.0008279
	22		.48	12.24	117.67	1.0200	.8171	0.0008333
	23		.52	12.16	116.13	1.0133	.8065	0.0008388
	24		.56	12.08	114.61	1.0067	.7959	0.0008444
	25		.60	12.00	113.10	1.0000	.7854	0.0008500
	26		.65	11.90	111.22	.9917	.7724	0.0008571
	27		.70	11.80	109.36	.9833	.7594	0.0008644
	28		.76	11.68	107.15	.9733	.7441	0.0008733
	29		.82	11.56	104.96	.9633	.7289	0.0008824
	30		.89	11.42	102.43	.9517	.7113	0.0008939

IVB-1 CAST IRON PIPE DIMENSIONS

Pipe Size Inches	Thickness Classes	Outside Diameter Inches	Wall Thickness Inches	Internal Diameter Inches	Internal Area Square Inches	Internal Diameter Feet	Internal Area Square Feet	ε/D ε=0.00085 Ft.
14	20	15.30	.43	14.44	163.77	1.2033	1.1373	0.0007064
	21		.48	14.34	161.51	1.1950	1.1216	0.0007113
	22		.51	14.28	160.16	1.1900	1.1122	0.0007143
	23		.55	14.20	158.36	1.1833	1.0998	0.0007183
	24		.59	14.12	156.59	1.1767	1.0874	0.0007224
	25		.64	14.02	154.38	1.1683	1.0721	0.0007275
	26		.69	13.92	152.18	1.1600	1.0568	0.0007328
	27		.75	13.80	149.57	1.1500	1.0387	0.0007391
	28		.81	13.68	146.98	1.1400	1.0207	0.0007456
	29		.87	13.56	144.41	1.1300	1.0029	0.0007522
	30		.94	13.42	141.45	1.1183	.9823	0.0007601
16	20	17.40	.46	16.48	213.31	1.3733	1.4813	0.0006189
	21		.50	16.40	211.24	1.3667	1.4669	0.0006220
	22		.54	16.32	209.18	1.3600	1.4527	0.0006250
	23		.58	16.24	207.14	1.3533	1.4385	0.0006281
	24		.63	16.14	204.60	1.3450	1.4208	0.0006320
	25		.68	16.04	202.07	1.3367	1.4033	0.0006359
	26		.73	15.94	199.56	1.3283	1.3858	0.0006399
	27		.79	15.82	196.56	1.3183	1.3650	0.0006448
	28		.85	15.70	193.59	1.3083	1.3444	0.0006497
	29		.92	15.56	190.16	1.2967	1.3205	0.0006555
	30		.99	15.42	186.75	1.2850	1.2969	0.0006615
18	20	19.50	0.50	18.50	268.80	1.5417	1.8667	.0005514
	21		0.54	18.42	266.48	1.5350	1.8506	.0005537
	22		0.58	18.34	264.17	1.5283	1.8345	.0005562
	23		0.63	18.24	261.30	1.5200	1.8146	.0005592
	24		0.68	18.14	258.44	1.5117	1.7947	.0005623
	25		0.73	18.04	255.60	1.5033	1.7750	.0005654
	26		0.79	17.92	252.21	1.4933	1.7515	.0005692
	27		0.85	17.80	248.85	1.4833	1.7281	.0005730
	28		0.92	17.66	244.95	1.4717	1.7010	.0005776
	29		0.99	17.52	241.08	1.4600	1.6742	.0005822
	30		1.07	17.36	236.70	1.4467	1.6437	.0005876
20	20	21.60	0.53	20.54	331.35	1.7117	2.3011	.0004966
	21		0.57	20.46	328.78	1.7050	2.2832	.0004985
	22		0.62	20.36	325.57	1.6967	2.2609	.0005010
	23		0.67	20.26	322.38	1.6883	2.2388	.0005035
	24		0.72	20.16	319.21	1.6800	2.2167	.0005060
	25		0.78	20.04	315.42	1.6700	2.1904	.0005090
	26		0.84	19.92	311.65	1.6600	2.1642	.0005120
	27		0.91	19.78	307.29	1.6483	2.1339	.0005157
	28		0.98	19.64	302.95	1.6367	2.1038	.0005193
	29		1.06	19.48	298.04	1.6233	2.0697	.0005236
	30		1.14	19.32	293.16	1.6100	2.0358	.0005280
24	20	25.80	0.58	24.64	476.84	2.0533	3.3114	.0004140
	21		0.63	24.54	472.98	2.0450	3.2846	.0004156
	22		0.68	24.44	469.13	2.0367	3.2578	.0004173
	23		0.73	24.34	465.30	2.0283	3.2312	.0004191
	24		0.79	24.22	460.72	2.0183	3.1995	.0004211
	25		0.85	24.10	456.17	2.0083	3.1678	.0004232
	26		0.92	23.96	450.88	1.9967	3.1311	.0004257
	27		0.99	23.82	445.63	1.9850	3.0946	.0004282
	28		1.07	23.66	439.66	1.9717	3.0532	.0004311
	29		1.16	23.48	433.00	1.9567	3.0069	.0004344
	30		1.25	23.30	426.38	1.9417	2.9610	.0004378

IVB-1 CAST IRON PIPE DIMENSIONS

Pipe Size Inches	Thickness Classes	Outside Diameter Inches	Wall Thickness Inches	Internal Diameter Inches	Internal Area Square Inches	Internal Diameter Feet	Internal Area Square Feet	ϵ/D ϵ=0.00085 Ft.
30	20	32.00	0.68	30.64	737.34	2.5533	5.1204	.0003329
	21		0.73	30.54	732.53	2.5450	5.0870	.0003340
	22		0.79	30.42	726.79	2.5350	5.0471	.0003353
	23		0.85	30.30	721.07	2.5250	5.0074	.0003366
	24		0.92	30.16	714.42	2.5133	4.9612	.0003382
	25		0.99	30.02	707.80	2.5017	4.9153	.0003398
	26		1.07	29.86	700.28	2.4883	4.8630	.0003416
	27		1.16	29.68	691.86	2.4733	4.8046	.0003437
	28		1.25	29.50	683.49	2.4583	4.7465	.0003458
	29		1.35	29.30	674.26	2.4417	4.6823	.0003481
	30		1.46	29.08	664.17	2.4233	4.6123	.0003508
36	20	38.30	0.75	36.80	1063.62	3.0667	7.3862	0.0002772
	21		0.81	36.68	1056.69	3.0567	7.3381	0.0002781
	22		0.87	36.56	1049.79	3.0467	7.2902	0.0002790
	23		0.94	36.42	1041.77	3.0350	7.2345	0.0002801
	24		1.02	36.26	1032.63	3.0217	7.1711	0.0002813
	25		1.10	36.10	1023.54	3.0083	7.1079	0.0002825
	26		1.19	35.92	1013.36	2.9933	7.0372	0.0002840
	27		1.29	35.72	1002.10	2.9766	6.9591	0.0002856
	28		1.39	35.52	990.91	2.9600	6.8813	0.0002872
	29		1.50	35.30	978.68	2.9417	6.7964	0.0002890
	30		1.62	35.06	965.41	2.9217	6.7043	0.0002909
42	20	44.50	0.83	42.84	1441.41	3.5700	10.0098	0.0002381
	21		0.90	42.70	1432.01	3.5583	9.9445	0.0002389
	22		0.97	42.56	1422.63	3.5467	9.8794	0.0002397
	23		1.05	42.40	1411.96	3.5333	9.8053	0.0002406
	24		1.13	42.24	1401.32	3.5200	9.7314	0.0002415
	25		1.22	42.06	1389.40	3.5050	9.6486	0.0002425
	26		1.32	41.86	1376.22	3.4883	9.5571	0.0002437
	27		1.43	41.64	1361.79	3.4700	9.4569	0.0002450
	28		1.54	41.46	1347.44	3.4517	9.3572	0.0002463
	29		1.66	41.18	1331.87	3.4317	9.2491	0.0002477
	30		1.79	40.92	1315.11	3.4100	9.1327	0.0002493
48	20	50.80	0.91	48.98	1884.20	4.0817	13.0847	0.0002032
	21		0.98	48.84	1873.45	4.0700	13.0100	0.0002088
	22		1.06	48.68	1861.19	4.0567	12.9249	0.0002095
	23		1.14	48.52	1848.98	4.0433	12.8401	0.0002102
	24		1.23	48.34	1835.28	4.0283	12.7450	0.0002110
	25		1.33	48.14	1820.13	4.0117	12.6398	0.0002119
	26		1.44	47.92	1803.53	3.9933	12.5245	0.0002129
	27		1.56	47.68	1785.51	3.9733	12.3993	0.0002139
	28		1.68	47.44	1767.58	3.9533	12.2749	0.0002150
	29		1.81	47.18	1748.26	3.9317	12.1407	0.0002162
	30		1.95	46.90	1727.57	3.9083	11.9970	0.0002175

IVB-2 DUCTILE IRON PIPE DIMENSIONS

Data obtained from **Handbook of Cast Iron Pipe,** Cast Iron Pipe Research Association, Oak Brook, Illinois 60521.

Pipe Size Inches	Thickness Classes	Outside Diameter Inches	Wall Thickness Inches	Internal Diameter Inches	Internal Area Square Inches	Internal Diameter Feet	Internal Area Square Feet	ϵ/D ϵ=0.00085 Ft.
3	2	3.96	0.28	3.40	9.08	0.2833	0.0631	0.0030000
	3		0.31	3.34	8.76	0.2783	0.0608	0.0030539
	4		0.34	3.28	8.45	0.2733	0.0587	0.0031098
	5		0.37	3.22	8.14	0.2683	0.0566	0.0031677
	6		0.40	3.16	7.84	0.2633	0.0545	0.0032278
4	2	4.80	0.29	4.22	13.99	0.3517	0.0971	0.0024171
	3		0.32	4.16	13.59	0.3467	0.0944	0.0024519
	4		0.35	4.10	13.20	0.3417	0.0917	0.0024878
	5		0.38	4.04	12.8190	0.3367	0.0890	0.0025248
	6		0.41	3.98	12.4410	0.3317	0.0864	0.0025628
6	2	6.90	0.31	6.28	30.97	0.5233	0.2151	0.0016242
	3		0.34	6.22	30.39	0.5183	0.2110	0.0016399
	4		0.37	6.16	29.80	0.5133	0.2070	0.0016558
	5		0.40	6.10	29.22	0.5083	0.2029	0.0016721
	6		0.43	6.04	28.65	0.5033	0.1990	0.0016887
8	2	9.05	0.33	8.39	55.29	0.6992	0.3839	0.0012157
	3		0.36	8.33	54.50	0.6942	0.3785	0.0012245
	4		0.39	8.27	53.72	0.6892	0.3730	0.0012334
	5		0.42	8.21	52.94	0.6842	0.3676	0.0012424
	6		0.45	8.15	52.17	0.6792	0.3623	0.0012515
10	2	11.10	0.35	10.40	84.95	0.8667	0.5899	0.0009808
	3		0.38	10.34	83.97	0.8617	0.5831	0.0009865
	4		0.41	10.28	83.00	0.8567	0.5764	0.0009922
	5		0.44	10.22	82.03	0.8517	0.5697	0.0009980
	6		0.47	10.16	81.07	0.8467	0.5630	0.0010039
12	2	13.20	0.37	12.46	121.93	1.0383	0.8468	0.0008186
	3		0.40	12.40	120.76	1.0333	0.8386	0.0008228
	4		0.43	12.34	119.60	1.0283	0.8305	0.0008266
	5		0.46	12.28	118.44	1.0233	0.8225	0.0008306
	6		0.49	12.22	117.28	1.0183	0.8145	0.0008347
14	1	15.30	0.36	14.58	166.96	1.2150	1.1594	0.0006996
	2		0.39	14.52	165.59	1.2100	1.1499	0.0007025
	3		0.42	14.46	164.22	1.2050	1.1404	0.0007054
	4		0.45	14.40	162.86	1.2000	1.1310	0.0007083
	5		0.48	14.34	161.51	1.1950	1.1216	0.0007113
	6		0.51	14.28	160.16	1.1900	1.1122	0.0007143
16	1	17.40	0.37	16.66	217.99	1.3883	1.5138	0.0006122
	2		0.40	16.60	216.42	1.3833	1.5029	0.0006145
	3		0.43	16.54	214.86	1.3783	1.4921	0.0006167
	4		0.46	16.48	213.31	1.3733	1.4813	0.0006189
	5		0.49	16.42	211.76	1.3683	1.4705	0.0006212
	6		0.52	16.36	210.21	1.3633	1.4598	0.0006235
18	1	19.50	0.38	18.74	275.82	1.5617	1.9154	0.0005443
	2		0.41	18.68	274.06	1.5567	1.9032	0.0005460
	3		0.44	18.62	272.30	1.5517	1.8910	0.0005478
	4		0.47	18.56	270.55	1.5467	1.8788	0.0005496
	5		0.50	18.50	268.80	1.5417	1.8667	0.0005514
	6		0.53	18.44	267.06	1.5367	1.8546	0.0005531
20	1	21.60	0.39	20.82	340.45	1.7350	2.3642	0.0004899
	2		0.42	20.76	338.49	1.7300	2.3506	0.0004913
	3		0.45	20.70	336.54	1.7250	2.3371	0.0004928
	4		0.48	20.64	334.59	1.7200	2.3235	0.0004942
	5		0.51	20.58	332.64	1.7150	2.3100	0.0004956
	6		0.54	20.52	330.71	1.7100	2.2966	0.0004971

IVB-2 DUCTILE IRON PIPE DIMENSIONS

Data obtained from **Handbook of Cast Iron Pipe**, Cast Iron Pipe Research Association, Oak Brook, Illinois 60521.

Pipe Size Inches	Thickness Classes	Outside Diameter Inches	Wall Thickness Inches	Internal Diameter Inches	Internal Area Square Inches	Internal Diameter Feet	Internal Area Square Feet	ϵ/D $\epsilon=0.00085$ Ft.
24	1	25.80	0.41	24.98	490.09	2.0817	3.4034	0.0004083
	2		0.44	24.92	487.73	2.0767	3.3871	0.0004093
	3		0.47	24.86	485.39	2.0717	3.3708	0.0004103
	4		0.50	24.80	483.05	2.0667	3.3545	0.0004113
	5		0.53	24.74	480.72	2.0617	3.3383	0.0004123
	6		0.56	24.68	478.39	2.0567	3.3221	0.0004133
30	1	32.00	0.43	31.14	761.60	2.5950	5.2889	0.0003276
	2		0.47	31.06	757.69	2.5883	5.2618	0.0003284
	3		0.51	30.98	753.79	2.5817	5.2347	0.0003292
	4		0.55	30.90	749.91	2.5750	5.2077	0.0003301
	5		0.59	30.82	746.03	2.5683	5.1808	0.0003310
	6		0.63	30.74	742.16	2.5617	5.1539	0.0003318
36	1	38.30	0.48	37.34	1095.06	3.1117	7.6046	0.0002732
	2		0.53	37.24	1089.20	3.1033	7.5639	0.0002739
	3		0.58	37.14	1083.36	3.0950	7.5233	0.0002746
	4		0.63	37.04	1077.54	3.0867	7.4829	0.0002754
	5		0.68	36.94	1071.73	3.0783	7.4425	0.0002761
	6		0.73	36.84	1065.93	3.0700	7.4023	0.0002769
42	1	44.50	0.53	43.44	1482.07	3.6200	10.2922	0.0002348
	2		0.59	43.32	1473.90	3.6100	10.2354	0.0002355
	3		0.65	43.20	1465.74	3.6000	10.1788	0.0002361
	4		0.71	43.08	1457.61	3.5900	10.1223	0.0002368
	5		0.77	42.96	1449.50	3.5800	10.0660	0.0002374
	6		0.83	42.84	1441.41	3.5700	10.0098	0.0002381
48	1	50.80	0.58	49.64	1935.32	4.1367	13.4397	0.0002055
	2		0.65	49.50	1924.42	4.1250	13.3640	0.0002061
	3		0.72	49.36	1913.55	4.1133	13.2886	0.0002066
	4		0.79	49.22	1902.71	4.1017	13.2133	0.0002072
	5		0.86	49.08	1891.90	4.0900	13.1382	0.0002078
	6		0.93	48.94	1881.13	4.0783	13.0634	0.0002084
54	1	57.10	0.65	55.80	2445.46	4.6500	16.9823	0.0001828
	2		0.73	55.64	2431.44	4.6367	16.8850	0.0001833
	4		0.89	55.32	2403.56	4.6100	16.6914	0.0001844
	6		1.05	55.00	2375.83	4.5833	16.4988	0.0001855

Characteristics of Piping Materials

Non-Ferrous Pipe and Tubing

IVC-1 ALUMINUM—ALLOY SEAMLESS PIPE AND EXTRUDED TUBE (ANSI H35.2; TABLES 12.1, 12.3, 16.7)

DIAMETERS, WALL THICKNESS, AND WEIGHTS OF ALUMINUM ALLOY PIPE

Nominal Pipe Size, in.	ANSI Schedule Number[a]	Outside Diameter, in.			Inside Diameter, in.	Wall Thickness, in.			Weight, lb per ft	
		Nom	Min[b]	Max[b]	Nom	Nom	Min	Max[c]	Nom[d]	Max
1/8	40	0.405	0.374	0.420	0.269	0.068	0.060	—	0.085	0.091
	80	0.405	0.374	0.420	0.215	0.095	0.083	—	0.109	0.117
1/4	40	0.540	0.509	0.555	0.364	0.088	0.077	—	0.147	0.159
	80	0.540	0.509	0.555	0.302	0.119	0.104	—	0.185	0.200
3/8	40	0.675	0.644	0.690	0.493	0.091	0.080	—	0.196	0.212
	80	0.675	0.644	0.690	0.423	0.126	0.110	—	0.256	0.276
1/2	5	0.840	0.809	0.855	0.710	0.065	0.053	0.077	0.186	—
	10	0.840	0.809	0.855	0.674	0.083	0.071	0.095	0.232	—
	40	0.840	0.809	0.855	0.622	0.109	0.095	—	0.294	0.318
	80	0.840	0.809	0.855	0.546	0.147	0.129	—	0.376	0.406
	160	0.840	0.809	0.855	0.466	0.188	0.164	—	0.453	0.489
3/4	5	1.050	1.019	1.065	0.920	0.065	0.053	0.077	0.237	—
	10	1.050	1.019	1.065	0.884	0.083	0.071	0.095	0.297	—
	40	1.050	1.019	1.065	0.824	0.113	0.099	—	0.391	0.422
	80	1.050	1.019	1.065	0.742	0.154	0.135	—	0.510	0.551
	160	1.050	1.019	1.065	0.612	0.219	0.192	—	0.672	0.726
1	5	1.315	1.284	1.330	1.185	0.065	0.053	0.077	0.300	—
	10	1.315	1.284	1.330	1.097	0.109	0.095	0.123	0.486	—
	40	1.315	1.284	1.330	1.049	0.133	0.116	—	0.581	0.627
	80	1.315	1.284	1.330	0.957	0.179	0.157	—	0.751	0.811
	160	1.315	1.284	1.330	0.815	0.250	0.219	—	0.984	1.062
1¼	5	1.660	1.629	1.675	1.530	0.065	0.053	0.077	0.383	—
	10	1.660	1.629	1.675	1.442	0.109	0.095	0.123	0.625	—
	40	1.660	1.629	1.675	1.380	0.140	0.122	—	0.786	0.849
	80	1.660	1.629	1.675	1.278	0.191	0.167	—	1.037	1.120
	160	1.660	1.629	1.675	1.160	0.250	0.219	—	1.302	1.406
1½	5	1.900	1.869	1.915	1.770	0.065	0.053	0.077	0.441	—
	10	1.900	1.869	1.915	1.682	0.109	0.095	0.123	0.721	—
	40	1.900	1.869	1.916	1.610	0.145	0.127	—	0.940	1.015
	80	1.900	1.869	1.916	1.500	0.200	0.175	—	1.256	1.357
	160	1.900	1.869	1.916	1.338	0.281	0.246	—	1.681	1.815
2	5	2.375	2.344	2.406	2.245	0.065	0.053	0.077	0.555	—
	10	2.375	2.344	2.406	2.157	0.109	0.095	0.123	0.913	—
	40	2.375	2.351	2.399	2.067	0.154	0.135	—	1.264	1.365
	80	2.375	2.351	2.399	1.939	0.218	0.191	—	1.737	1.876
	160	2.375	2.351	2.399	1.687	0.344	0.301	—	2.581	2.788
2½	5	2.875	2.844	2.906	2.709	0.083	0.071	0.095	0.856	—
	10	2.875	2.844	2.906	2.635	0.120	0.105	0.135	1.221	—
	40	2.875	2.846	2.904	2.469	0.203	0.178	—	2.004	2.164
	80	2.875	2.846	2.904	2.323	0.276	0.242	—	2.650	2.862
	160	2.875	2.846	2.904	2.125	0.375	0.328	—	3.464	3.741
3	5	3.500	3.469	3.531	3.334	0.083	0.071	0.095	1.048	—
	10	3.500	3.469	3.531	3.260	0.120	0.105	0.135	1.498	—
	40	3.500	3.465	3.535	3.068	0.216	0.189	—	2.621	2.830
	80	3.500	3.465	3.535	2.900	0.300	0.262	—	3.547	3.830
	160	3.500	3.465	3.535	2.624	0.438	0.383	—	4.955	5.351

IVC-1 ALUMINUM—ALLOY SEAMLESS PIPE AND EXTRUDED TUBE (ANSI H35.2: TABLE 16.7)

DIAMETERS, WALL THICKNESS, AND WEIGHTS OF ALUMINUM ALLOY PIPE (CONT'D)

Nominal Pipe Size, in.	ANSI Schedule Number[a]	Outside Diameter, in.			Inside Diameter, in.	Wall Thickness, in.			Weight, lb per ft	
		Nom	Min[b]	Max[b]	Nom	Nom	Min	Max[c]	Nom[d]	Max
3½	5	4.000	3.969	4.031	3.834	0.083	0.071	0.095	1.201	—
	10	4.000	3.969	4.031	3.760	0.120	0.105	0.135	1.720	—
	40	4.000	3.960	4.040	3.548	0.226	0.198	—	3.151	3.403
	80	4.000	3.960	4.040	3.364	0.318	0.278	—	4.326	4.672
4	5	4.500	4.469	4.531	4.334	0.083	0.071	0.095	1.354	—
	10	4.500	4.469	4.531	4.260	0.120	0.105	0.135	1.942	—
	40	4.500	4.455	4.545	4.026	0.237	0.207	—	3.733	4.031
	80	4.500	4.455	4.545	3.826	0.337	0.295	—	5.183	5.598
	120	4.500	4.455	4.545	3.624	0.438	0.383	—	6.573	7.099
	160	4.500	4.455	4.545	3.438	0.531	0.465	—	7.786	8.409
5	5	5.563	5.532	5.625	5.345	0.109	0.095	0.123	2.196	—
	10	5.563	5.532	5.625	5.295	0.134	0.117	0.151	2.688	—
	40	5.563	5.507	5.619	5.047	0.258	0.226	—	5.057	5.461
	80	5.563	5.507	5.619	4.813	0.375	0.328	—	7.188	7.763
	120	5.563	5.507	5.619	4.563	0.500	0.438	—	9.353	10.10
	160	5.563	5.507	5.619	4.313	0.625	0.547	—	11.40	12.31
6	5	6.625	6.594	6.687	6.407	0.109	0.095	0.123	2.624	—
	10	6.625	6.594	6.687	6.357	0.134	0.117	0.151	3.213	—
	40	6.625	6.559	6.691	6.065	0.280	0.245	—	6.564	7.089
	80	6.625	6.559	6.691	5.761	0.432	0.378	—	9.884	10.67
	120	6.625	6.559	6.691	5.501	0.562	0.492	—	12.59	13.60
	160	6.625	6.559	6.691	5.187	0.719	0.692	—	15.69	16.94
8	5	8.625	8.594	8.718	8.407	0.109	0.095	0.123	3.429	—
	10	8.625	8.594	8.718	8.329	0.148	0.130	0.166	4.635	—
	20	8.625	8.539	8.711	8.125	0.250	0.219	—	7.735	8.354
	30	8.625	8.539	8.711	8.071	0.277	0.242	—	8.543	9.227
	40	8.625	8.539	8.711	7.981	0.322	0.282	—	9.878	10.67
	60	8.625	8.539	8.711	7.813	0.406	0.355	—	12.33	13.31
	80	8.625	8.539	8.711	7.625	0.500	0.438	—	15.01	16.21
	100	8.625	8.539	8.711	7.437	0.594	0.520	—	17.62	19.03
	120	8.625	8.539	8.711	7.187	0.719	0.629	—	21.00	22.68
	140	8.625	8.539	8.711	7.001	0.812	0.710	—	23.44	25.31
	160	8.625	8.539	8.711	6.813	0.906	0.793	—	25.84	27.90
10	5	10.750	10.719	10.843	10.482	0.134	0.117	0.151	5.256	—
	10	10.750	10.719	10.843	10.420	0.165	0.144	0.186	6.453	—
	20	10.750	10.642	10.858	10.250	0.250	0.219	—	9.698	10.47
	30	10.750	10.642	10.858	10.136	0.307	0.269	—	11.84	12.79
	40	10.750	10.642	10.858	10.020	0.365	0.319	—	14.00	15.12
	60	10.750	10.642	10.858	9.750	0.500	0.438	—	18.93	20.45
	80	10.750	10.642	10.858	9.562	0.594	0.520	—	22.29	24.07
	100	10.750	10.642	10.858	9.312	0.719	0.629	—	26.65	28.78
12	5	12.750	12.719	12.843	12.438	0.156	0.136	0.176	7.258	—
	10	12.750	12.719	12.843	12.390	0.180	0.158	0.202	8.359	—
	20	12.750	12.622	12.878	12.250	0.250	0.219	—	11.55	12.47
	30	12.750	12.622	12.878	12.090	0.330	0.289	—	15.14	16.35
	40	12.750	12.622	12.878	11.938	0.406	0.355	—	18.52	20.00
	60	12.750	12.622	12.878	11.626	0.562	0.492	—	22.31	24.33
	80	12.750	12.622	12.878	11.374	0.688	0.602	—	30.66	33.11

[a] Schedule 40 is also designated as "standard" pipe and schedule 80 is also designated as "extra heavy" pipe.
[b] For schedules 5 and 10 these values apply to mean outside diameter.
[c] For schedules other than 5 and 10, maximum wall thickness is controlled by weight tolerance.
[d] Based on a density of 0.098 lb per cu in., the density of 6061, and 6063 alloys, and on nominal dimensions and plain ends. For 3003 alloy multiply by 1.01; for 5083, 5086, and 5456 alloys multiply by 0.098; for 5454 alloy multiply by 0.099.

IVC-2 ALUMINUM ALLOY PIPE (ANSI B35.2; TABLE 16.7) OIL AND GAS TRANSMISSION AND DISTRIBUTION PIPING SYSTEMS

NORMAL SIZES, DIMENSIONS, AND WEIGHTS[a, b] OF ALUMINUM ALLOY PIPE

Nominal Pipe Size, in.	Schedule Number[c]	Outside Diameter, in.			Inside Diameter, in., Nominal	Wall Thickness, in.			Weight per ft,[a] lb	
		Nominal	Min[d]	Max[d]		Nominal	Min[d]	Max[d]	Nominal	Max[d]
1	2	3	4	5	6	7	8	9	10	11
2	5	2.375	2.344	2.406	2.245	0.065	0.053	0.077	0.555	—
	10	2.375	2.344	2.406	2.157	0.109	0.095	0.123	0.913	—
	40	2.375	2.351	2.399	2.067	0.154	0.135	—	1.264	1.365
	80	2.375	2.351	2.399	1.939	0.218	0.191	—	1.737	1.876
2½	5	2.875	2.844	2.906	2.709	0.083	0.071	0.095	0.856	—
	10	2.875	2.844	2.906	2.635	0.120	0.105	0.135	1.221	—
	40	2.875	2.846	2.904	2.469	0.203	0.178	—	2.004	2.164
	80	2.875	2.846	2.904	2.323	0.276	0.242	—	2.650	2.862
3	5	3.500	3.469	3.531	3.334	0.083	0.071	0.095	1.048	—
	10	3.500	3.469	3.531	3.260	0.120	0.105	0.135	1.498	—
	40	3.500	3.465	3.535	3.068	0.216	0.189	—	2.621	2.830
	80	3.500	3.465	3.535	2.900	0.300	0.262	—	3.547	3.830
4	5	4.500	4.469	4.531	4.334	0.083	0.071	0.095	1.354	—
	10	4.500	4.469	4.531	4.260	0.120	0.105	0.135	1.942	—
	40[e]	4.500	4.455	4.545	4.026	0.237	0.207	—	3.733	4.031
	80[g]	4.500	4.455	4.545	3.826	0.337	0.295	—	5.183	5.598
	120	4.500	4.455	4.545	3.624	0.438	0.383	—	6.573	7.099
	160	4.500	4.455	4.545	3.438	0.531	0.465	—	7.786	8.409
5	5	5.563	5.532	5.625	5.345	0.109	0.095	0.123	2.196	—
	10	5.563	5.532	5.625	5.295	0.134	0.117	0.151	2.688	—
	40[e]	5.563	5.507	5.619	5.047	0.258	0.226	—	5.057	5.461
	80[g]	5.563	5.507	5.619	4.813	0.375	0.328	—	7.188	7.763
	120	5.563	5.507	5.619	4.563	0.500	0.438	—	9.353	10.10
	160	5.563	5.507	5.619	4.313	0.625	0.547	—	11.40	12.31
6	5	6.625	6.594	6.687	6.407	0.109	0.095	0.123	2.264	—
	10	6.625	6.594	6.687	6.357	0.134	0.117	0.151	3.213	—
	40[e]	6.625	6.559	6.691	6.065	0.280	0.245	—	6.564	7.080
	80[g]	6.625	6.559	6.691	5.761	0.432	0.378	—	9.884	10.67
	120	6.625	6.559	6.691	5.501	0.562	0.492	—	12.59	13.60
	160	6.625	6.559	6.691	5.187	0.719	0.692	—	15.69	16.94
8	5	8.625	8.594	8.718	8.407	0.109	0.095	0.123	3.429	—
	10	8.625	8.594	8.718	8.329	0.148	0.130	0.166	4.635	—
	20	8.625	8.539	8.711	8.125	0.250	0.219	—	7.735	8.354
	30	8.625	8.539	8.711	8.071	0.277	0.242	—	8.543	9.227
	40[e]	8.625	8.539	8.711	7.981	0.322	0.282	—	9.878	10.67
	60	8.625	8.539	8.711	7.813	0.406	0.355	—	12.33	13.31
	80[g]	8.625	8.539	8.711	7.625	0.500	0.438	—	15.01	16.21
	100	8.625	8.539	8.711	7.437	0.594	0.520	—	17.62	19.03
	120	8.625	8.539	8.711	7.187	0.719	0.629	—	22.00	22.68
	140	8.625	8.539	8.711	7.001	0.812	0.710	—	23.44	25.31
	160	8.625	8.539	8.711	6.813	0.906	0.793	—	25.84	27.90
10	5	10.750	10.719	10.843	10.482	0.134	0.117	0.151	5.256	—
	10	10.750	10.719	10.843	10.420	0.165	0.144	0.186	6.453	—
	20	10.750	10.642	10.858	10.250	0.250	0.219	—	9.698	10.47
	30	10.750	10.642	10.858	10.136	0.307	0.269	—	11.84	12.79
	40[e]	10.750	10.642	10.858	10.020	0.365	0.319	—	14.00	15.12
	60	10.750	10.642	10.858	9.750	0.500	0.438	—	18.93	20.45
	80[g]	10.750	10.642	10.858	9.562	0.594	0.520	—	23.29	24.07
	100	10.750	10.642	10.858	9.312	0.719	0.629	—	26.65	28.78

IVC-2 ALUMINUM ALLOY PIPE (ASTM B345-88 & ANSI B35.2) OIL AND GAS TRANSMISSION AND DISTRIBUTION PIPING SYSTEMS

NORMAL SIZES, DIMENSIONS, AND WEIGHTS[a, b] OF ALUMINUM ALLOY PIPE

(ANSI B35.2—TABLE 16.7)

Nominal Pipe Size, in.	Schedule Number[c]	Outside Diameter, in.			Inside Diameter, in., Nominal	Wall Thickness, in.			Weight per ft,[a] lb	
		Nominal	Min[d]	Max[d]		Nominal	Min[d]	Max[d]	Nominal	Max[d]
1	2	3	4	5	6	7	8	9	10	11
12	5	12.750	12.719	12.843	12.438	0.156	0.136	0.176	7.258	—
	10	12.750	12.719	12.843	12.390	0.180	0.158	0.202	8.359	—
	20	12.750	12.622	12.878	12.250	0.250	0.219	—	11.55	12.47
	30	12.750	12.622	12.878	12.090	0.330	0.289	—	15.14	16.35
	40[e]	12.750	12.622	12.878	11.938	0.406	0.355	—	18.52	20.00
	60	12.750	12.622	12.878	11.626	0.562	0.492	—	22.31	24.33
	80[g]	12.750	12.622	12.878	11.374	0.688	0.602	—	30.66	33.11

(ASTM B345-88—TABLE 4)

Nominal Pipe Size, in.	Schedule Number[c]	Outside Diameter, in.			Inside Diameter, in., Nominal	Wall Thickness, in.			Weight per ft,[a] lb	
		Nominal	Min[d]	Max[d]		Nominal	Min[d]	Max[d]	Nominal	Max[d]
1	2	3	4	5	6	7	8	9	10	11
14[f]	10	14.000	—	—	—	0.250	—	—	12.70	—
	20	14.000	—	—	—	0.312	—	—	15.78	—
	30	14.000	—	—	—	0.375	—	—	18.88	—
	40[e]	14.000	—	—	—	0.438	—	—	21.95	—
	60	14.000	—	—	—	0.593	—	—	29.42	—
	80[g]	14.000	—	—	—	0.750	—	—	36.71	—
16[f]	10	16.000	—	—	—	0.250	—	—	14.55	—
	20	16.000	—	—	—	0.312	—	—	18.08	—
	30	16.000	—	—	—	0.375	—	—	21.65	—
	40[e]	16.000	—	—	—	0.500	—	—	28.63	—
	60	16.000	—	—	—	0.656	—	—	37.19	—
	80[g]	16.000	—	—	—	0.844	—	—	47.26	—
18[f]	40	18.000	—	—	—	0.562	—	—	36.21	—
20[f]	40	20.000	—	—	—	0.594	—	—	42.59	—

[a] Based on density of 0.098 lb per cu in., the density of alloys 1060, 6061, and 6063 determined according to the equation: (outside diameter − wall thickness) × wall thickness × 3.6945 = weight in lb per linear ft.
For alloys 5086 and 5456, multiply by 0.980; for alloys 5050 and 5052 by 0.990, and for alloy 3003 by 1.010.
[b] Other pipe sizes with outside diameters listed in Table 2 of ANSI Standard for Wrought-Steel and Wrought-Iron Pipe (B36.10M-1985) may be considered covered by this specification if agreed upon between the manufacturer and the purchaser.
[c] USA Standard Code for Pressure Piping (ASTM B345—Table 4).
[d] Based on standard tolerances for pipe.
[e] Also designated as "Standard" pipe.
[f] Dimensional and weight tolerances for 14, 16, 18, and 20-in. pipe are subject to special inquiry to the manufacturer.
[g] Also designated as "Extra Heavy" or "Extra Strong" pipe.

IVC-3 ALUMINUM COILED TUBING (ASTM B210-88) INSTRUMENT LINES, OIL LINES, REFRIGERATION SERVICES

TUBE CLASSIFICATION[a] AND SIZES

Outside Diameter, in.	Wall Thickness, in.
	Type A
1/8	0.025
3/16	0.028
1/4	0.032
5/16	0.035
3/8	0.035
7/16	0.035
1/2	0.035
9/16	0.042
5/8	0.042
3/4	0.049
7/8	0.049
1	0.049
1 1/8	0.049
1 1/4	0.058
1 3/8	0.058
1 1/2	0.065
1 5/8	0.065
1 3/4	0.065
2	0.072

[a] Type A tubes are designed for use where operating pressures are relatively low and strength is not of prime consideration.

Reprinted with permission of
American Society for Testing and Materials

PERMISSIBLE VARIATIONS IN WALL THICKNESS

Specified Wall[a] Thickness, in.	Allowable Deviation of Mean Wall Thickness[b] from Specified Wall Thickness, plus or minus, in.
0.010 to 0.035	0.002
0.036 to 0.049	0.003
0.050 to 0.083	0.004
0.084 to 0.109	0.005

[a] Intermediate wall thickness shall be rounded off to the third decimal place in accordance with the Recommended Practices for Designating Significant Places in Specified Limiting Values (ASTM Designation: E 29).[3]
[b] The mean wall thickness is determined by the average of two measurements taken opposite each other.

IVC-4 SEAMLESS COPPER PIPE AND SEAMLESS RED BRASS PIPE (ASTM B42-88 AND ASTM B43-88)

STANDARD DIMENSIONS, WEIGHTS, AND TOLERANCES

NOTE—ALL TOLERANCES PLUS AND MINUS EXCEPT AS OTHERWISE INDICATED

Standard Pipe Size, in.	Nominal Outside Diameter, in.	Average Outside Diameter Tolerance,[a] in. All Minus	Nominal Wall Thickness and Tolerance, in.				Theoretical Weight lb/ft			
			Regular		Extra Strong		Copper		Red Brass	
			Wall Thickness	Tolerance[b]	Wall Thickness	Tolerance[b]	Regular	Extra Strong	Regular	Extra Strong
1/8	0.405	0.004	0.062	0.004	0.100	0.006	0.259	0.371	0.253	0.363
1/4	0.540	0.004	0.082	0.005	0.123	0.007	0.457	0.625	0.447	0.611
3/8	0.675	0.005	0.090	0.005	0.127	0.007	0.641	0.847	0.627	0.829
1/2	0.840	0.005	0.107	0.006	0.149	0.008	0.955	1.25	0.934	1.23
3/4	1.050	0.006	0.114	0.006	0.157	0.009	1.30	1.71	1.27	1.67
1	1.315	0.006	0.126	0.007	0.182	0.010	1.82	2.51	1.78	2.46
1 1/4	1.660	0.006	0.146	0.008	0.194	0.010	2.69	3.46	2.63	3.39
1 1/2	1.900	0.006	0.150	0.008	0.203	0.011	3.20	4.19	3.13	4.10
2	2.375	0.008	0.156	0.009	0.221	0.012	4.22	5.80	4.12	5.67
2 1/2	2.875	0.008	0.187	0.010	0.280	0.015	6.12	8.85	5.99	8.66
3	3.500	0.010	0.219	0.012	0.304	0.016	8.76	11.8	8.56	11.6
3 1/2	4.000	0.010	0.250	0.013	0.321	0.017	11.4	14.4	11.2	14.1
4	4.500	0.012	0.250	0.014	0.341	0.018	12.9	17.3	12.7	16.9
5	5.562	0.014	0.250	0.014	0.375	0.019	16.2	23.7	15.8	23.2
6	6.625	0.016	0.250	0.014	0.437	0.027	19.4	32.9	19.0	32.2
8	8.625	0.020	0.312	0.022	0.500	0.035	31.6	49.5	30.9	48.4
10	10.750	0.022	0.365	0.030	0.500	0.040	46.2	62.4	45.2	61.1
12	12.750	0.024	0.375	0.030	—	—	56.5	—	55.3	—

[a] The average outside diameter of a tube is the average of the maximum and minimum outside diameters as determined at any one cross section of the pipe.
[b] Maximum deviaiton at any one point.

IVC-5 THREADLESS COPPER PIPE (ASTM B302-88) SEAMLESS, DEOXIDIZED FOR BRAZED JOINT ASSEMBLY

DIMENSIONS AND WEIGHTS, AND TOLERANCES IN DIAMETER AND WALL THICKNESS FOR COPPER THREADLESS PIPE (TP) SIZES

Nominal or Standard Pipe Size, in.	Nominal Dimensions, in.			Cross-Sectional Area of Bore, sq. in. (cm²)	Nominal Weight lb per ft (kg/m)	Tolerances, in. (mm)	
	Outside Diameter	Inside Diameter	Wall Thickness			Average Outside Diameter,[a] all minus	Wall Thickness, plus and minus
¼	0.540 (13.7)	0.410 (10.4)	0.065 (1.65)	0.132 (0.752)	0.376 (.559)	0.004 (0.10)	0.0035 (0.090)
⅜	0.675 (17.1)	0.545 (13.8)	0.065 (1.65)	0.233 (1.50)	0.483 (.719)	0.004 (0.10)	0.004 (0.10)
½	0.840 (21.3)	0.710 (18.0)	0.065 (1.65)	0.396 (2.55)	0.613 (.912)	0.005 (0.13)	0.004 (0.10)
¾	1.050 (26.7)	0.920 (23.4)	0.065 (1.65)	0.665 (4.29)	0.780 (1.16)	0.005 (0.13)	0.004 (0.10)
1	1.315 (33.4)	1.185 (30.1)	0.065 (1.65)	1.10 (7.10)	0.989 (1.47)	0.005 (0.13)	0.004 (0.10)
1¼	1.660 (42.2)	1.530 (38.9)	0.065 (1.65)	1.84 (11.9)	1.26 (1.87)	0.006 (0.15)	0.004 (0.10)
1½	1.900 (48.3)	1.770 (45.0)	0.065 (1.65)	2.46 (15.8)	1.45 (2.16)	0.006 (0.15)	0.004 (0.10)
2	2.375 (60.3)	2.245 (52.0)	0.065 (1.65)	3.96 (25.5)	1.83 (2.52)	0.007 (0.18)	0.006 (0.15)
2½	2.875 (73.0)	2.745 (69.7)	0.065 (1.65)	5.92 (38.2)	2.22 (3.30)	0.007 (0.18)	0.006 (0.15)
3	3.500 (68.9)	3.334 (84.7)	0.083 (2.11)	8.73 (56.3)	3.45 (5.13)	0.008 (0.20)	0.007 (0.18)
3½	4.000 (102)	3.810 (86.3)	0.095 (3.41)	11.4 (73.5)	4.52 (6.73)	0.008 (0.20)	0.007 (0.18)
4	4.500 (114)	4.286 (109)	0.107 (3.72)	14.4 (92.9)	5.72 (8.51)	0.010 (0.25)	0.009 (0.23)
5	5.562 (141)	5.298 (135)	0.132 (3.40)	22.0 (142)	8.73 (13.0)	0.012 (0.30)	0.010 (0.25)
6	6.625 (165)	6.309 (160)	0.158 (4.00)	31.3 (202)	12.4 (12.5)	0.014 (0.36)	0.010 (0.25)
8	8.625 (219)	8.215 (205)	0.205 (5.21)	53.0 (342)	21.0 (31.2)	0.018 (0.46)	0.014 (0.36)
10	10.750 (273)	10.238 (260)	0.256 (6.50)	82.3 (531)	32.7 (48.7)	0.018 (0.46)	0.016 (0.41)
12	12.750 (324)	12.124 (300)	0.313 (7.93)	115 (742)	47.4 (20.5)	0.018 (0.46)	0.020 (0.20)

[a] The average outside diameter of a tube is the average of the maximum and minimum outside diameters, as determined at any one cross-section of the tube.

ROUNDNESS TOLERANCES

t/D, Ratio of Nominal Wall Thickness to Nominal Outside Diameter	Diameter Roundness Tolerances,[a] percent of Outside Diameter (Expressed to the Nearest 0.001 in. nearest 0.01 mm.)
0.01 to 0.03 incl.	1.5
Over 0.03 to 0.05, incl.	1.0
Over 0.05 to 0.10, incl.	0.8
Over 0.10	0.7

[a] The deviation from roundness is measured as the difference between major and minor outside diameters as determined at any one cross-section of the tube.

IVC-6 SEAMLESS COPPER WATER TUBE (ANSI H23.1-67) PLUMBING AND FLUID CONVEYANCE TUBING

DIMENSIONS, WEIGHTS, AND TOLERANCES IN DIAMETER AND WALL THICKNESS FOR STANDARD COPPER WATER TUBE SIZES

(ALL TOLERANCES IN THIS TABLE ARE PLUS AND MINUS EXCEPT AS OTHERWISE INDICATED)

Standard Size, in.	Nominal Outside Diameter, in.	Average Outside Diameter[a] Tolerance in.		Nominal Wall Thickness and Tolerances, in.						Theoretical Weight, lb per ft		
				Type K		Type L		Type M				
		Annealed	Drawn	Wall Thickness	Tolerance[b]	Wall Thickness	Tolerance[b]	Wall Thickness	Tolerance[b]	Type K	Type L	Type M
1/4	0.375	0.002	0.001	0.035	0.004	0.030	0.0035	[c]	[c]	0.145	0.126	[c]
3/8	0.500	0.0025	0.001	0.049	0.004	0.035	0.0035	0.025	0.0025	0.269	0.198	0.145
1/2	0.625	0.0025	0.001	0.049	0.004	0.040	0.0035	0.028	0.0025	0.344	0.285	0.204
5/8	0.750	0.0025	0.001	0.049	0.004	0.042	0.0035	[c]	[c]	0.418	0.362	[c]
3/4	0.875	0.003	0.001	0.065	0.0045	0.045	0.004	0.032	0.003	0.641	0.455	0.328
1	1.125	0.0035	0.0015	0.065	0.0045	0.050	0.004	0.035	0.0035	0.839	0.655	0.465
1 1/4	1.375	0.004	0.0015	0.065	0.0045	0.055	0.0045	0.042	0.0035	1.04	0.884	0.682
1 1/2	1.625	0.0045	0.002	0.072	0.005	0.060	0.0045	0.049	0.004	1.36	1.14	0.940
2	2.125	0.005	0.002	0.083	0.007	0.070	0.006	0.058	0.006	2.06	1.75	1.46
2 1/2	2.625	0.005	0.002	0.095	0.007	0.080	0.006	0.065	0.006	2.93	2.48	2.03
3	3.125	0.005	0.002	0.109	0.007	0.090	0.007	0.072	0.006	4.00	3.33	2.68
3 1/2	3.625	0.005	0.002	0.120	0.008	0.100	0.007	0.083	0.007	5.12	4.29	3.58
4	4.125	0.005	0.002	0.134	0.010	0.110	0.009	0.095	0.009	6.51	5.38	4.66
5	5.125	0.005	0.002	0.160	0.010	0.125	0.010	0.109	0.009	9.67	7.61	6.66
6	6.125	0.005	0.002	0.192	0.012	0.140	0.011	0.122	0.010	13.9	10.2	8.92
8	8.125	0.006	+0.002 −0.004	0.271	0.016	0.200	0.014	0.170	0.014	25.9	19.3	16.5
10	10.125	0.008	+0.002 −0.006	0.338	0.018	0.250	0.016	0.212	0.015	40.3	30.1	25.6
12	12.125	0.008	+0.002 −0.006	0.405	0.020	0.280	0.018	0.254	0.016	57.8	40.4	36.7

[a] The average outside diameter of a tube is the average of the maximum and minimum outside diameter, as determined at any one cross section of the tube.
[b] Maximum deviation at any one point.
[c] Indicates that the material is not generally available or that no tolerance has been established.

IVC-7 WROUGHT SEAMLESS COPPER AND COPPER ALLOY TUBE (ANSI 23.4-67)

APPLICABLE PRODUCTS, UNLESS OTHERWISE SPECIFIED:

Title of Product Specification	ASTM Designation[a]
Seamless Copper Tube, Bright Annealed	B 68
Seamless Copper Tube	B 75
Seamless Brass Tube	B 135
Seamless Copper-Nickel Pipe and Tube	B 466

[a] These designations refer to the latest issue of the respective specification which appears in this publication.

Characteristics of Piping Materials
Non-Ferrous Pipe and Tubing

IVC-7 WROUGHT SEAMLESS COPPER AND COPPER ALLOY TUBE (ANSI 23.4-67)

PREFERRED SIZES OF ROUND SEAMLESS COPPER AND COPPER-ALLOY TUBE[a, b]

APPLICABLE TO ASTM DESIGNATIONS B 75 AND B 135

[X indicates the preferred sizes]

Outside Diameter, in. (mm)	Wall Thickness, in. (mm)																		
	0.010 (0.254)	0.013 (0.330)	0.016 (0.406)	0.020 (0.508)	0.025 (0.635)	0.032 (0.813)	0.040 (1.00)	0.049 (1.24)	0.065 (1.65)	0.083 (2.11)	0.100 (2.54)	1/8 (3.18)	5/32 (3.96)	3/16 (4.78)	1/4 (6.35)	5/16 (7.92)	3/8 (9.52)	1/2 (12.7)	5/8 (15.9)
1/8 (3.18)	X	X	X	X	X	X	X	—	—	—	—	—	—	—	—	—	—	—	—
3/16 (4.78)	X	X	X	X	X	X	X	X	—	—	—	—	—	—	—	—	—	—	—
1/4 (6.35)	—	—	X	X	X	X	X	X	X	X	—	—	—	—	—	—	—	—	—
5/16 (7.92)	—	—	X	X	X	X	X	X	X	X	X	—	—	—	—	—	—	—	—
3/8 (9.52)	—	—	X	X	X	X	X	X	X	X	X	X	—	—	—	—	—	—	—
1/2 (12.7)	—	—	X	X	X	X	X	X	X	X	X	X	X	—	—	—	—	—	—
5/8 (15.9)	—	—	X	X	X	X	X	X	X	X	X	X	X	—	—	—	—	—	—
3/4 (19.1)	—	—	X	X	X	X	X	X	X	X	X	X	X	—	—	—	—	—	—
7/8 (22.3)	—	—	X	X	X	X	X	X	X	X	X	X	X	X	—	—	—	—	—
1 (25.4)	—	—	—	X	X	X	X	X	X	X	X	X	X	X	X	—	—	—	—
1 1/4 (31.8)	—	—	—	X	X	X	X	X	X	X	X	X	X	X	X	X	—	—	—
1 1/2 (38.1)	—	—	—	X	X	X	X	X	X	X	X	X	X	X	X	X	X	—	—
1 3/4 (44.4)	—	—	—	X	X	X	X	X	X	X	X	X	X	X	X	X	X	—	—
2 (50.8)	—	—	—	—	X	X	X	X	X	X	X	X	X	X	X	X	X	X	—
2 1/4 (57.2)	—	—	—	—	X	X	X	X	X	X	X	X	X	X	X	X	X	X	—
2 1/2 (63.5)	—	—	—	—	X	X	X	X	X	X	X	X	X	X	X	X	X	X	X
2 3/4 (69.8)	—	—	—	—	X	X	X	X	X	X	X	X	X	X	X	X	X	X	X
3 (76.2)	—	—	—	—	—	X	X	X	X	X	X	X	X	X	X	X	X	X	X
3 1/2 (88.9)	—	—	—	—	—	X	X	X	X	X	X	X	X	X	X	X	X	X	X
4 (102)	—	—	—	—	—	—	X	X	X	X	X	X	X	X	X	X	X	X	X
4 1/2 (114)	—	—	—	—	—	—	X	X	X	X	X	X	X	X	X	X	X	X	X
5 (127)	—	—	—	—	—	—	—	X	X	X	X	X	X	X	X	X	X	X	X
5 1/2 (140)	—	—	—	—	—	—	—	X	X	X	X	X	X	X	X	X	X	X	X
6 (152)	—	—	—	—	—	—	—	—	X	X	X	X	X	X	X	X	X	X	X
7 (178)	—	—	—	—	—	—	—	—	X	X	X	X	X	X	X	X	X	X	X
8 (203)	—	—	—	—	—	—	—	—	—	X	X	X	X	X	X	X	X	X	X
9 (229)	—	—	—	—	—	—	—	—	—	X	X	X	X	X	X	X	X	X	X
10 (254)	—	—	—	—	—	—	—	—	—	—	X	X	X	X	X	X	X	X	X
11 (279)	—	—	—	—	—	—	—	—	—	—	X	X	X	X	X	X	X	X	X
12 (305)	—	—	—	—	—	—	—	—	—	—	—	X	X	X	X	X	X	X	X

[a] In conformance with the Simplified Practice Recommendations R 235–48 for Copper and Copper-Alloy Round Seamless Tube issued by the U.S. Department of Commerce.
[b] This tube is not necessarily available in all alloys in the full range of sizes shown.

Characteristics of Piping Materials
Non-Ferrous Pipe and Tubing

IVC-7 WROUGHT SEAMLESS COPPER AND COPPER ALLOY TUBE (ANSI 23.4-67) CONTINUED

WALL THICKNESS TOLERANCES FOR COPPER AND COPPER-ALLOY TUBE

(Applicable to Specifications B 68, B 75, and B 135)

NOTE—*Maximum Deviation at Any Point:* The following tolerances are plus and minus; if tolerances all plus or all minus are desired, double the values given.

Wall Thickness, in. (mm)	Outside Diameter, in.[a] (mm)						
	1/32 (0.792) to 1/8 (3.18), incl	Over 1/8 (3.18) to 5/8 (15.9), incl	Over 5/8 (15.9) to 1 (25.4), incl	Over 1 (25.4) to 2 (50.8), incl	Over 2 (50.8) to 4 (102), incl	Over 4 (102) to 7 (213), incl	Over 7 (213) to 10 (254), incl
Under 0.018 (0.457)	0.002 (0.051)	0.001 (0.025)	0.0015 (0.038)	0.002 (0.051)	—	—	—
0.018 (0.457), incl to 0.025 (0.635)	0.003 (0.076)	0.002 (0.051)	0.002 (0.051)	0.0025 (0.064)	—	—	—
0.025 (0.635), incl to 0.035 (0.889)	0.003 (0.076)	0.0025 (0.064)	0.0025 (0.064)	0.003 (0.076)	0.004 (0.10)	—	—
0.035 (0.889), incl to 0.058 (1.47)	0.003 (0.076)	0.003 (0.076)	0.0035 (0.089)	0.0035 (0.089)	0.005 (0.13)	0.007 (0.18)	—
0.058 (1.47), incl to 0.083 (2.11)		0.0035 (0.089)	0.004 (0.10)	0.004 (0.10)	0.006 (0.15)	0.008 (0.20)	0.010 (0.25)
0.083 (2.11), incl to 0.120 (3.03)		0.004 (0.10)	0.005 (0.13)	0.005 (0.13)	0.007 (0.18)	0.009 (0.23)	0.011 (0.28)
0.120 (3.03), incl to 0.165 (4.19)		0.005 (0.13)	0.006 (0.15)	0.006 (0.15)	0.008 (0.20)	0.010 (0.25)	0.012 (0.30)
0.165 (4.19), incl to 0.220 (5.59)	—	0.007 (0.18)	0.0075 (0.19)	0.008 (0.20)	0.010 (0.25)	0.012 (0.30)	0.014 (0.36)
0.220 (5.59), incl to 0.284 (7.21)	—	—	0.009 (0.23)	0.010 (0.25)	0.012 (0.30)	0.014 (0.36)	0.016 (0.41)
0.284 (7.21), incl to 0.380 (9.65)	—	—	0.011 (0.28)	0.012 (0.30)	0.014 (0.36)	0.016 (0.41)	0.018 (0.46)
0.380 (9.65) and over	—	—	—	5 percent	5 percent	6 percent	6 percent

[a] When round tube is ordered by outside and inside diameters, the maximum plus and minus deviation of the wall thickness from the nominal at any point shall not exceed the values given in the table by more than 50 percent.

IVC-7 WROUGHT SEAMLESS COPPER AND COPPER ALLOY TUBE (ANSI 23.4-67) CONTINUED

AVERAGE DIAMETER TOLERANCES FOR COPPER AND COPPER-ALLOY TUBE

(Applicable to Specifications B 68, B 75, and B 135)

Specified Diameter, in. (mm)	Diameter to Which Tolerance Applies	Tolerance, plus and minus, in. (mm)
Up to ⅛ (3.18) incl	inside or outside	0.002 (0.051)
Over ⅛ (3.18) to ⅝ (15.9), incl	inside or outside	0.002 (0.051)
Over ⅝ (15.9) to 1 (25.4), incl	inside or outside	0.0025 (0.064)
Over 1 (25.4) to 2 (50.8), incl	inside or outside	0.003 (0.076)
Over 2 (50.8) to 3 (76.2), incl	inside or outside	0.004 (0.10)
Over 3 (76.2) to 4 (102), incl	inside or outside	0.005 (0.13)
Over 4 (102) to 5 (127), incl	inside or outside	0.006 (0.15)
Over 5 (127) to 6 (152), incl	inside or outside	0.007 (0.18)
Over 6 (152) to 8 (203), incl	inside or outside	0.008 (0.20)
Over 8 (203) to 10 (254), incl	inside or outside	0.010 (0.25)

IVC-8 COPPER CAPILLARY TUBE, HARD-DRAWN (ASTM B360)

STANDARD DIMENSIONS AND RESIDUE LIMITS OF INTERIOR SURFACE FOR CAPILLARY TUBES.

Outside Diameter, in. (mm)	Inside Diameter, in. (mm)	Mean Wall Thickness, in. (mm)	Cross-Sectional Area of Tube Bore, sq in. (mm²)	Weight Per Foot, lb (k/m)	Maximum Allowable Residue, grams per linear foot (g/linear m)
0.072 (1.83)	0.026 (0.660)	0.023 (0.584)	5.309×10^{-4} (0.343)	0.01373 (0.0204)	0.00020 (0.000656)
0.072 (1.83)	0.028 (0.711)	0.022 (0.558)	6.158 (0.397)	0.01340 (0.0199)	0.00021 (0.000689)
0.081 (2.06)	0.031 (0.787)	0.025 (0.635)	7.548 (0.487)	0.01705 (0.0254)	0.00023 (0.000754)
0.081 (2.06)	0.033 (0.838)	0.024 (0.606)	8.553 (0.552)	0.01666 (0.0248)	0.00025 (0.000820)
0.087 (2.21)	0.036 (0.914)	0.0255 (0.648)	10.18 (0.657)	0.01910 (0.0284)	0.00027 (0.000886)
0.087 (2.21)	0.039 (0.991)	0.024 (0.606)	11.95 (0.771)	0.01842 (0.0239)	0.00029 (0.000951)
0.093 (2.36)	0.042 (1.07)	0.0255 (0.648)	13.85 (0.8)	0.02096 (0.0312)	0.00032 (0.00105)
0.097 (2.47)	0.046 (1.17)	0.0255 (0.648)	16.62 (1.07)	0.02221 (0.0331)	0.00035 (0.00115)
0.099 (2.51)	0.049 (1.24)	0.025 (0.635)	18.86 (1.22)	0.02253 (0.0335)	0.00037 (0.00121)
0.106 (2.69)	0.054 (1.37)	0.026 (0.660)	22.90 (1.44)	0.02533 (0.0337)	0.00041 (0.00134)
0.112 (2.84)	0.059 (1.50)	0.0265 (0.673)	27.34 (1.76)	0.02760 (0.0411)	0.00044 (0.00144)
0.125 (3.18)	0.064 (1.63)	0.0305 (0.775)	32.17 (2.07)	0.03511 (0.0522)	0.00048 (0.00157)
0.125 (3.18)	0.070 (1.78)	0.0275 (0.698)	38.48 (2.48)	0.03266 (0.0486)	0.00053 (0.00174)
0.125 (3.18)	0.075 (1.91)	0.025 (0.635)	44.18 (2.85)	0.03045 (0.0454)	0.00057 (0.00187)
0.145 (3.68)	0.080 (2.03)	0.0325 (0.826)	50.27 (3.24)	0.04453 (0.0663)	0.00060 (0.00197)
0.145 (3.68)	0.085 (2.16)	0.030 (0.762)	56.74 (3.66)	0.04202 (0.0625)	0.00064 (0.00210)
0.145 (3.68)	0.090 (2.29)	0.0275 (0.698)	63.62 (4.10)	0.03936 (0.0586)	0.00068 (0.00223)
0.160 (4.06)	0.100 (2.54)	0.030 (0.762)	78.54 (5.07)	0.04750 (0.0707)	0.00075 (0.00246)
0.160 (4.06)	0.110 (2.79)	0.025 (0.635)	95.03 (6.13)	0.04111 (0.0611)	0.00083 (0.00272)
0.188 (4.78)	0.120 (3.03)	0.034 (0.864)	113.1 (7.29)	0.06377 (0.0949)	0.00090 (0.00295)
0.188 (4.78)	0.130 (3.30)	0.029 (0.737)	132.7 (8.56)	0.05616 (0.0836)	0.00098 (0.00321)
0.200 (5.08)	0.145 (3.68)	0.0275 (0.698)	165.1 (10.7)	0.05779 (0.0860)	0.00109 (0.00358)
0.220 (5.59)	0.160 (4.06)	0.030 (0.762)	201.1 (13.0)	0.06943 (0.103)	0.00121 (0.00397)
0.240 (6.10)	0.175 (4.45)	0.0325 (0.826)	240.5 (15.5)	0.08107 (0.121)	0.00132 (0.00433)

IVC-9 SEAMLESS COPPER—NICKEL PIPE AND TUBE (ASTM B466 - 86 & B466M - 86)

ALLOYS		
Copper Alloy Nos.	Nominal Composition, percent	
	Copper	Nickel
C 70400	95	5
C 70600	90	10
C 71000	80	20
C 71500	70	30
C 72200	60	40

PREFERRED WALL THICKNESSES OF DRAWN SEAMLESS PIPE, BASED ON SPS DIAMETERS

SPS, in.	Outside Diameter		Wall Thicknesses														
			Specials										Regular		Extra Strong	
	in.	mm	in.	mm	in.	mm	in.	mm	in.	mm	in.	mm	in.	mm	in.	mm
1/8	0.405	10.3	0.058	1.47	—	—	—	—	—	—	—	—	0.062	1.57	0.100	2.54
1/4	0.540	13.7	0.065	1.65	0.072	1.83	—	—	—	—	—	—	0.082	2.08	0.123	3.12
3/8	0.675	17.1	0.065	1.65	0.072	1.83	0.095	2.41	0.148	3.76	—	—	0.090	2.29	0.127	3.23
1/2	0.840	21.3	0.065	1.65	0.072	1.83	0.120	3.03	0.203	5.16	—	—	0.107	2.72	0.149	3.78
3/4	1.050	26.7	0.065	1.65	0.083	2.11	0.148	3.76	0.238	6.05	—	—	0.114	2.90	0.157	3.99
1	1.315	33.4	0.065	1.65	0.095	2.41	0.203	5.16	0.340	8.64	—	—	0.126	3.20	0.182	4.62
1 1/4	1.660	42.2	0.072	1.83	0.095	2.41	0.120	3.03	0.220	5.59	0.380	9.65	0.146	3.71	0.194	4.93
1 1/2	1.900	48.3	0.072	1.83	0.109	2.77	0.134	3.40	0.250	6.35	0.425	10.8	0.150	3.81	0.203	5.16
2	2.375	60.3	0.083	2.11	0.120	3.03	0.165	4.19	0.340	8.64	0.520	13.2	0.156	3.96	0.221	5.61
2 1/2	2.875	73.0	0.083	2.11	0.134	3.40	0.203	5.16	0.380	9.65	—	—	0.187	4.75	0.280	7.11
3	3.500	88.9	0.095	2.41	0.165	4.19	0.250	6.35	0.458	11.6	—	—	0.219	5.56	0.304	7.72
3 1/2	4.000	102	0.095	2.41	0.180	4.57	0.284	7.21	—	—	—	—	0.250	6.35	0.321	8.15
4	4.500	114	0.109	2.77	0.203	5.16	0.340	8.64	—	—	—	—	0.250	6.35	0.341	8.66
5	5.562	141	0.125	3.18	0.220	5.59	0.425	10.8	—	—	—	—	0.250	6.35	0.375	9.52
6	6.625	168	0.134	3.40	0.259	6.58	0.457	11.6	—	—	—	—	0.250	6.35	0.437	11.1

Note—The values stated in U.S. customary units are to be regarded as the standard. The metric equivalents of U.S. customary units may be approximate.

IVC-9 SEAMLESS COPPER—NICKEL PIPE AND TUBE (ASTM B466M-86) (CONTINUED)

WALL THICKNESS TOLERANCES

NOTE—*Maximum Deviation of Any Point:* The following tolerances are plus and minus; if tolerances all plus or all minus are desired, double the values given.

Wall Thickness, in. (mm)	Outside Diameter,[A] in. (mm)						
	1/32 to 1/8 (0.792 to 3.18) incl	Over 1/8 to 5/8 (3.18 to 15.9) incl	Over 5/8 to 1 (15.9 to 25.4) incl	Over 1 to 2 (25.4 to 50.8) incl	Over 2 to 4 (50.8 to 102) incl	Over 4 to 7 (102 to 178) incl	Over 7 to 10 (178 to 254) incl
Up to 0.017 (0.432) incl	0.0025 (0.064)	0.0015 (0.038)	0.002 (0.051)	0.0025 (0.064)	—	—	—
Over 0.017 (0.432) to 0.024 (0.610) incl	0.004 (0.10)	0.0025 (0.064)	0.0025 (0.064)	0.003 (0.076)	—	—	—
Over 0.024 (0.610) to 0.034 (0.864) incl	0.004 (0.10)	0.003 (0.076)	0.003 (0.076)	0.004 (0.10)	0.005 (0.13)	—	—
Over 0.034 (0.864) to 0.057 (1.45) incl	0.004 (0.10)	0.004 (0.10)	0.0045 (0.11)	0.0045 (0.11)	0.0065 (0.17)	0.009 (0.23)	—
Over 0.057 (1.45) to 0.082 (2.08) incl	—	0.0045 (0.11)	0.005 (0.13)	0.005 (0.13)	0.0075 (0.19)	0.010 (0.25)	0.013 (0.33)
Over 0.082 (2.08) to 0.119 (3.02) incl	—	0.005 (0.13)	0.0065 (0.17)	0.0065 (0.17)	0.009 (0.23)	0.011 (0.28)	0.014 (0.36)
Over 0.119 (3.02) to 0.164 (4.17) incl	—	0.007 (0.18)	0.007 (0.18)	0.0075 (0.19)	0.010 (0.25)	0.013 (0.33)	0.015 (0.38)
Over 0.164 (4.17) to 0.219 (5.56) incl	—	—	0.009 (0.23)	0.010 (0.25)	0.012 (0.30)	0.015 (0.38)	0.018 (0.46)
Over 0.219 (5.56) to 0.283 (7.19) incl	—	—	0.012 (0.30)	0.013 (0.33)	0.015 (0.38)	0.018 (0.46)	0.020 (0.51)
Over 0.283 (7.19) to 0.379 (9.62) incl	—	—	—	0.015 (0.38)	0.018 (0.46)	0.020 (0.51)	0.023 (0.58)
Over 0.379 (9.62)	—	—	—	6[B]	6[B]	8[B]	8[B]

[A]When tube is ordered by outside and inside diameters, the maximum plus and minus deviation of the wall thickness from the nominal at any point shall not exceed the values given in this table by more than 50%.

[B]Percent of the specified wall thickness expressed to the nearest 0.001 in. (0.025 mm).

AVERAGE DIAMETER[A] TOLERANCES

Specified Diameter, in. (mm)	Tolerance Applies to	Tolerances, plus and minus,[B] in. (mm) for Tubes of Copper Alloy UNS Nos. C70400, C70600, C71000, C71500, and C72200
Up to 1/8 (3.18), incl	inside diameter	0.003 (0.076)
Up to 1/8 (3.18), incl	outside diameter	0.0025 (0.064)
Over 1/8 (3.18) to 5/8 (15.9), incl	inside or outside	0.0025 (0.064)
Over 5/8 (15.9) to 1 (25.4), incl	inside or outside	0.003 (0.076)
Over 1 (25.4) to 2 (50.8), incl	inside or outside	0.004 (0.10)
Over 2 (50.8) to 3 (76.2), incl	inside or outside	0.005 (0.13)
Over 3 (76.2) to 4 (102), incl	inside or outside	0.006 (0.15)
Over 4 (102) to 5 (127), incl	inside or outside	0.008 (0.20)
Over 5 (127) to 6 (152), incl	inside or outside	0.009 (0.23)
Over 6 (152) to 8 (203), incl	inside or outside	0.010 (0.25)
Over 8 (203) to 10 (254), incl	inside or outside	0.013 (0.33)

[A]The average outside or inside diameter of a tube is the average of the maximum and minimum outside diameters, or of the maximum and minimum inside diameters, whichever is applicable, as determined at any one cross section of the tube.

[B]If tolerances all plus or all minus are desired, double the values given.

Characteristics of Piping Materials
Plastic Line Pipe

IVD-1 API STD 5LR (TENTATIVE STANDARD) THERMOSETTING RESIN LINE PIPE GLASS FIBER REINFORCED

TABLE 6.1
DIMENSIONS AND WEIGHTS

Nominal Size, Inches	Outside Diameter, Inches	Inside Diameter, Inches	Total Wall Thickness, Inches	Min. Reinforced Wall Thickness, Inches* (t)	Nominal Weight, Lbs./Ft.	Process of Manufacture
2	2.375	2.235	.070	.060	.40	FW
		2.135	.120	.100	.65	FW
		1.995	.190	.166	1.19	FW
		1.975	.200	.140	.83	CC
		1.875	.250	.170	1.25	CC
		1.775	.300	.220	1.40	CC
2½	2.875	2.735	.070	.060	.51	FW
		2.495	.190	.166	1.32	FW
		2.475	.200	.140	1.20	CC
		2.375	.250	.170	1.46	CC
		2.275	.300	.220	1.70	CC
3	3.500	3.360	.070	.060	.60	FW
		3.340	.080	.070	.65	FW
		3.260	.120	.100	1.00	FW
		3.100	.200	.140	1.40	CC
		3.000	.250	.170	1.75	CC
		2.900	.300	.220	2.10	CC
4	4.500	4.360	.070	.060	.80	FW
		4.340	.080	.070	.80	FW
		4.320	.090	.079	.90	FW
		4.260	.120	.100	1.30	FW
		4.100	.200	.140	1.77	CC
		4.000	.250	.170	2.40	CC
		3.900	.300	.220	2.90	CC
6	6.625	6.405	.110	.100	1.70	FW
		6.305	.160	.140	2.60	FW
		6.125	.250	.170	3.50	CC
		6.025	.300	.220	4.15	CC
		5.825	.400	.300	5.40	CC
8	8.625	8.355	.135	.125	2.00	FW
		8.265	.180	.160	3.80	FW
		8.125	.250	.170	4.50	CC
		8.025	.300	.220	5.40	CC
		7.825	.400	.300	7.00	CC
10	10.750	10.410	.170	.150	4.20	FW
		10.310	.220	.200	5.83	FW
		10.250	.250	.170	5.60	CC
		10.150	.300	.220	6.70	CC
		9.950	.400	.320	8.75	CC
12	12.750	12.350	.200	.175	6.00	FW
		12.310	.220	.200	7.00	FW
		12.250	.250	.170	6.70	CC
		12.150	.300	.220	8.00	CC
		11.950	.400	.320	10.40	CC

DIMENSIONAL TOLERANCES

Nominal Size, Inches	Outside Diameter, Percent	Total Wall Thickness, Percent
2	±1	+20 — 12.5*
2½	±1	+20 — 12.5*
3	±1	+20 — 12.5*
4	±1	+20 — 12.5*
6	±1	+15 — 12.5*
8	±1	+15 — 12.5*
10	±1	+15 — 12.5*
12	±1	+15 — 12.5*

IVD-2 API STD 5LP (TENTATIVE STANDARDS) THERMOPLASTIC LINE PIPE EXTRUDED POLYVINYL CHLORIDE (PVC)

DIMENSIONS, WEIGHTS AND TOLERANCES

Nominal size, in.	Schedule	Outside Dia. and Tolerances, In.	Minimum Wall Thickness, In.	Wall Thickness, Tolerance, In.	Out of Roundness Max. & Min. In.	Nominal Weight lb/100 ft
½	Sch 80	0.840 ±1 0.004	0.147	+0.020	±0.008	20.0
	Sch 40		0.109	+ 0.020	±0.008	15.7
	SDR 11		0.076	+ 0.018	±0.008	11.4
	SDR 13.5		0.062	+ 0.015	±0.008	9.5
	SDR 17		0.062	+ 0.015	±0.008	9.5
	SDR 21		0.062	+ 0.015	±0.015	9.5
¾	Sch 80	1.050 ± 0.004	0.154	+ 0.020	±0.010	27.4
	Sch 40		0.113	+ 0.020	±0.010	21.0
	SDR 11		0.095	+ 0.021	±0.010	18.0
	SDR 13.5		0.090	+ 0.020	±0.010	17.1
	SDR 17		0.090	+ 0.020	±0.010	17.1
	SDR 21		0.090	+ 0.020	±0.015	17.1
1	Sch 80	1.315 ± 0.005	0.179	+ 0.021	±0.010	40.2
	Sch 40		0.133	+ 0.020	±0.010	31.0
	SDR 11		0.119	+ 0.026	±0.010	28.2
	SDR 13.5		0.097	+ 0.020	±0.010	23.4
	SDR 17		0.090	+ 0.020	±0.010	21.8
	SDR 21		0.090	+ 0.020	±0.015	21.8
1¼	Sch 80	1.660 ± 0.005	0.191	+ 0.023	±0.012	55.4
	Sch 40		0.140	+ 0.020	±0.012	42.0
	SDR 11		0.151	+ 0.026	±0.012	45.0
	SDR 13.5		0.123	+ 0.020	±0.012	37.3
	SDR 17		0.098	+ 0.020	±0.012	30.2
	SDR 21		0.090	+ 0.020	±0.015	27.9
1½	Sch 80	1.900 ± 0.006	0.200	+ 0.024	±0.012	67.1
	Sch 40		0.145	+ 0.020	±0.012	50.4
	SDR 11		0.173	+ 0.026	±0.012	59.2
	SDR 13.5		0.141	+ 0.020	±0.012	49.1
	SDR 17		0.112	+ 0.020	±0.012	39.7
	SDR 21		0.090	+ 0.020	±0.015	32.3
2	Sch 80	2.375 ± 0.006	0.218	+ 0.026	±0.012	92.9
	Sch 40		0.154	+ 0.020	±0.012	67.6
	SDR 11		0.216	+ 0.026	±0.012	92.1
	SDR 13.5		0.176	+ 0.021	±0.012	76.5
	SDR 17		0.140	+ 0.020	±0.012	61.8
	SDR 21		0.113	+ 0.020	±0.030	50.5
2½	Sch 80	2.875 ± 0.007	0.276	+ 0.033	±0.015	142
	Sch 40		0.203	+ 0.024	±0.015	108
	SDR 11		0.261	+ 0.031	±0.015	135
	SDR 13.5		0.213	+ 0.025	±0.015	112
	SDR 17		0.169	+ 0.020	±0.015	90.6
	SDR 21		0.137	+ 0.020	±0.030	74.3
3	Sch 80	3.500 ± 0.008	0.300	+ 0.036	±0.015	189
	Sch 40		0.216	+ 0.026	±0.015	140
	SDR 11		0.318	+ 0.038	±0.015	199
	SDR 13.5		0.259	+ 0.031	±0.015	165
	SDR 17		0.206	+ 0.025	±0.015	134
	SDR 21		0.167	+ 0.020	±0.030	110

IVD-2 API STD 5LP (TENTATIVE STANDARDS) THERMOPLASTIC LINE PIPE EXTRUDED POLYVINYL CHLORIDE (PVC)

DIMENSIONS, WEIGHTS AND TOLERANCES (CONT'D)

Nominal size, in.	Schedule	Outside Dia. and Tolerances, In.	Minimum Wall Thickness, In.	Wall Thickness, Tolerance, In.	Out of Roundness Max. & Min. In.	Nominal Weight lb/100 ft
3½	Sch 80	4.000 ± 0.008	0.318	+ 0.038	±0.015	232
	Sch 40		0.226	+ 0.027	±0.050	169
	SDR 11		0.363	+ 0.044	±0.015	262
	SDR 13.5		0.296	+ 0.035	±0.015	217
	SDR 17		0.236	+ 0.028	±0.015	176
	SDR 21		0.190	+ 0.023	±0.050	143
4	Sch 80	4.500 ± 0.009	0.337	+ 0.040	±0.015	277
	Sch 40		0.237	+ 0.028	±0.050	200
	SDR 11		0.409	+ 0.049	±0.015	331
	SDR 13.5		0.333	+ 0.040	±0.015	274
	SDR 17		0.264	+ 0.032	±0.015	221
	SDR 21		0.214	+ 0.026	±0.050	181
5	Sch 80	5.563 ± 0.010	0.375	+ 0.045	±0.030	387
	Sch 40		0.258	+ 0.031	±0.050	272
	SDR 11		0.506	+ 0.061	±0.030	509
	SDR 13.5		0.413	+ 0.050	±0.030	423
	SDR 17		0.328	+ 0.039	±0.030	342
	SDR 21		0.265	+ 0.032	±0.050	279
6	Sch 80	6.625 ± 0.011	0.432	+ 0.052	±0.035	530
	Sch 40		0.280	+ 0.034	±0.050	352
	SDR 11		0.603	+ 0.068	±0.035	719
	SDR 13.5		0.491	+ 0.059	±0.035	597
	SDR 17		0.390	+ 0.047	±0.035	482
	SDR 21		0.316	+ 0.038	±0.050	395
	SDR 26		0.255	+ 0.031	±0.055	322
	SDR 32.5		0.204	+ 0.024	±0.060	259
8	Sch 80	8.625 ± 0.015	0.500	+ 0.060	±0.075	803
	Sch 40	8.625 ± 0.015	0.322	+ 0.039	±0.075	529
	Sch 11	8.625 ± 0.013	0.785	+ 0.094	±0.040	1217
	SDR 13.5	8.625 ± 0.013	0.639	+ 0.077	±0.040	1009
	SDR 17	8.625 ± 0.013	0.508	+ 0.061	±0.045	815
	SDR 21	8.625 ± 0.013	0.410	+ 0.049	±0.060	666
	SDR 26	8.625 ± 0.013	0.332	+ 0.040	±0.080	545
	SDR 32.5	8.625 ± 0.013	0.265	+ 0.032	±0.120	438
10	Sch 80	10.750 ± 0.015	0.593	+ 0.071	±0.075	1191
	Sch 40		0.365	+ 0.044	±0.075	750
	Sch 11		0.978	+ 0.117	±0.050	1890
	SDR 13.5		0.797	+ 0.096	±0.050	1569
	SDR 17		0.633	+ 0.076	±0.050	1267
	SDR 21		0.511	+ 0.061	±0.070	1035
	SDR 26		0.413	+ 0.050	±0.100	844
	SDR 32.5		0.331	+ 0.040	±0.120	682
12	Sch 80	12.750 ± 0.015	0.687	+ 0.082	±0.075	1639
	Sch 40	12.750 ± 0.015	0.406	+ 0.049	±0.075	991
	SDR 11	12.750 ± 0.017	1.160	+ 0.139	±0.050	2659
	SDR 13.5	12.750 ± 0.017	0.945	+ 0.113	±0.050	2206
	SDR 17	12.750 ± 0.017	0.745	+ 0.090	±0.050	1769
	SDR 21	12.750 ± 0.017	0.608	+ 0.073	±0.070	1460
	SDR 26	12.750 ± 0.017	0.490	+ 0.059	±0.100	1188
	SDR 32.5	12.750 ± 0.017	0.392	+ 0.047	±0.140	958

IVE-1 150-LB STEEL PIPE FLANGES AND FLANGED FITTINGS

FLANGE

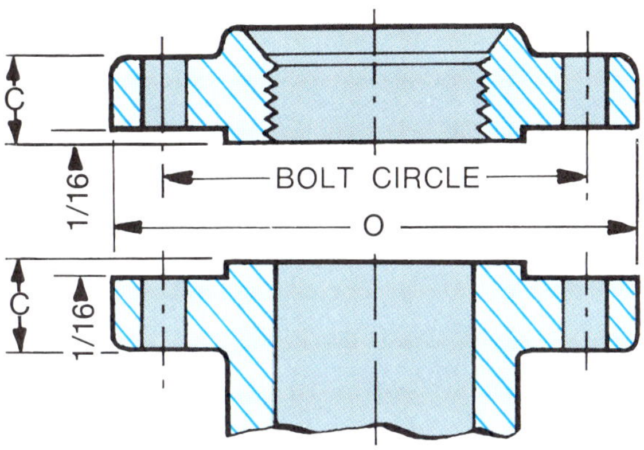

FLANGE FITTING

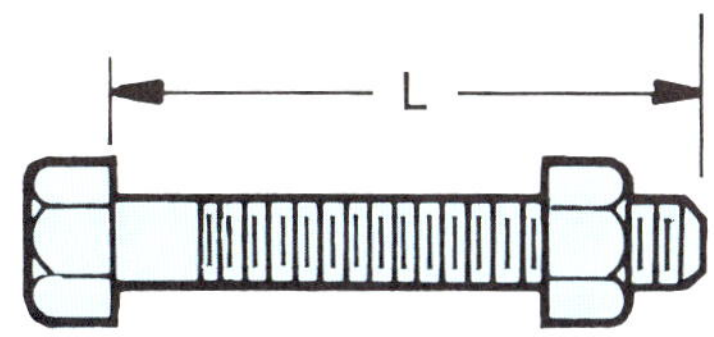

MACHINE BOLT WITH NUT

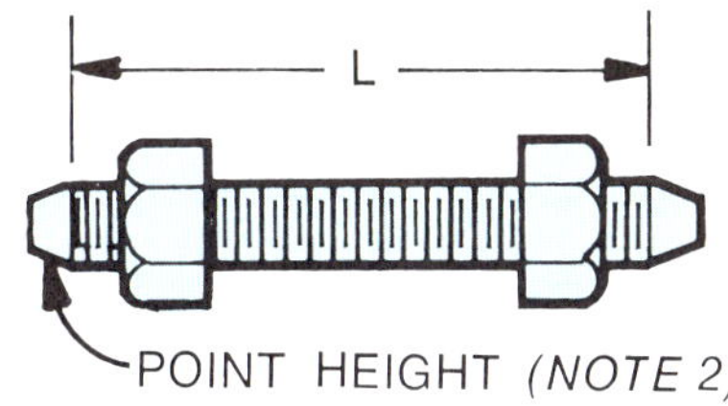

STUD BOLT WITH NUTS

TEMPLATES FOR DRILLING STEEL FLANGES AND FLANGED FITTINGS

Nominal Pipe Size	Outside Diameter of Flange O	Flange Thicknesses[4] (Min)		Drilling[1]				Length of Bolts[3] L		
								Stud Bolts[2]		Machine Bolts
		Flange C	Fitting C	Diameter of Bolt Circle	Diameter of Bolt Holes	Number of Bolts	Diameter of Bolts	0.006 In. Raised Face	Ring Joint	0.006 In. Raised Face
½	3½	7/16	—	2⅜	⅝	4	½	2.25	—	2.00
¾	3⅞	½	—	2¾	⅝	4	½	2.50	—	2.00
1	4¼	9/16	7/16	3⅛	⅝	4	½	2.50	3.00	2.25
1¼	4⅝	⅝	½	3½	⅝	4	½	2.75	3.25	2.25
1½	5	11/16	9/16	3⅞	⅝	4	½	2.75	3.25	2.50
2	6	¾	⅝	4¾	¾	4	⅝	3.25	3.75	2.75
2½	7	⅞	11/16	5½	¾	4	⅝	3.50	4.00	3.00
3	7½	15/16	¾	6	¾	4	⅝	3.50	4.00	3.00
3½	8½	15/16	13/16	7	¾	8	⅝	3.50	4.00	3.00
4	9	15/16	15/16	7½	¾	8	⅝	3.50	4.00	3.00
5	10	15/16	15/16	8½	⅞	8	¾	3.75	4.25	3.25
6	11	1	1	9½	⅞	8	¾	4.00	4.50	3.25
8	13½	1⅛	1⅛	11¾	⅞	8	¾	4.25	4.75	3.50
10	16	13/16	13/16	14¼	1	12	⅞	4.50	5.00	4.00
12	19	1¼	1¼	17	1	12	⅞	4.75	5.25	4.00
14	21	1⅜	1⅜	18¾	1⅛	12	1	5.25	5.75	4.50
16	23½	1 7/16	1 7/16	21¼	1⅛	16	1	5.25	5.75	4.50
18	25	1 9/16	1 9/16	22¾	1¼	16	1⅛	5.75	6.25	5.00
20	27½	1 11/16	1 11/16	25	1¼	20	1⅛	6.25	6.75	5.50
24	32	1⅞	1⅞	29½	1⅜	20	1¼	6.75	7.25	6.00

All dimensions given in inches.
[1]Bolt holes straddle the center lines.
[2]Length of stud-bolts do not include the height of the points.
[3]Bolt lengths not shown in table are determined in accordance with ANSI B16.5-1988
[4]Flange fittings not available in ½ and ¾ sizes.

IVE-1 300-LB STEEL PIPE FLANGES AND FLANGED FITTINGS

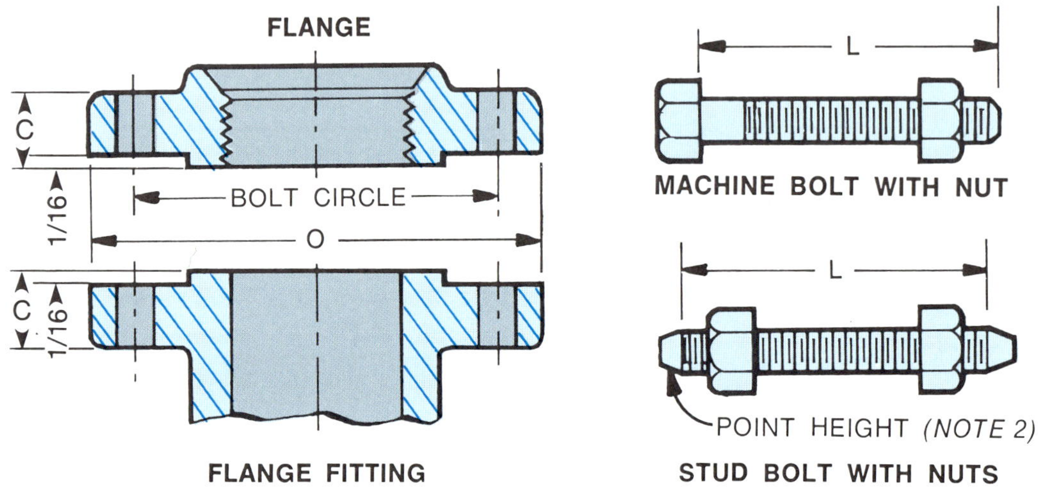

TEMPLATES FOR DRILLING STEEL FLANGES AND FLANGED FITTINGS

Nominal Pipe Size	Outside Diameter of Flange O	Drilling[1]					Length of Bolts[3] L		
							Stud Bolts[2]		Machine Bolts
		Thickness of Flange[4] (Min) C	Diameter of Bolt Circle	Diameter of Bolt Holes	Number of Bolts	Diameter of Bolts	0.006 In. Raised Face	Ring Joint	0.006 In. Raised Face
½	3¾	9/16	2⅝	⅝	4	½	2½	3	2¼
¾	4⅝	⅝	3¼	¾	4	⅝	3	3½	2½
1	4⅞	11/16	3½	¾	4	⅝	3	3½	2½
1¼	5¼	¾	3⅞	¾	4	⅝	3¼	3¾	2¾
1½	6⅛	13/16	4½	⅞	4	¾	3½	4	3
2	6½	⅞	5	¾	8	⅝	3½	4	3
2½	7½	1	5⅞	⅞	8	¾	4	4½	3¼
3	8¼	1⅛	6⅝	⅞	8	¾	4¼	4¾	3½
3½	9	1 3/16	7¼	⅞	8	¾	4¼	5	3¾
4	10	1¼	7⅞	⅞	8	¾	4½	5	3¾
5	11	1⅜	9¼	⅞	8	¾	4¾	5¼	4¼
6	12½	1 7/16	10⅝	⅞	12	¾	4¾	5½	4¼
8	15	1⅝	13	1	12	⅞	5½	6	4¾
10	17½	1⅞	15¼	1⅛	16	1	6¼	6¾	5½
12	20½	2	17¾	1¼	16	1⅛	6¾	7¼	5¾
14	23	2⅛	20¼	1¼	20	1⅛	7	7½	6¼
16	25½	2¼	22½	1⅜	20	1¼	7½	8	6½
18	28	2⅜	24¾	1⅜	24	1¼	7¾	8¼	6¾
20	30½	2½	27	1⅜	24	1¼	8	8¾	7¼
24	36	2¾	32	1⅝	24	1½	9	10	8

All dimensions given in inches.
[1]Bolt holes straddle the center lines.
[2]Length of stud-bolts does not include the height of the points.
[3]Bolt lengths not shown in table are determined in accordance with ANSI B16.5-1988.
[4]Flanged fittings not available in ½ and ¾ sizes.

IVE-1 400-LB STEEL PIPE FLANGES AND FLANGED FITTINGS

FLANGE

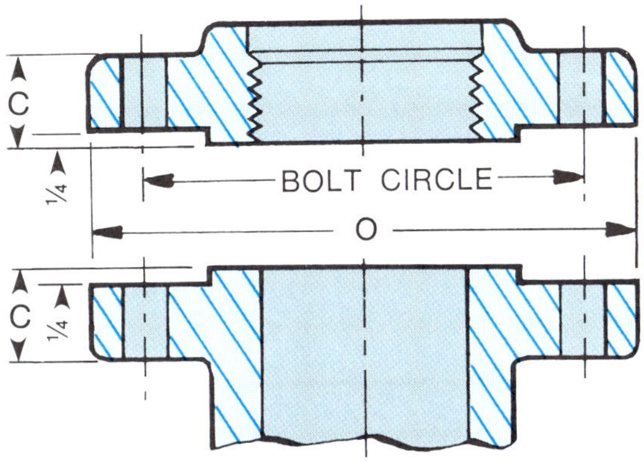

FLANGE FITTING

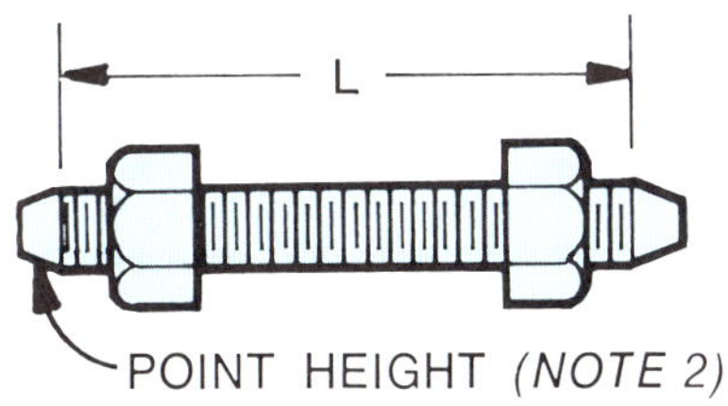

STUD BOLT WITH NUTS

TEMPLATES FOR DRILLING STEEL FLANGES AND FLANGED FITTINGS

Nominal Pipe Size	Outside Diameter of Flange O	Thickness of Flange C (Min)	Drilling[1]				Length[2, 3] of Stud Bolts L		
			Diameter of Bolt Circle	Diameter of Bolt Holes	Number of Bolts	Diameter of Bolts	¼-In. Raised Face	Male & Female also Tongue & Groove	Ring Joint
½	Use 600-lb dimensions in these sizes.								
¾									
1									
1¼									
1½									
2									
2½									
3									
3½									
4	10	1⅜	7⅞	1	8	⅞	5½	5¼	5½
5	11	1½	9¼	1	8	⅞	5¾	5½	5¾
6	12½	1⅝	10⅝	1	12	⅞	6	5¾	6
8	15	1⅞	13	1⅛	12	1	6¾	6½	6¾
10	17½	2⅛	15¼	1¼	16	1⅛	7½	7¼	7½
12	20½	2¼	17¾	1⅜	16	1¼	8	7¾	8
14	23	2⅜	20¼	1⅜	20	1¼	8¼	8	8¼
16	25½	2½	22½	1½	20	1⅜	8¾	8½	8¾
18	28	2⅝	24¾	1½	24	1⅜	9	8¾	9
20	30½	2¾	27	1⅝	24	1½	9½	9¼	9¾
24	36	3	32	1⅞	24	1¾	10½	10¼	11

All dimensions given in inches.
[1]Bolt holes straddle the center lines.
[2]Length of stud-bolts does not include the height of the points.
[3]Bolt lengths not shown in table are determined in accordance with ANSI B16.5-1988.

IVE-1 600-LB STEEL PIPE FLANGES AND FLANGED FITTINGS

FLANGE

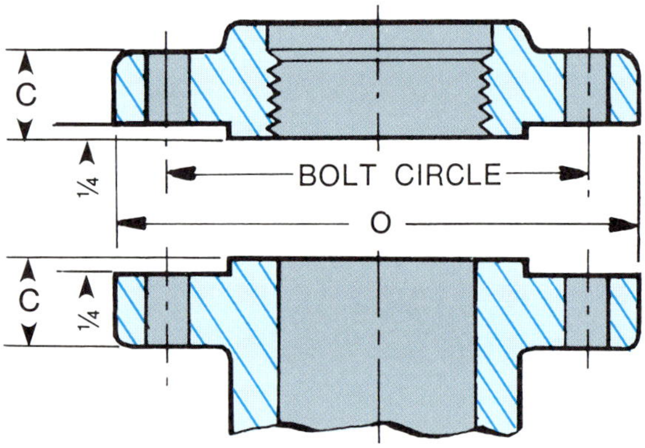

FLANGE FITTING

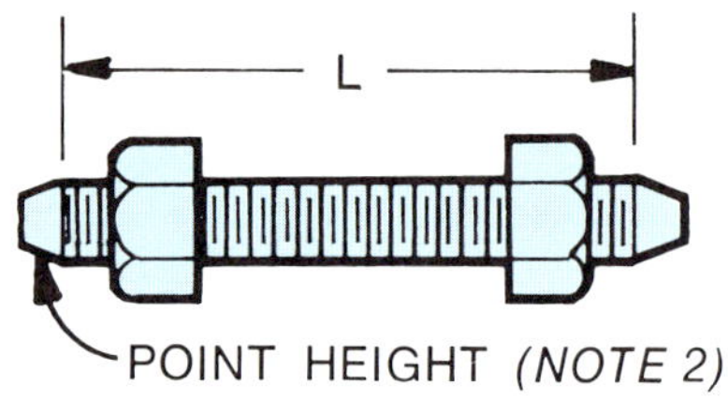

STUD BOLT WITH NUTS

TEMPLATES FOR DRILLING STEEL FLANGES AND FLANGED FITTINGS

Nominal Pipe Size	Outside Diameter of Flange O	Thickness of Flange C (Min)	Drilling[1]				Length[2, 3] of Stud Bolts L		
			Diameter of Bolt Circle	Diameter of Bolt Holes	Number of Bolts	Diameter of Bolts	¼-In. Raised Face	Male & Female also Tongue & Groove	Ring Joint
½	3¾	9/16	2⅝	⅝	4	½	3	2¾	3
¾	4⅝	⅝	3¼	¾	4	⅝	3½	3¼	3½
1	4⅞	11/16	3½	¾	4	⅝	3½	3¼	3½
1¼	5¼	13/16	3⅞	¾	4	⅝	3¾	3½	3¾
1½	6⅛	⅞	4½	⅞	4	¾	4¼	4	4¼
2	6½	1	5	¾	8	⅝	4¼	4	4¼
2½	7½	1⅛	5⅞	⅞	8	¾	4¾	4½	4¾
3	8¼	1¼	6⅝	⅞	8	¾	5	4¾	5
3½	9	1⅜	7¼	1	8	⅞	5½	5¼	5½
4	10¾	1½	8½	1	8	⅞	5¾	5½	5¾
5	13	1¾	10½	1⅛	8	1	6½	6¼	6½
6	14	1⅞	11½	1⅛	12	1	6¾	6½	6¾
8	16½	2 3/16	13¾	1¼	12	1⅛	7½	7¼	7¾
10	20	2½	17	1⅜	16	1¼	8½	8¼	8½
12	22	2⅝	19¼	1⅜	20	1¼	8¾	8½	8¾
14	23¾	2¾	20¾	1½	20	1⅜	9¼	9	9¼
16	27	3	23¾	1⅝	20	1½	10	9¾	10
18	29¼	3¼	25¾	1¾	20	1⅝	10¾	10½	10¾
20	32	3½	28½	1¾	24	1⅝	11¼	11	11½
24	37	4	33	2	24	1⅞	13	12¾	13¼

All Dimensions given in inches.
[1]Bolt holes straddle the center lines.
[2]Length of stud-bolts does not include the height of the points.
[3]Bolt lengths not shown in table are determined in accordance with ANSI B16.5-1988.

IVE-1 900-LB STEEL PIPE FLANGES AND FLANGED FITTINGS

FLANGE

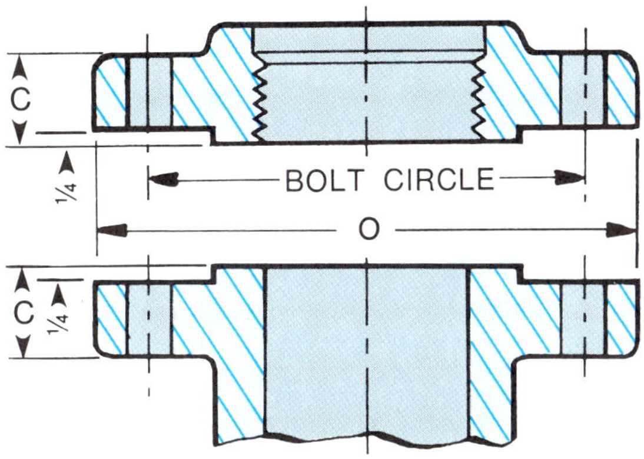

FLANGE FITTING

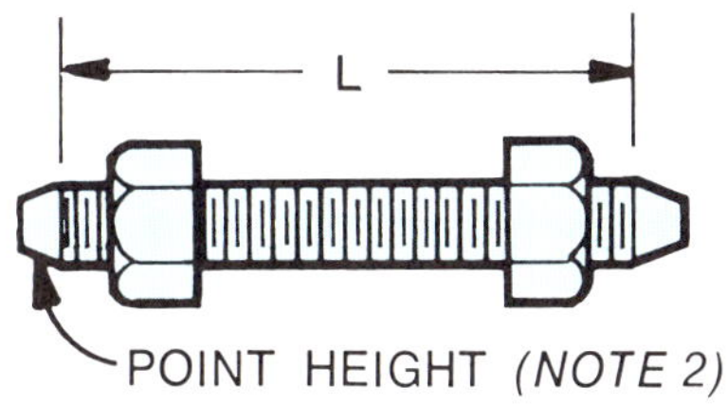

STUD BOLT WITH NUTS

TEMPLATES FOR DRILLING STEEL FLANGES AND FLANGED FITTINGS

Nominal Pipe Size	Outside Diameter of Flange O	Thickness of Flange C (Min)	Drilling[1]				Length[2, 3] of Stud Bolts L		
			Diameter of Bolt Circle	Diameter of Bolt Holes	Number of Bolts	Diameter of Bolts	¼-In. Raised Face	Male & Female also Tongue & Groove	Ring Joint
½ ¾ 1 1¼ 1½ 2 2½	Use 1500-lb dimensions in these sizes.								
3	9	1½	7½	1	8	⅞	5¾	5½	5¾
4	11½	1¾	9¼	1¼	8	1⅛	6¾	6½	6¾
5	13¾	2	11	1⅜	8	1¼	7½	7¼	7½
6	15	2 3/16	12½	1¼	12	1⅛	7½	7¼	7¾
8	18½	2½	15½	1½	12	1⅜	8¾	8½	8¾
10	21½	2¾	18½	1½	16	1⅜	9¼	9	9¼
12	24	3⅛	21	1½	20	1⅜	10	9¾	10
14	25¼	3⅜	22	1⅝	20	1½	10¾	10½	11
16	27¾	3½	24¼	1¾	20	1⅝	11¼	11	11½
18	31	4	27	2	20	1⅞	12¾	12½	13¼
20	33¼	4¼	29½	2⅛	20	2	13¾	13½	14¼
24	41	5½	35½	2⅝	20	2½	17¼	17	18

All dimensions given in inches.
[1]Bolt holes straddle the center lines.
[2]Length of stud-bolts does not include the height of the points.
[3]Bolt lengths not shown in table are determined in accordance with ANSI B16.5-1988.

IVE-1 1500-LB STEEL PIPE FLANGES AND FLANGED FITTINGS

FLANGE

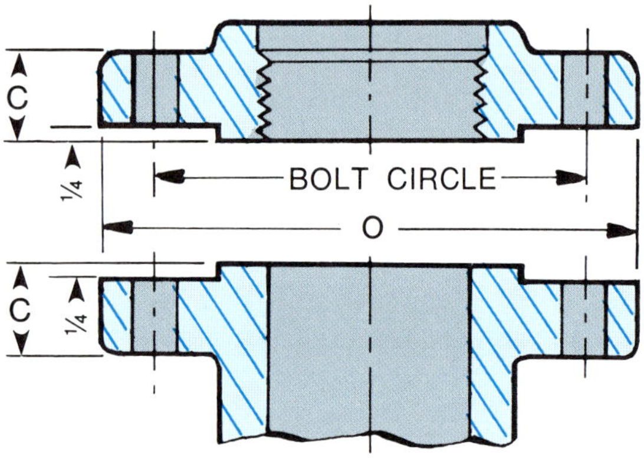

FLANGE FITTING

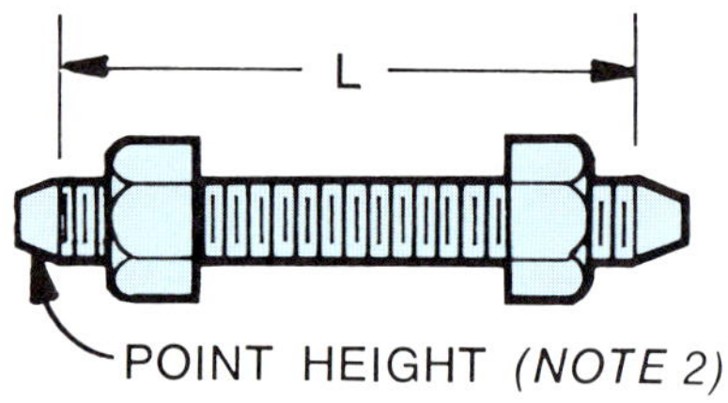

STUD BOLT WITH NUTS

TEMPLATES FOR DRILLING STEEL FLANGES AND FLANGED FITTINGS

Nominal Pipe Size	Outside Diameter of Flange O	Thickness of Flange C (Min)	Drilling[1]				Length[2, 3] of Stud Bolts L		
			Diameter of Bolt Circle	Diameter of Bolt Holes	Number of Bolts	Diameter of Bolts	¼-In. Raised Face	Male & Female also Tongue & Groove	Ring Joint
½	4¾	⅞	3¼	⅞	4	¾	4½	4	4½
¾	5⅛	1	3½	⅞	4	¾	4½	4¼	4½
1	5⅞	1⅛	4	1	4	⅞	5	4¾	5
1¼	6¼	1⅛	4⅜	1	4	⅞	5	4¾	5
1½	7	1¼	4⅞	1⅛	4	1	5½	5¼	5½
2	8½	1½	6½	1	8	⅞	5¾	5½	5¾
2½	9⅝	1⅝	7½	1⅛	8	1	6¼	6	6¼
3	10½	1⅞	8	1¼	8	1⅛	7	6¾	7
4	12¼	2⅛	9½	1⅜	8	1¼	7¾	7½	7¾
5	14¾	2⅞	11½	1⅝	8	1½	9¾	9½	9¾
6	15½	3¼	12½	1½	12	1⅜	10½	10	10½
8	19	3⅝	15½	1¾	12	1⅝	11½	11¼	12¾
10	23	4¼	19	2	12	1⅞	13¼	13	13½
12	26½	4⅞	22½	2⅛	16	2	14¾	14½	15¼
14	29½	5¼	25	2⅜	16	2¼	16	15¾	16¾
16	32½	5¾	27¾	2⅝	16	2½	17½	17¼	18½
18	36	6⅜	30½	2⅞	16	2¾	19½	19¼	20¾
20	38¾	7	32¾	3⅛	16	3	21½	21	22¼
24	46	8	39	3⅝	16	3½	24¼	24	25½

All Dimensions given in inches.
[1]Bolt holes straddle the center lines.
[2]Length of stud-bolts does not include the height of the points.
[3]Bolt lengths not shown in table are determined in accordance with ANSI B16.5-1988.

IVE-1 2500-LB STEEL PIPE FLANGES AND FLANGED FITTINGS

FLANGE

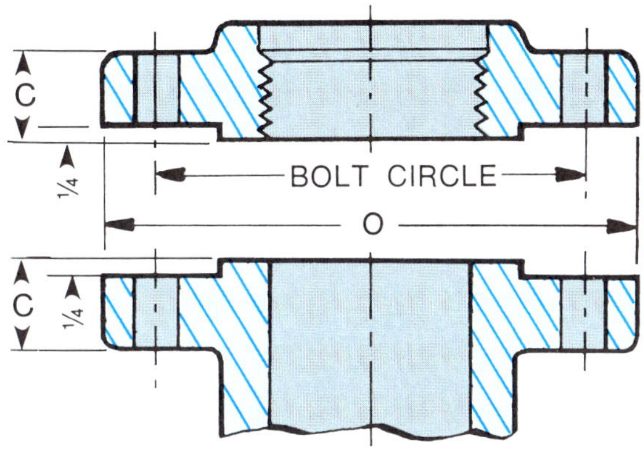

FLANGE FITTING

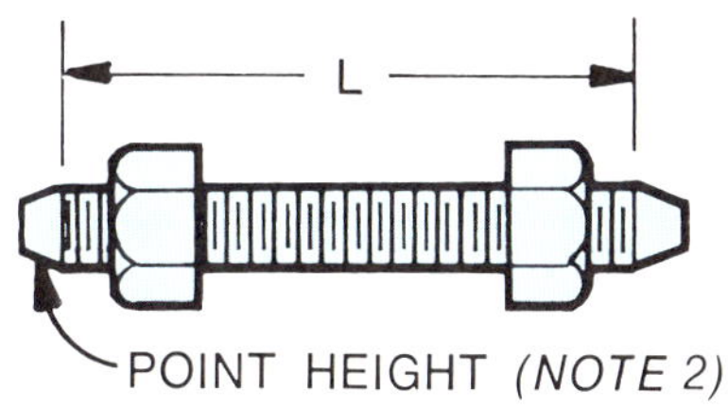

STUD BOLT WITH NUTS

TEMPLATES FOR DRILLING STEEL FLANGES AND FLANGED FITTINGS

Nominal Pipe Size	Outside Diameter of Flange O	Thickness of Flange C (Min)	Drilling[1]				Length[2, 3] of Stud Bolts L		
			Diameter of Bolt Circle	Diameter of Bolt Holes	Number of Bolts	Diameter of Bolts	¼-In. Raised Face	Male & Female also Tongue & Groove	Ring Joint
½	5¼	1 3/16	3½	⅞	4	¾	4¾	4½	4¾
¾	5½	1¼	3¾	⅞	4	¾	5	4¾	5
1	6¼	1⅜	4¼	1	4	⅞	5½	5¼	5½
1¼	7¼	1½	5⅛	1⅛	4	1	6	5¾	6
1½	8	1¾	5¾	1¼	4	1⅛	6¾	6½	6¾
2	9¼	2	6¾	1⅛	8	1	7	6¾	7
2½	10½	2¼	7¾	1¼	8	1⅛	7¾	7½	8
3	12	2⅝	9	1⅜	8	1¼	8¾	8½	9
4	14	3	10¾	1⅝	8	1½	10	9¾	10¼
5	16½	3⅝	12¾	1⅞	8	1¾	11¾	11½	12¼
6	19	4¼	14½	2⅛	8	2	13½	13¼	14
8	21¾	5	17¼	2⅛	12	2	15	14¾	15½
10	26½	6½	21¼	2⅝	12	2½	19¼	19	20
12	30	7¼	24⅜	2⅞	12	2¾	21¼	21	22

All Dimensions given in inches.
[1]Bolt holes straddle the center lines.
[2]Length of stud-bolts does not linclude the height of the points.
[3]Bolt lengths not shown in table are determined in accordance with ANSI B16.5-1988.

IVE-2 ANSI B16.5-1988 STEEL FLANGE FACINGS[1] 150 AND 300 LB

[1] See IVE-3 for dimensions of facings (other than ring-joint) and IVE-4 for ring-joint facing.

[2] For small male and female joints care should be taken in the use of these dimensions to insure that the inside diameter of fitting or pipe is small enough to permit sufficient bearing surface to prevent the crushing of the gasket. (See IV-3). This applies particularly on lines where the joint is made on the end of the pipe. Screwed companion flanges for small male and female joints are furnished with plain face and are threaded with ANSI Standard Locknut Thread (NPSL).

[3] See IVE-4.

[4] See IVE-3.

[5] Large male and female faces are not applicable to 150-lb class because of potential dimensional conflicts.

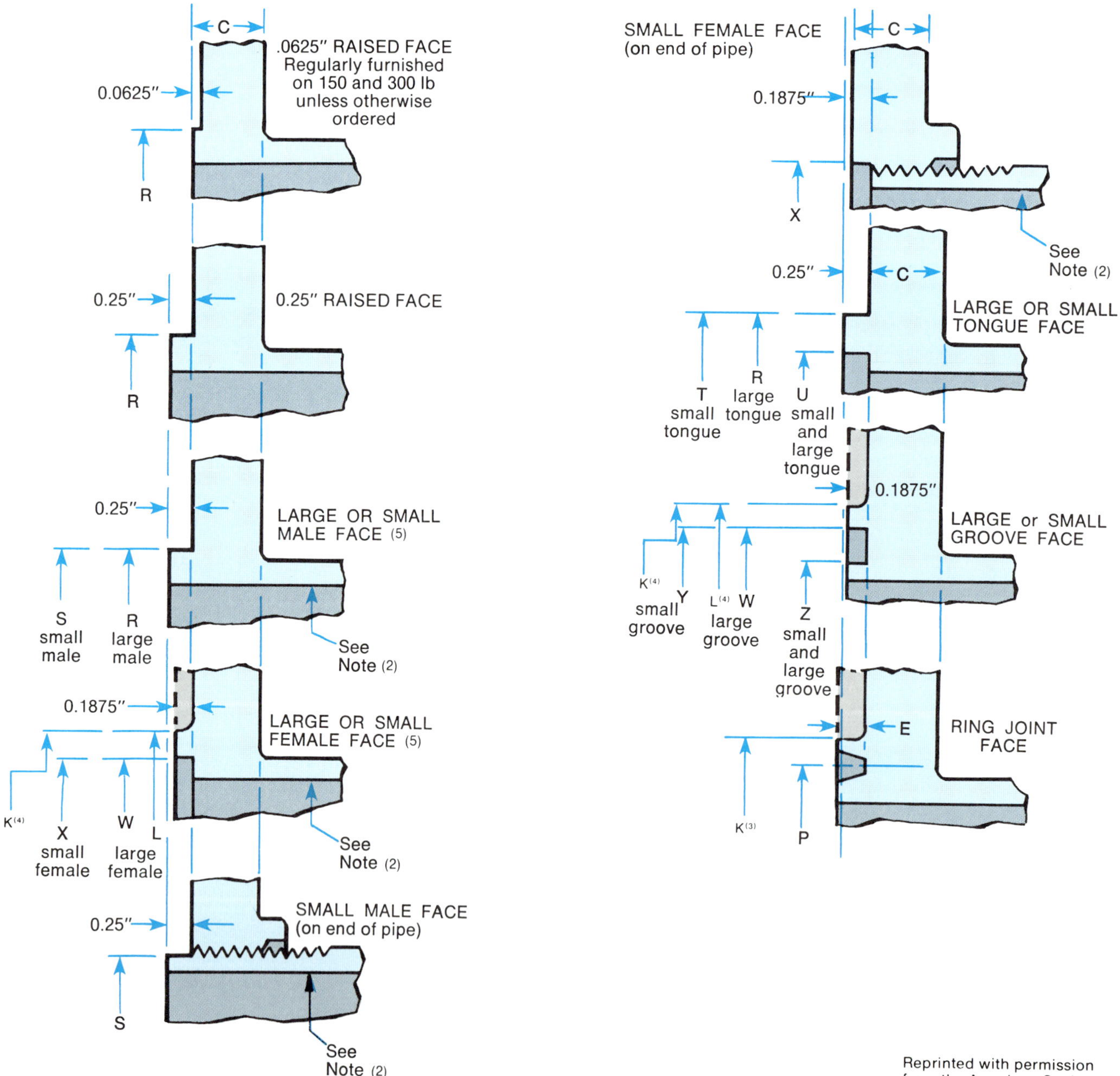

Reprinted with permission from the American Society of Mechanical Engineers, N.Y., N.Y.

IVE-2 STEEL FLANGE FACINGS (Cont.)
400 LB AND HIGHER

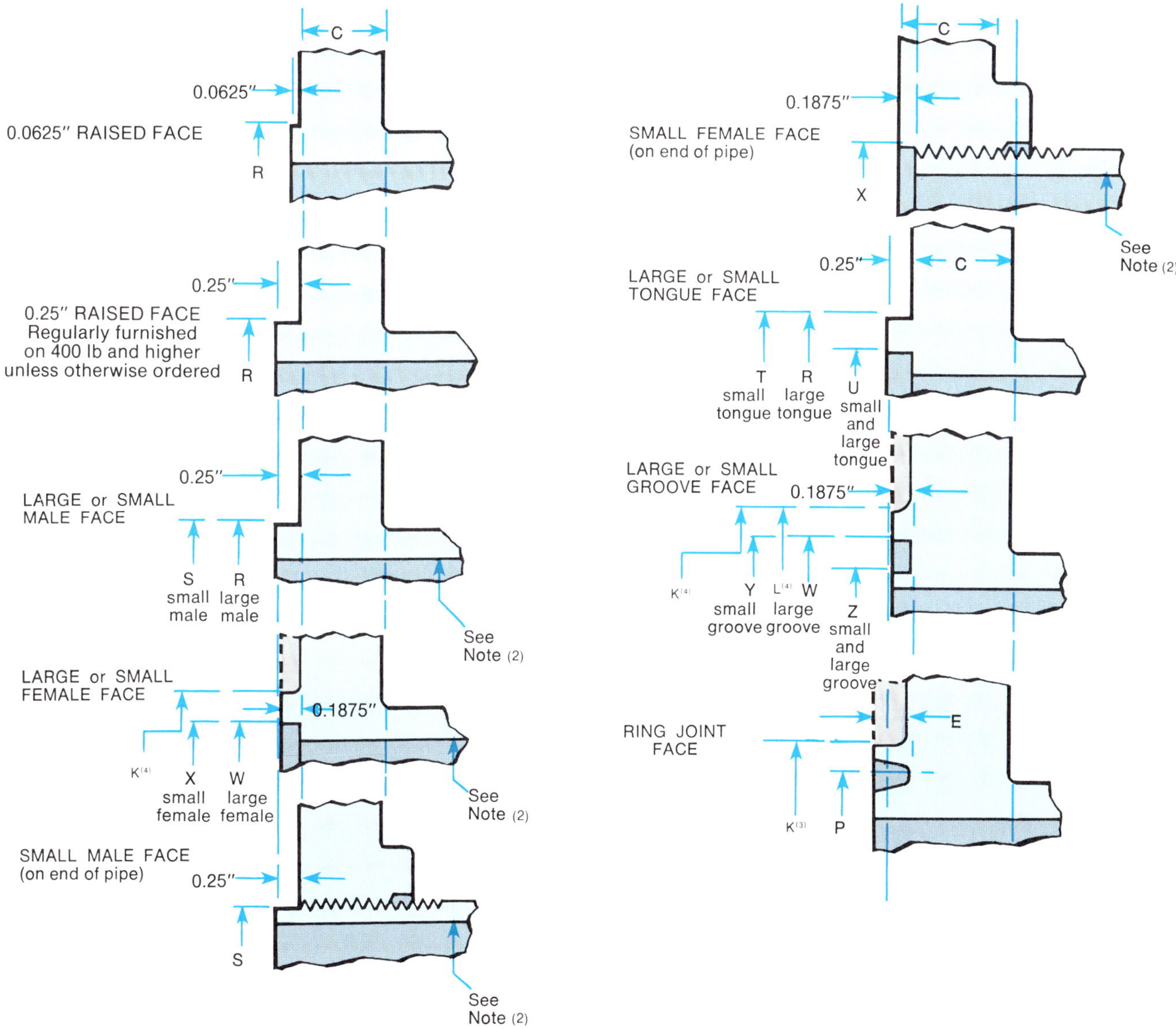

Reprinted with permission
from the American Society
of Mechanical Engineers,
N.Y., N.Y.

IVE-2 FLANGE FACINGS
LAPPED JOINTS

[1] See IVE-3 for dimensions of facings (other than ring-joint) and IVE-4 for ring-joint facing.

[2] *For Lapped Joints,* facings shall be furnished as follows:
Raised Face. Finished height of face shall be no less than pipe wall thickness.

Large Male and Female. Finished height of male shall be no less than wall thickness of pipe used or ¼ in., whichever is greater. Thickness of lap remaining after machining female face shall be no less than wall thickness of pipe used.

Tongue and Groove. Thickness of lap remaining after machining tongue or groove face shall be no less than wall thickness of pipe used.

Ring-Joint. Thickness of lap remaining after machining ring-groove shall be no less than wall thickness of pipe used.

[3] The outside diameters of lap for ring joints are shown in IVE-4, dimension K. The outside diameters of laps for large female, large tongue and groove and small tongue and groove are shown in IVE-3. Small male and female are not used with lapped joints.

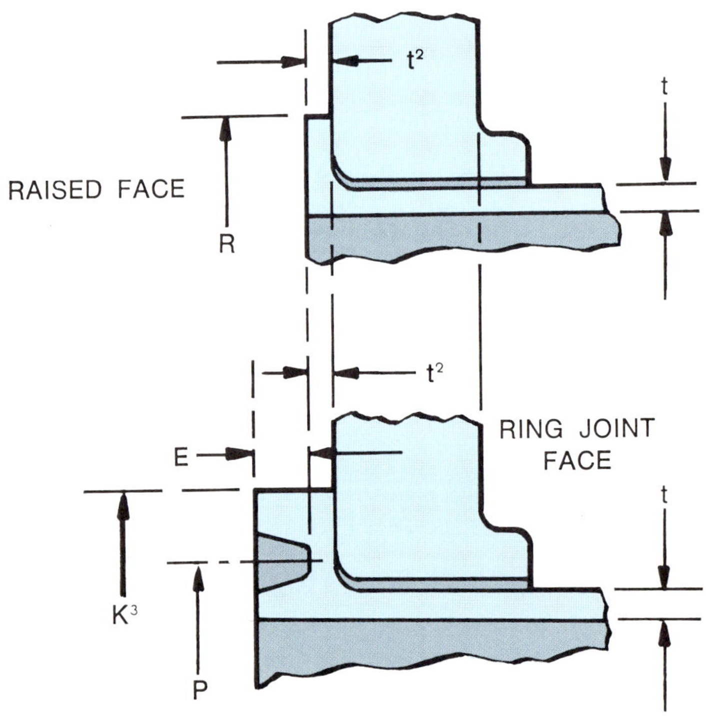

Characteristics of Piping Materials

Non-Ferrous Pipe and Tubing

IVE-3 DIMENSIONS OF FACINGS[1,2,3,7] (OTHER THAN RING JOINTS) (ALL PRESSURES)

Nominal Pipe Size	Outside Diameter: Raised Face Large Male and Large Tongue	Outside Diameter: Small Male[9]	Outside Diameter: Small Tongue	Inside Diameter of Large and Small Tongue	Inside Diameter of Small Male	Outside Diameter: Large Female and Large Groove	Outside Diameter: Small Female[9]	Outside Diameter: Small Groove	Inside Diameter of Large and Small Groove	Height: Raised Face	Height: Large and Small Male and Tongue	Depth of Groove or Female	Minimum[8,10] Outside Diameter of Raised Portion: Small Female and Groove	Minimum[8,10] Outside Diameter of Raised Portion: Large Female and Groove	Nominal Pipe Size
	R	S	T	U		W	X	Y	Z				K	L	
1/2	1 3/8	23/32	1 3/8	1	See Note 9	1 7/16	25/32	1 7/16	15/16	See Note 4 & 7	See Note 5 & 7	See Note 6 & 7	1 3/4	1 13/16	1/2
3/4	1 11/16	15/16	1 11/16	1 5/16		1 3/4	1	1 3/4	1 1/4				2 1/16	2 1/8	3/4
1	2	1 3/16	1 7/8	1 1/2		2 1/16	1 1/4	1 15/16	1 7/16				2 1/4	2 7/16	1
1 1/4	2 1/2	1 1/2	2 1/4	1 7/8		2 9/16	1 9/16	2 5/16	1 13/16				2 5/8	2 15/16	1 1/4
1 1/2	2 7/8	1 3/4	2 1/2	2 1/8		2 15/16	1 13/16	2 9/16	2 1/16				2 7/8	3 5/16	1 1/2
2	3 5/8	2 1/4	3 1/4	2 7/8		3 11/16	2 5/16	3 5/16	2 13/16				3 5/8	4 1/16	2
2 1/2	4 1/8	2 11/16	3 3/4	3 3/8		4 3/16	2 3/4	3 13/16	3 5/16				4 1/8	4 9/16	2 1/2
3	5	3 5/16	4 5/8	4 1/4		5 1/16	3 3/8	4 11/16	4 3/16				5	5 7/16	3
3 1/2	5 1/2	3 13/16	5 1/8	4 3/4		5 9/16	3 7/8	5 3/16	4 11/16				5 1/2	5 15/16	3 1/2
4	6 3/16	4 5/16	5 11/16	5 3/16		6 1/4	4 3/8	5 3/4	5 1/8				6 3/16	6 5/8	4
5	7 5/16	5 3/8	6 13/16	6 5/16		7 3/8	5 7/16	6 7/8	6 1/4				7 5/16	7 3/4	5
6	8 1/2	6 3/8	8	7 1/2		8 9/16	6 7/16	8 1/16	7 7/16				8 1/2	8 15/16	6
8	10 5/8	8 3/8	10	9 3/8		10 11/16	8 7/16	10 1/16	9 5/16				10 5/8	11 1/16	8
10	12 3/4	10 1/2	12	11 1/4		12 13/16	10 9/16	12 1/16	11 3/16				12 3/4	13 3/16	10
12	15	12 1/2	14 1/4	13 1/2		15 1/16	12 9/16	14 5/16	13 7/16				15	15 7/16	12
14	16 1/4	13 3/4	15 1/2	14 3/4		16 5/16	13 13/16	15 9/16	14 11/16				16 1/4	16 11/16	14
16	18 1/2	15 3/4	17 5/8	16 3/4		18 9/16	15 13/16	17 11/16	16 11/16				18 1/2	18 15/16	16
18	21	17 3/4	20 1/8	19 1/4		21 1/16	17 13/16	20 3/16	19 3/16				21	21 7/16	18
20	23	19 3/4	22	21		23 1/16	19 13/16	22 1/16	20 15/16				23	23 7/16	20
24	27 1/4	23 3/4	26 1/4	25 1/4		27 5/16	23 13/16	26 5/16	25 3/16				27 1/4	27 11/16	24

All dimensions are given in inches.

1 For facing requirements of flanges, see IVE-2.

2 For facing requirements for lapped joints—see IVE-2.

3 Facing Tolerances.—Inside and Outside Diameter of large and small tongue and groove and females, ±1/64 In.
Outside Diameter, 1/16 in. raised face, ±1/32 in.
Outside Diameter, 1/4 in. raised face, ±1/64 in.

4 Height of raised face is either 1/16 in. or 1/4 in.

5 Height of large and small male and tongue is 1/4 in.

6 Depth of groove or female is 3/16 in.

7 See IVE-2 for thickness and outside diameters of laps.

8 Raised portion or full face may be furnished unless otherwise specified on order.

9 For small male and female joints care should be taken in the use of these dimensions to insure that the inside diameter of fitting or pipe is small enough to permit sufficient bearing surface to prevent the crushing of the gasket. This applies particularly on lines where the joint is made on the end of the pipe. Inside diameter of fitting should match inside diameter of pipe as specified by purchaser. Screwed companion flanges for small male and female joints are furnished with plain face and are threaded with USA Standard Locknut Thread (NPSL).

10 Large male and female faces are not applicable to 150-lb class because of potential dimensional conflicts.

IVE-4 DIMENSIONS OF RING JOINT FACINGS

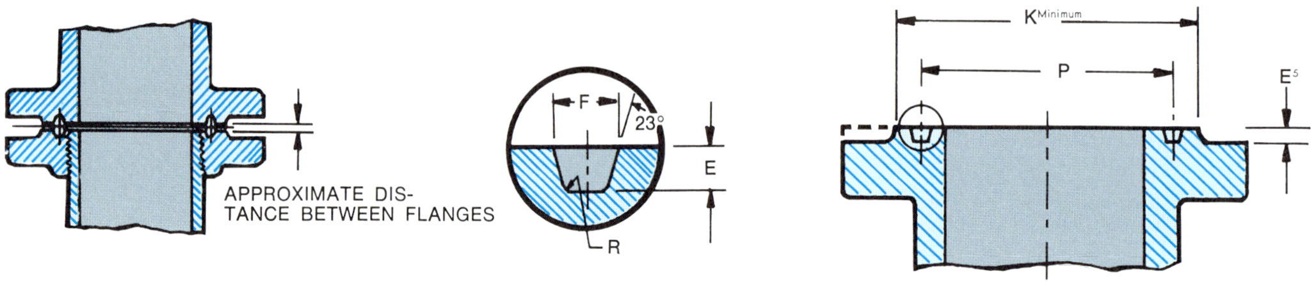

DIMENSIONS OF RING-JOINT FACINGS[1,2] (ALL PRESSURES)

Nominal Pipe Size							Groove Number	Groove Dimensions				Diameter of Raised Portion K					Approximate Distance Between Flanges						
150 Psi	300 Psi	400 Psi[3]	600 Psi	900 Psi[4]	1500 Psi	2500 Psi		Pitch Diameter P	Depth E	Width F	Radius at Bottom R	150 Psi	300 400 600 Psi	900 Psi	1500 Psi	2500 Psi	150 Psi	300 Psi	400 Psi	600 Psi	900 Psi	1500 Psi	2500 Psi
—	1/2	—	1/2	—	—	—	R11	1 11/32	7/32	9/32	1/32	—	2	—	—	—	—	1/8	—	1/8	—	—	—
—	—	—	—	—	1/2	—	12	1 9/16	1/4	11/32	1/32	—	—	—	2 3/8	—	—	—	—	—	—	5/32	—
—	3/4	—	3/4	—	—	1/2	13	1 11/16	1/4	11/32	1/32	—	2 1/2	—	—	2 9/16	—	5/32	—	5/32	—	—	5/32
—	—	—	—	—	3/4	—	14	1 3/4	1/4	11/32	1/32	—	—	—	2 5/8	—	—	—	—	—	—	5/32	—
1	—	—	—	—	—	—	15	1 7/8	1/4	11/32	1/32	2 1/2	—	—	—	—	5/32	—	—	—	—	—	—
—	1	—	1	—	1	3/4	16	2	1/4	11/32	1/32	—	2 3/4	—	2 13/16	2 7/8	—	5/32	—	5/32	—	5/32	5/32
1 1/4	—	—	—	—	—	—	17	2 1/4	1/4	11/32	1/32	2 7/8	—	—	—	—	5/32	—	—	—	—	—	—
—	1 1/4	—	1 1/4	—	1 1/4	1	18	2 3/8	1/4	11/32	1/32	—	3 1/8	—	3 3/16	3 1/4	—	5/32	—	5/32	—	5/32	5/32
1 1/2	—	—	—	—	—	—	19	2 9/16	1/4	11/32	1/32	3 1/4	—	—	—	—	5/32	—	—	—	—	—	—
—	1 1/2	—	1 1/2	—	1 1/2	—	20	2 11/16	1/4	11/32	1/32	—	3 9/16	—	3 5/8	—	—	5/32	—	5/32	—	5/32	—
—	—	—	—	—	—	1 1/4	21	2 27/32	5/16	15/32	1/32	—	—	—	—	4	—	—	—	—	—	—	1/8
2	—	—	—	—	—	—	22	3 1/4	1/4	11/32	1/32	4	—	—	—	—	5/32	—	—	—	—	—	—
—	2	—	2	—	—	1 1/2	23	3 1/4	5/16	15/32	1/32	—	4 1/4	—	—	4 1/2	—	7/32	—	3/16	—	—	1/8
—	—	—	—	—	2	—	24	3 3/4	5/16	15/32	1/32	—	—	—	4 7/8	—	—	—	—	—	—	1/8	—
2 1/2	—	—	—	—	—	—	25	4	1/4	11/32	1/32	4 3/4	—	—	—	—	5/32	—	—	—	—	—	—
—	2 1/2	—	2 1/2	—	—	2	26	4	5/16	15/32	1/32	—	5	—	—	5 1/4	—	7/32	—	3/16	—	—	1/8
—	—	—	—	—	2 1/2	—	27	4 1/4	5/16	15/32	1/32	—	—	—	5 3/8	—	—	—	—	—	—	1/8	—
—	—	—	—	—	—	2 1/2	28	4 3/8	3/8	17/32	1/16	—	—	—	—	5 7/8	—	—	—	—	—	—	1/8
3	—	—	—	—	—	—	29	4 1/2	1/4	11/32	1/32	5 1/4	—	—	—	—	5/32	—	—	—	—	—	—
—	†	—	†	—	—	—	30	4 5/8	5/16	15/32	1/32	—	—	—	—	—	—	—	—	—	—	—	—
—	†3	—	†3	3	—	—	31	4 7/8	5/16	15/32	1/32	—	5 3/4	6 1/8	—	—	—	7/32	—	3/16	5/32	—	—
—	—	—	—	—	—	3	32	5	3/8	17/32	1/16	—	—	—	—	6 5/8	—	—	—	—	—	—	1/8
3 1/2	—	—	—	—	—	—	33	5 3/16	1/4	11/32	1/32	6 1/16	—	—	—	—	5/32	—	—	—	—	—	—
—	3 1/2	—	3 1/2	—	—	—	34	5 3/16	5/16	15/32	1/37	—	6 1/4	—	—	—	—	7/32	—	3/16	—	—	—
—	—	—	—	—	3	—	35	5 3/8	5/16	15/32	1/32	—	—	—	6 5/8	—	—	—	—	—	—	1/8	—
4	—	—	—	—	—	—	36	5 7/8	1/4	11/32	1/32	6 3/4	—	—	—	—	5/32	—	—	—	—	—	—
—	4	4	4	4	—	—	37	5 7/8	5/16	15/32	1/32	—	6 7/8	7 1/8	—	—	—	7/32	7/32	3/16	5/32	—	—
—	—	—	—	—	—	4	38	6 3/16	7/16	21/32	1/16	—	—	—	—	8	—	—	—	—	—	—	5/32
—	—	—	—	—	4	—	39	6 3/8	5/16	15/32	1/32	—	—	—	7 5/8	—	—	—	—	—	—	1/8	—
5	—	—	—	—	—	—	40	6 3/4	1/4	11/32	1/32	7 5/8	—	—	—	—	5/32	—	—	—	—	—	—
—	5	5	5	5	—	—	41	7 1/8	5/16	15/32	1/32	—	8 1/4	8 1/2	—	—	—	7/32	7/32	3/16	5/32	—	—
—	—	—	—	—	—	5	42	7 1/2	1/2	25/32	1/16	—	—	—	—	9 1/2	—	—	—	—	—	—	5/32
6	—	—	—	—	—	—	43	7 5/8	1/4	11/32	1/32	8 5/8	—	—	—	—	5/32	—	—	—	—	—	—
—	—	—	—	—	5	—	44	7 5/8	5/16	15/32	1/32	—	—	—	9	—	—	—	—	—	—	1/8	—
—	6	6	6	6	—	—	45	8 5/16	5/16	15/32	1/32	—	9 1/2	9 1/2	—	—	—	7/32	7/32	3/16	5/32	—	—

IVE-4 DIMENSIONS OF RING JOINT FACINGS

DIMENSIONS OF RING-JOINT FACINGS[1,2] (ALL PRESSURES) (CONT'D)

Nominal Pipe Size							Groove Number	Groove Dimensions				Diameter of Raised Portion K					Approximate Distance Between Flanges						
150 Psi	300 Psi	400 Psi[3]	600 Psi	900 Psi	1500 Psi	2500 Psi		Pitch Diameter P	Depth E	Width F	Radius at Bottom R	150 Psi	300 400 600 Psi	900 Psi	1500 Psi	2500 Psi	150 Psi	300 Psi	400 Psi	600 Psi	900 Psi	1500 Psi	2500 Psi
—	—	—	—	—	6	—	46	8 5/16	3/8	17/32	1/16	—	—	—	9 3/4	—	—	—	—	—	—	1/8	—
—	—	—	—	—	—	6	47	9	1/2	25/32	1/16	—	—	—	—	11	—	—	—	—	—	—	5/32
8	—	—	—	—	—	—	48	9 3/4	1/4	11/32	1/32	10 3/4	—	—	—	—	5/32	—	—	—	—	—	—
—	8	8	8	8	—	—	49	10 5/8	5/16	15/32	1/32	—	11 7/8	12 1/8	—	—	—	7/32	7/32	3/16	5/32	—	—
—	—	—	—	—	8	—	50	10 5/8	7/16	21/32	1/16	—	—	—	12 1/2	—	—	—	—	—	—	5/32	—
—	—	—	—	—	—	8	51	11	9/16	29/32	1/16	—	—	—	—	13 3/8	—	—	—	—	—	—	3/16
10	—	—	—	—	—	—	52	12	1/4	11/32	1/32	13	—	—	—	—	5/32	—	—	—	—	—	—
—	10	10	10	10	—	—	53	12 3/4	5/16	15/32	1/32	—	14	14 1/4	—	—	—	7/32	7/32	3/16	5/32	—	—
—	—	—	—	—	10	—	54	12 3/4	7/16	21/32	1/16	—	—	—	14 5/8	—	—	—	—	—	—	5/32	—
—	—	—	—	—	—	10	55	13 1/2	11/16	1 3/16	3/32	—	—	—	—	16 3/4	—	—	—	—	—	—	1/4
12	—	—	—	—	—	—	56	15	1/4	11/32	1/32	16	—	—	—	—	5/32	—	—	—	—	—	—
—	12	12	12	12	—	—	57	15	5/16	15/32	1/32	—	16 1/4	16 1/2	—	—	—	7/32	7/32	3/16	5/32	—	—
—	—	—	—	—	12	—	58	15	9/16	29/32	1/16	—	—	—	17 1/4	—	—	—	—	—	—	3/16	—
14	—	—	—	—	—	—	59	15 5/8	1/4	11/32	1/32	16 3/4	—	—	—	—	1/8	—	—	—	—	—	—
—	—	—	—	—	—	12	60	16	11/16	1 5/16	3/32	—	—	—	—	19 1/2	—	—	—	—	—	—	5/16
—	14	14	14	—	—	—	61	16 1/2	5/16	15/32	1/32	—	18	—	—	—	—	7/32	7/32	3/16	—	—	—
—	—	—	—	14	—	—	62	16 1/2	7/16	21/32	1/16	—	—	18 3/8	—	—	—	—	—	—	5/32	—	—
—	—	—	—	—	14	—	63	16 1/2	5/8	1 1/16	3/32	—	—	—	19 1/4	—	—	—	—	—	—	7/32	—
16	—	—	—	—	—	—	64	17 7/8	1/4	11/32	1/32	19	—	—	—	—	1/8	—	—	—	—	—	—
—	16	16	16	—	—	—	65	18 1/2	5/16	15/32	1/32	—	20	—	—	—	—	7/32	7/32	3/16	—	—	—
—	—	—	—	16	—	—	66	18 1/2	7/16	21/32	1/16	—	—	20 5/8	—	—	—	—	—	—	5/32	—	—
—	—	—	—	—	16	—	67	18 1/2	11/16	1 3/16	3/32	—	—	—	21 1/2	—	—	—	—	—	—	5/16	—
16	—	—	—	—	—	—	68	20 3/8	1/4	11/32	1/32	21 1/2	—	—	—	—	1/8	—	—	—	—	—	—
—	18	18	18	—	—	—	69	21	5/16	15/32	1/32	—	22 5/8	—	—	—	—	7/32	7/32	3/16	—	—	—
—	—	—	—	18	—	—	70	21	1/2	25/32	1/16	—	—	23 3/8	—	—	—	—	—	—	3/16	—	—
—	—	—	—	—	18	—	71	21	11/16	1 3/16	3/32	—	—	—	24 1/8	—	—	—	—	—	—	5/16	—
20	—	—	—	—	—	—	72	22	1/4	11/32	1/32	23 1/2	—	—	—	—	1/8	—	—	—	—	—	—
—	20	20	20	—	—	—	73	23	3/8	17/32	1/16	—	25	—	—	—	—	7/32	7/32	3/16	—	—	—
—	—	—	—	20	—	—	74	23	1/2	25/32	1/16	—	—	25 1/2	—	—	—	—	—	—	3/16	—	—
—	—	—	—	—	20	—	75	23	11/16	1 5/16	3/32	—	—	—	26 1/2	—	—	—	—	—	—	3/8	—
24	—	—	—	—	—	—	76	26 1/2	1/4	11/32	1/32	28	—	—	—	—	1/8	—	—	—	—	—	—
—	24	24	24	—	—	—	77	27 1/4	7/16	21/32	1/16	—	29 1/2	—	—	—	—	1/4	1/4	7/32	—	—	—
—	—	—	—	24	—	—	78	27 1/4	5/8	1 1/16	3/32	—	—	30 3/8	—	—	—	—	—	—	7/32	—	—
—	—	—	—	—	24	—	79	27 1/4	13/16	1 7/16	3/32	—	—	—	31 1/4	—	—	—	—	—	—	7/16	—

All dimensions given in inches.

† For ring joints with lapped flanges in the 300, and 600 lb standards, ring and groove number R30 are used instead of R31.

[1] For facing requirements for flanges, see IVE-2.

[2] For facing requirements for lapped joints see IVE-2.

[3] Use 600 lb in sizes ½ to 3½ for 400 lb pressure.

[4] Use 1500 lb in sizes ½ to 2½ for 900 lb pressure.

[5] Height of raised portion is equal to the depth of groove dimensions "E" but is not subject to the tolerance for "E". Former full-face contour may be used.

TOLERANCES

E (depth) $^{+1/64}_{-0}$
F (width) ±.008
P (pitch diam) ±.005
R (radius at bottom) Max
23° (angle) ±½°

Characteristics of Piping Materials

Flanges

IVE-5 API TYPE 6B FLANGES FOR 2000-PSI MAXIMUM WORKING PRESSURE

All dimensions in inches unless otherwise noted.

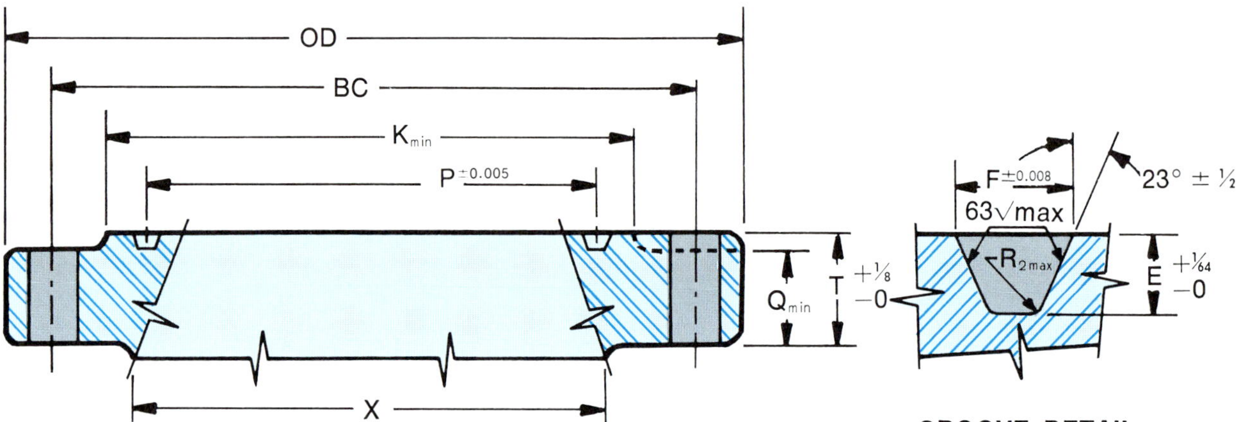

FLANGE SECTION
INTEGRAL FLANGE

GROOVE DETAIL

Reprinted with permission from the American Petroleum Institute.

	Basic Flange Dimensions				Bolting Dimensions				Ring-Joint Groove and Flange Facing Dimensions				
Nominal Size of Flange	Outside Diameter of Flange OD	Total Thickness of Flange T	Basic Thickness of Flange Q	Diameter of Hub X	Diameter of Bolt Circle BC	Number of Bolts	Diameter of Bolts	Length of Stud Bolts L_{SSB}**	Ring Number R or RX	Pitch Diameter of Type R Ring & Groove P	Width of Groove F	Depth of Groove E	Diameter of Raised Face K
1 1/2*	6 1/8	1 1/8	7/8	2 3/4	4 1/2	4	3/4	4 1/4	20	2 11/16	11/32	1/4	3 9/16
2	6 1/2	1 5/16	1	3 5/16	5	8	5/8	4 1/2	23	3 1/4	15/32	5/16	4 1/4
2 1/2	7 1/2	1 7/16	1 1/8	3 15/16	5 7/8	8	3/4	5	26	4	15/32	5/16	5
3	8 1/4	1 9/16	1 1/4	4 5/8	6 5/8	8	3/4	5 1/4	31	4 7/8	15/32	5/16	5 3/4
4	10 3/4	1 13/16	1 1/2	6	8 1/2	8	7/8	6	37	5 7/8	15/32	5/16	6 7/8
5*	13	2 1/16	1 3/4	7 7/16	10 1/2	8	1	6 3/4	41	7 1/8	15/32	5/16	8 1/4
6	14	2 3/16	1 7/8	8 3/4	11 1/2	12	1	7	45	8 5/16	15/32	5/16	9 1/2
8	16 1/2	2 1/2	2 3/16	10 3/4	13 3/4	12	1 1/8	8	49	10 5/8	15/32	5/16	11 7/8
10	20	2 13/16	2 1/2	13 1/2	17	16	1 1/4	8 3/4	53	12 3/4	15/32	5/16	14
12	22	2 15/16	2 5/8	15 3/4	19 1/4	20	1 1/4	9	57	15	15/32	5/16	16 1/4
16	27	3 5/16	3	19 1/2	23 3/4	20	1 1/2	10 1/4	65	18 1/2	15/32	5/16	20
18*	29 1/4	3 9/16	3 1/4	21 1/2	25 3/4	20	1 5/8	11	69	21	15/32	5/16	22 5/8
20	32	3 7/8	3 1/2	24	28 1/2	24	1 5/8	11 3/4	73	23	17/32	3/8	25

* These sizes inactive; available on special order only.

**L_{SSB} includes thickness of two nuts but not either point height.

IVE-5 API TYPE 6B FLANGES FOR 2000-PSI MAXIMUM WORKING PRESSURE (CONTINUED)

HUB AND BORE DIMENSIONS					
Nominal Size of Flange	Hub Length Threaded Line-Pipe Flange L_L	Hub Length Threaded Casing Flange L_C	Hub Length Welding-Neck Line-Pipe Flange L_N	Neck Diameter Welding-Neck Line-Pipe Flange H_L	Maximum Bore of Welding Neck Flange J_L
1½*	1½	—	3	1.90	1.610
2	1¾	—	3 3/16	2.38	2.067
2½	1 15/16	—	3 7/16	2.88	2.469
3	2⅛	—	3 9/16	3.50	3.068
4	2 7/16	3½	4 5/16	4.50	4.026
5*	2 11/16	4	4 13/16	5.56	4.813
6	2 15/16	4½	4 15/16	6.63	5.761
8	3 5/16	5	5 9/16	8.63	7.813
10	3 11/16	5¼	6 5/16	10.75	9.750
12	3 15/16	3 15/16	—	—	—
16	4½	4½	—	—	—
18*	4 15/16	4 15/16	—	—	—
20	5⅜	5⅜	—	—	—

*These sizes inactive; available on special order only.

REQUIREMENTS

1. Bolt holes shall be drilled ⅛ in. larger in diameter than the specified bolt diameter.
2. The contour of the flange face outside the K diameter is optional with the manufacturer unless raised or full face is specified on the purchase order.
3. Ring-groove radius R_1 shall be 1/32 in. for groove widths 11/32 and 15/32; 1/16 in. for width 17/32.
4. The diameter and depth of counterbore for threaded flanges shall be as specified in IVE-11.
5. The bore J_L of welding-neck flanges shall be as specified on the purchase order.

   NOTE: *Bore diameter should be the same as ID of pipe to be used, but shall not exceed values of J_L shown.*
6. The wall thickness of welding-neck flanges shall be not less than 87½ per cent of the nominal wall thickness of the pipe to which the flange is to be attached.
7. The welding end of welding-neck flanges shall be cylindrical or shall have a maximum draft of 7 deg. The length shall be sufficient to insure a sound weld, but in no case shall be less than ¼ in. See IVE-8.
8. The welding bevel of welding-neck flanges shall be as shown in IVE-8.

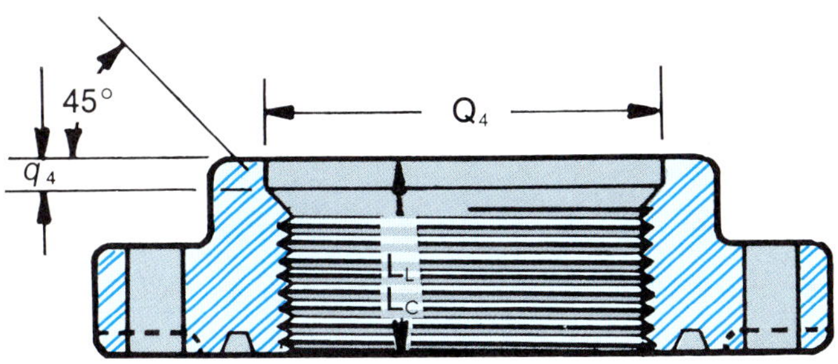

THREADED FLANGE

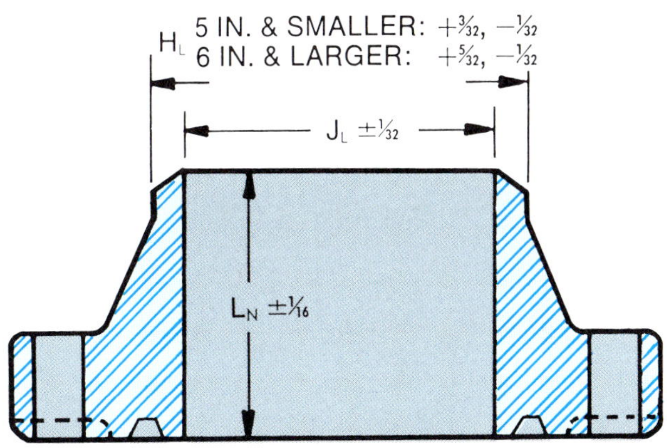

WELDING-NECK LINE-PIPE FLANGE

IVE-5 API TYPE 6B FLANGES FOR 3000-PSI MAXIMUM WORKING PRESSURE

All dimensions in inches unless otherwise noted.

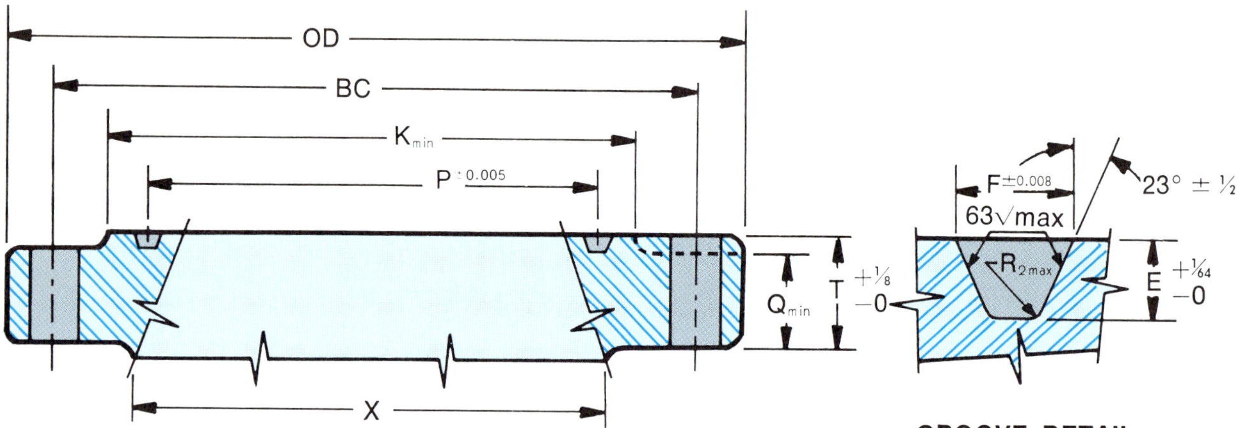

FLANGE SECTION INTEGRAL FLANGE

GROOVE DETAIL

Basic Flange Dimensions					Bolting Dimensions				Ring-Joint Groove and Flange Facing Dimensions				
Nominal Size of Flange	Outside Diameter of Flange *OD*	Total Thickness of Flange *T*	Basic Thickness of Flange *Q*	Diameter of Hub *X*	Diameter of Bolt Circle *BC*	Number of Bolts	Diameter of Bolts	Length of Stud Bolts L_{SSB}**	Ring Number R or RX	Pitch Diameter of Type R Ring & Groove *P*	Width of Groove *F*	Depth of Groove *E*	Diameter of Raised Face *K*
1½*	7	1½	1¼	2¾	4⅞	4	1	5½	20	2 11/16	11/32	¼	3⅝
2	8½	1 13/16	1½	4⅛	6½	8	⅞	6	24	3¾	15/32	5/16	4⅞
2½	9⅝	1 15/16	1⅝	4⅞	7½	8	1	6½	27	4¼	15/32	5/16	5⅜
3	9½	1 13/16	1½	5	7½	8	⅞	6	31	4⅞	15/32	5/16	6⅛
4	11½	2 1/16	1¾	6¼	9¼	8	1⅛	7	37	5⅞	15/32	5/16	7⅛
5*	13¾	2 5/16	2	7½	11	8	1¼	7¾	41	7⅛	15/32	5/16	8½
6	15	2½	2 3/16	9¼	12½	12	1⅛	8	45	8 5/16	15/32	5/16	9½
8	18½	2 13/16	2½	11¾	15½	12	1⅜	9	49	10⅝	15/32	5/16	12⅛
10	21½	3 1/16	2¾	14½	18½	16	1⅜	9½	53	12¾	15/32	5/16	14¼
12	24	3 7/16	3⅛	16½	21	20	1⅜	10¼	57	15	15/32	5/16	16½
16	27¾	3 15/16	3½	20	24¼	20	1⅝	11¾	66	18½	21/32	7/16	20⅝
18*	31	4½	4	22¼	27	20	1⅞	13¾	70	21	25/32	½	23⅜
20	33¾	4¾	4¼	24½	29½	20	2	14½	74	23	25/32	½	25½

*These sizes inactive; available on special order only.
**L_{SSB} includes thickness of two nuts but not either point height.

NOTE: *Except for bore of welding neck flanges, dimensions for sizes 1½ in. to 2½ in. inclusive are identical with 5000 psi flanges in Table.*

IVE-5 API TYPE 6B FLANGES FOR 3000-PSI MAXIMUM WORKING PRESSURE (CONTINUED)

HUB AND BORE DIMENSIONS						
Nominal Size of Flange	Hub Length Threaded Line-Pipe Flange L_L	Hub Length Threaded Casing Flange L_C	Hub Length Tubing Flange L_T	Hub Length Welding-Neck Line-Pipe Flange L_N	Neck Diameter Welding-Neck Line-Pipe Flange H_L	Maximum Bore of Welding Neck Flange J_L
1½*	2	—	2	3½	1.90	1.500
2	2 9/16	—	2 9/16	4 5/16	2.38	1.939
2½	2 13/16	—	2 13/16	4 7/16	2.88	2.323
3	2 7/16	—	2 15/16	4 5/16	3.50	2.900
4	3 1/16	3½	3½	4 13/16	4.50	3.826
5*	3 7/16	4	—	5 5/16	5.56	4.813
6	3 11/16	4½	—	5 13/16	6.63	5.761
8	4 5/16	5	—	6 11/16	8.63	7.439
10	4 9/16	5¼	—	7 9/16	10.75	9.314
12	4 15/16	4 15/16	—	—	—	—
16	5 11/16	5 11/16	—	—	—	—
18*	6½	6½	—	—	—	—
20	6¾	6¾	—	—	—	—

*These sizes inactive; available on special order only.

Reprinted with permission of the American Petroleum Institute.

REQUIREMENTS

1. Bolt holes shall be drilled 1/8 in. larger in diameter than the specified bolt diameter.
2. The contour of the flange face outside the K diameter is optional with the manufacturer unless raised or full face is specified on the purchase order.
3. Ring-groove radius R_2 shall be 1/32 in. for groove widths 11/32 and 15/32; and 1/16 in. for widths 21/32 and 25/32.
4. The diameter and depth of counterbore for threaded flanges shall be as specified in IVE-11.
5. The bore J_L of welding-neck flanges shall be as specified on the purchase order.

   NOTE: *Bore diameter should be the same as ID of pipe to be used, but shall not exceed values of J_L shown.*
6. The wall thickness of welding-neck flanges shall be not less than 87½ per cent of the nominal wall thickness of the pipe to which the flange is to be attached.
7. The welding end of welding-neck flanges shall be cylindrical or shall have a maximum draft of 7 deg. The length shall be sufficient to insure a sound weld, but in no case shall be less than ¼ in. See IVE-8.
8. The welding bevel of welding-neck flanges shall be as shown in IVE-8.

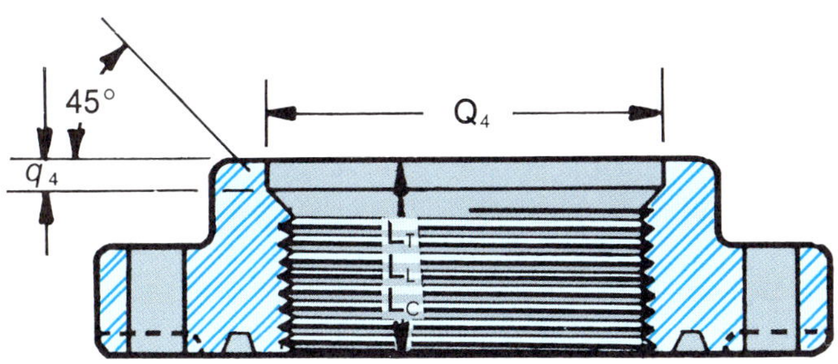

THREADED FLANGE

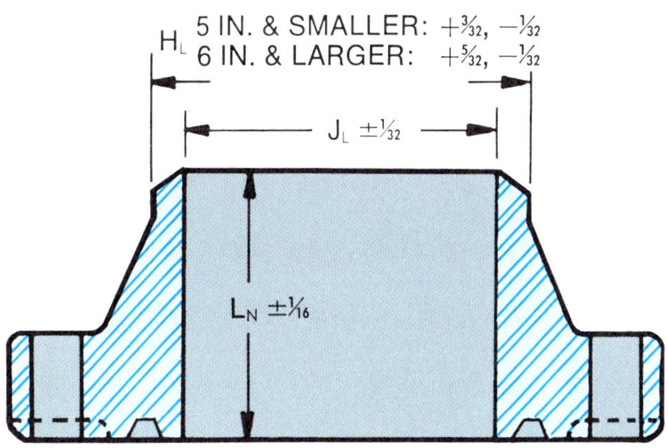

WELDING-NECK LINE-PIPE FLANGE

IVE-5 API TYPE 6B FLANGES FOR 5000-PSI MAXIMUM WORKING PRESSURE

All dimensions in inches unless otherwise noted.

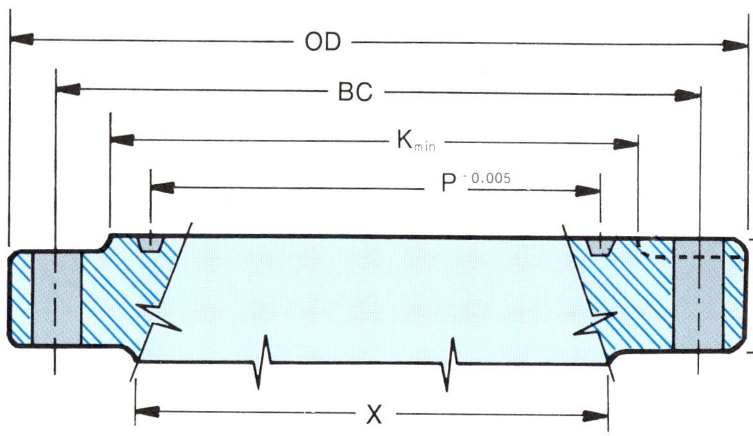

FLANGE SECTION
INTEGRAL FLANGE

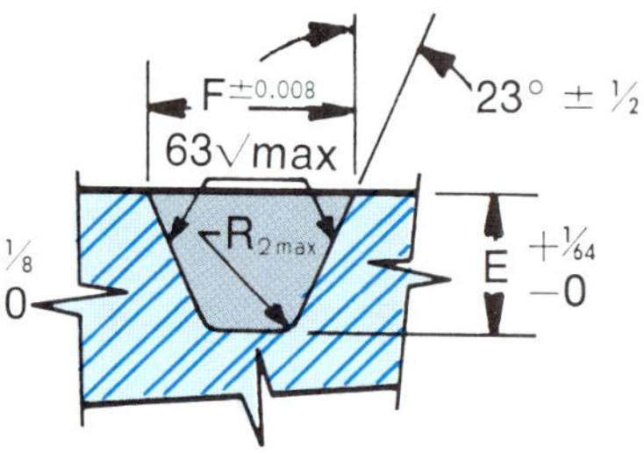

GROOVE DETAIL

Reprinted with permission of the American Petroleum Institute.

Basic Flange Dimensions					Bolting Dimensions				Ring-Joint Groove and Flange Facing Dimension				
Nominal Size of Flange	Outside Diameter of Flange OD	Total Thickness of Flange T	Basic Thickness of Flange Q	Diameter of Hub X	Diameter of Bolt Circle BC	Number of Bolts	Diameter of Bolts	Length of Stud Bolts L_{SSB}***	Ring Number R or RX	Pitch Diameter of Type R Ring & Groove P	Width of Groove F	Depth of Groove E	Diameter of Raised Face K
1½*	7	1½	1¼	2¾	4⅞	4	1	5½	20	2 11/16	11/32	¼	3⅝
2	8½	1 13/16	1½	4⅛	6½	8	⅞	6	24	3¾	15/32	5/16	4⅞
2½	9⅝	1 15/16	1⅝	4⅞	7½	8	1	6½	27	4¼	15/32	5/16	5⅜
3	10½	2 3/16	1⅞	5¼	8	8	1⅛	7¼	35	5⅜	15/32	5/16	6⅝
4	12¼	2 7/16	2⅛	6⅜	9½	8	1¼	8	39	6⅜	15/32	5/16	7⅝
5*	14¾	3 3/16	2⅞	7¾	11½	8	1½	10	44	7⅝	15/32	5/16	9
6	15½	3⅝	3¼	9	12½	12	1⅜	10¾	46	8 5/16	17/32	⅜	9¾
8	19	4 1/16	3⅝	11½	15½	12	1⅝	12	50	10⅝	21/32	7/16	12½
10	23	4 11/16	4¼	14½	19	12	1⅞	13¾	54	12¾	21/32	7/16	14⅝
13⅝**	—	—	—	—	—	—	—	—	—	—	—	—	—
16¾**	—	—	—	—	—	—	—	—	—	—	—	—	—

*These sizes inactive; available on special order only.
**See IVE-6 for dimension details on these sizes.
***L_{SSB} includes thickness of two nuts but not of either point height.

NOTE 1: *Except for bore welding-neck flanges, dimensions for sizes 1½ in. to 2½ in. inclusive are identical with 3,000 psi flanges in Table.*

IVE-5 API TYPE 6B FLANGES FOR 5000-PSI MAXIMUM WORKING PRESSURE (CONTINUED)

HUB AND BORE DIMENSIONS						
Nominal Size of Flange	Hub Length Threaded Line-Pipe Flange L_L	Hub Length Threaded Casing Flange L_C	Hub Length Tubing Flange L_T	Hub Length Welding-Neck Line-Pipe Flange L_N	Neck Diameter Welding-Neck Line-Pipe Flange H_L	Maximum Bore of Welding Neck Flange J_L
1½*	2	—	2	3½	1.90	1.337
2	2 9/16	—	2 9/16	4 5/16	2.38	1.689
2½	2 13/16	—	2 13/16	4 7/16	2.88	2.125
3	3 3/16	—	3 3/16	4 15/16	3.50	2.624
4	3 7/8	3 7/8	3 7/8	5 3/16	4.50	3.438
5*	4 7/16	4 7/16	—	6 7/16	5.56	4.313
6	5 1/16	5 1/16	—	7 1/8	6.63	5.189
8	6 1/16	6 1/16	—	8 13/16	8.63	6.813
10	6 11/16	6 11/16	—	10 7/16	10.75	8.500
13 5/8**	—	—	—	—	—	—
16¾**	—	—	—	—	—	—

*These sizes inactive; available on special order only.
**See IVE-6 for dimension details on these sizes.

Reprinted with permission of the American Petroleum Institute.

REQUIREMENTS

1. Bolt holes shall be drilled ⅛ in. larger in diameter than the specified bolt diameter.
2. The contour of the flange face outside the *K* diameter is optional with the manufacturer unless raised or full face is specified on the purchase order.
3. Ring-groove radius R^2 shall be 1/32 in. for groove widths 11/32 and 15/32, 1/16 in. for widths 17/32 and 21/32.
4. The diameter and depth of counterbore for threaded flanges shall be as specified in IVE-11.
5. The bore J_L of welding-neck flanges shall be as specified on the purchase order.

   NOTE: *Bore diameter should be the same as ID of pipe to be used, but shall not exceed values of J_L shown.*
6. The wall thickness of welding-neck flanges shall be not less than 87½ per cent of the nominal wall thickness of the pipe to which the flange is to be attached.
7. The welding end of welding-neck flanges shall be cylindrical or shall have a maximum draft of 7 deg. The length shall be sufficient to insure a sound weld, but in no case shall be less than ¼ in. See IVE-8.
8. The welding bevel of welding-neck flanges shall be as shown in Fig. IVE-8.

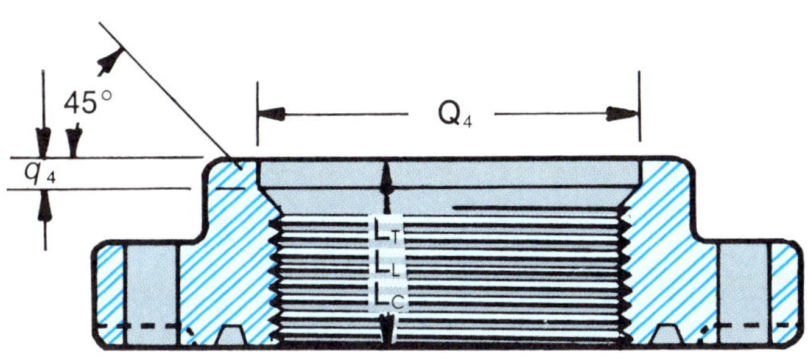

THREADED FLANGE

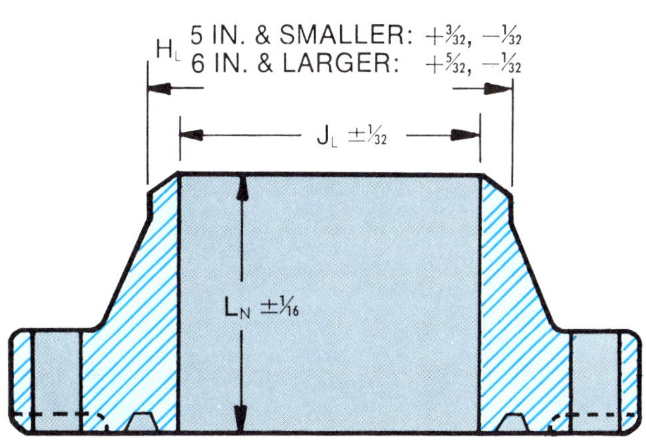

WELDING-NECK LINE-PIPE FLANGE

IVE-6 API TYPE 6BX INTEGRAL FLANGES FOR 5,000, 10,000 AND 15,000 PSI MAXIMUM WORKING PRESSURE

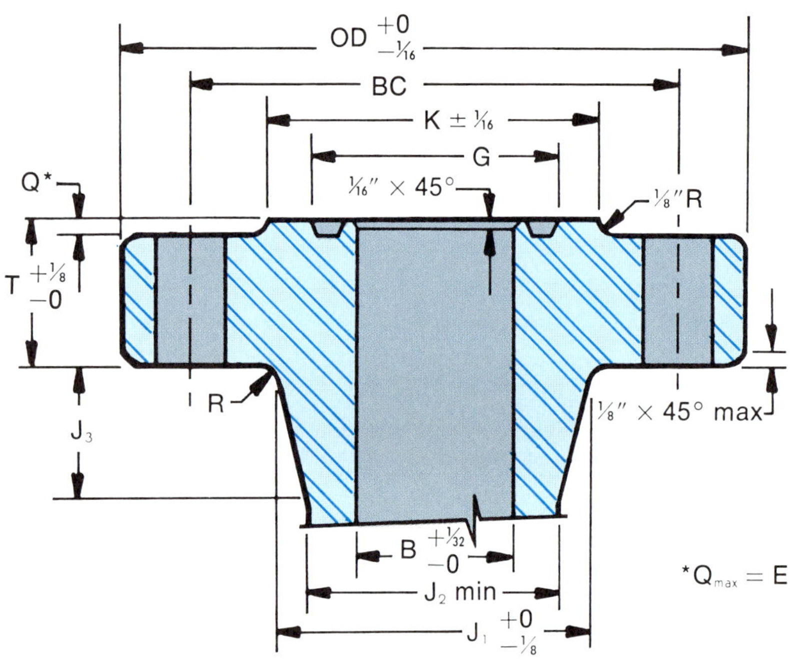

FLANGE SECTION

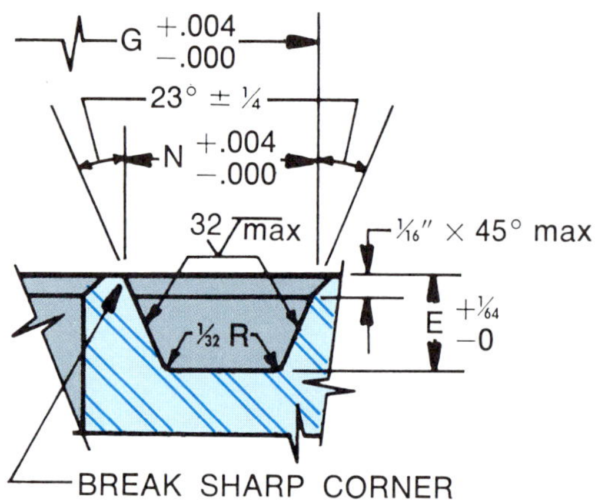

GROOVE DETAIL

$^{*}Q_{max} = E$, $Q_{min} = 1/8$

IVE-6 API TYPE 6BX INTEGRAL FLANGES FOR 5,000, 10,000 AND 15,000 PSI MAXIMUM WORKING PRESSURE (CONTINUED)

Basic Flange Dimensions							Bolting Dimensions				Facing and Groove Dimensions				
Nominal Size and Bore B	Outside Dia OD	Total Thickness T	Large Dia of Hub J_1	Small Dia of Hub J_2	Length of Hub J_3	Radius at Hub R	Dia of Bolt Circle BC	Number of Bolts	Dia of Bolts	Length of Stud Bolts L_{SSB}***	Raised Face Dia K	Groove OD G	Width of Groove N	Depth of Groove E	Ring Number
5,000 PSI															
13 5/8	26 1/2	4 7/16	18 15/16	16 11/16	4 1/2	5/8	23 1/4	16	1 5/8	12 1/2	18	16.063	0.786	9/16	BX-160
16 3/4†	30 3/8	5 1/8	21 7/8	20 3/4	3	3/4	26 5/8	16	1 7/8	14 1/2	21 1/16	18.832	0.705	21/64	BX-162
10,000 PSI															
1 11/16**	7 3/16	1 21/32	3 5/16	2 13/32	1 27/32	3/8	5 9/16	8	3/4	5	4	2.893	0.450	7/32	BX-150
1 13/16	7 3/8	1 21/32	3 1/2	2 9/16	1 29/32	3/8	5 3/4	8	3/4	5	4 1/8	3.062	0.466	7/32	BX-151
2 1/16	7 7/8	1 47/64	3 15/16	2 15/16	2 1/32	3/8	6 1/4	8	3/4	5 1/4	4 3/8	3.395	0.498	15/64	BX-152
2 9/16	9 1/8	2 1/64	4 3/4	3 5/8	2 1/4	3/8	7 1/4	8	7/8	6	5 3/16	4.046	0.554	17/64	BX-153
3 1/16	10 5/8	2 19/64	5 19/32	4 11/32	2 1/2	3/8	8 1/2	8	1	6 3/4	6	4.685	0.606	19/64	BX-154
4 1/16	12 7/16	2 49/64	7 3/16	5 3/4	2 7/8	3/8	10 3/16	8	1 1/8	8	7 9/32	5.930	0.698	21/64	BX-155
7 1/16	18 7/8	4 1/16	11 7/8	10	3 3/4	5/8	15 7/8	12	1 1/2	11 1/4	11 7/8	9.521	0.921	7/16	BX-156
9	21 3/4	4 7/8	14 3/4	12 7/8	3 11/16	5/8	18 3/4	16	1 1/2	13	14 1/8	11.774	1.039	1/2	BX-157
11	25 3/4	5 9/16	17 3/4	15 3/4	4 1/16	5/8	22 1/4	16	1 3/4	15	16 7/8	14.064	1.149	9/16	BX-158
13 5/8	30 1/4	6 5/8	21 3/4	19 1/2	4 1/2	5/8	26 1/2	20	1 7/8	17 1/4	20 3/8	17.033	1.279	5/8	BX-159
16 3/4	34 5/16	6 5/8	25 13/16	23 11/16	3	3/4	30 9/16	24	1 7/8	17 1/2	22 11/16	18.832	0.705	21/64	BX-162
15,000 PSI															
1 11/16**	7 5/8	1 3/4	3 11/16	2 11/16	1 7/8	3/8	6	8	3/4	5 1/4	3 13/16	2.893	0.450	7/32	BX-150
1 13/16	8 3/16	1 25/32	3 27/32	2 13/16	1 7/8	3/8	6 5/16	8	7/8	5 1/2	4 3/16	3.062	0.466	7/32	BX-151
2 1/16	8 3/4	2	4 3/8	3 1/4	2 1/8	3/8	6 7/8	8	7/8	6	4 1/2	3.395	0.498	15/64	BX-152
2 9/16	10	2 1/4	5 1/16	3 15/16	2 1/4	3/8	7 7/8	8	1	6 3/4	5 1/4	4.046	0.554	17/64	BX-153
3 1/16	11 5/16	2 17/32	6 1/16	4 13/16	2 1/2	3/8	9 1/16	8	1 1/8	7 1/2	6 1/16	4.685	0.606	19/64	BX-154
7 1/16	19 7/8	4 11/16	12 13/16	10 7/8	3 5/8	5/8	16 7/8	16	1 1/2	12 3/4	12	9.521	0.921	7/16	BX-156

*Dimension Q may be omitted on studded connections.

**This flange is inactive; available on special order only.

†This flange was adopted June 1969 and shall be marked with both the working pressure (5000 WP) and the test pressure (10,000 TP) in addition to other marking requirements.

NOTE 1: *Bolt holes shall be drilled 1/8 in. larger in diameter than the bolt size.*

***L_{SSB} includes thickness of two nuts but not of either point height.

IVE-7 API TYPE 6BX WELDING NECK FLANGES FOR 10,000 AND 15,000 PSI MAXIMUM WORKING PRESSURE

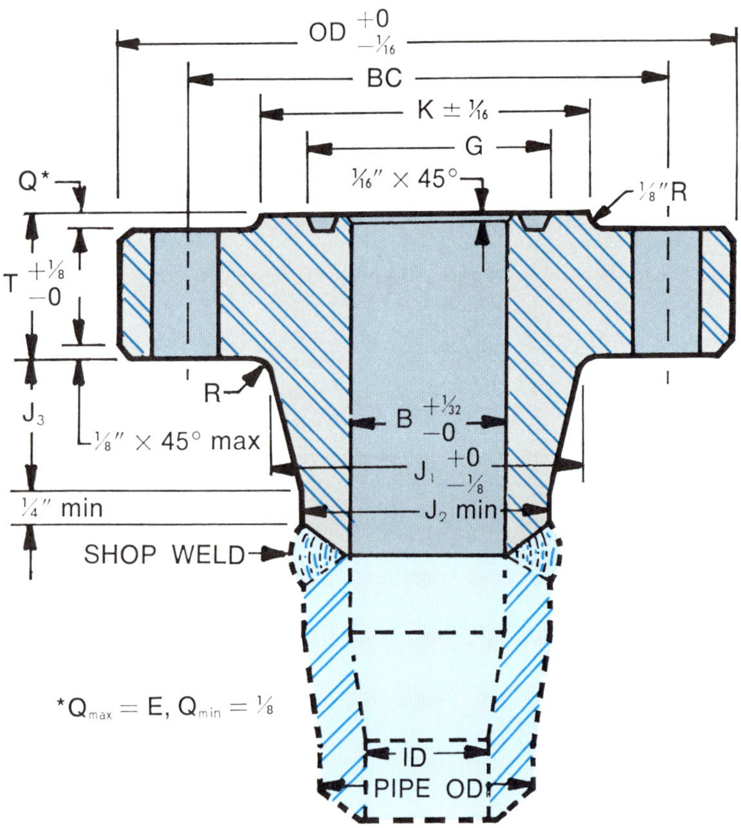

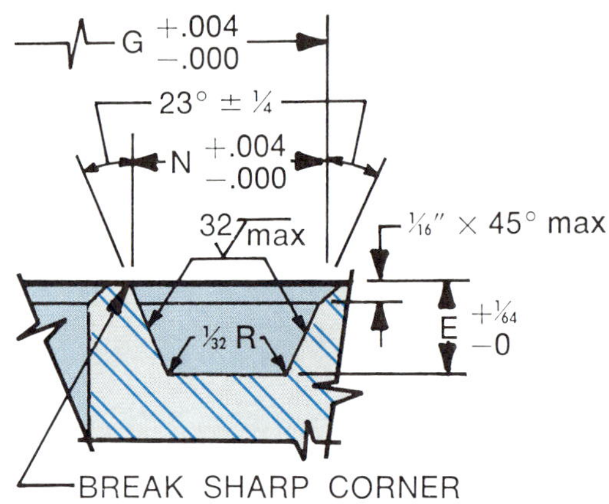

GROOVE DETAIL

IVE-7 API TYPE 6BX WELDING NECK FLANGES FOR 10,000 AND 15,000 PSI MAXIMUM WORKING PRESSURE (CONTINUED)

All dimensions in inches unless otherwise noted.

REQUIREMENTS

1. Due to the difficulty of field welding API Type 2 and 3 material, from which these flanges are made, a transition piece may be shop welded to the basic flange and the weld properly heat treated. This transition piece shall be made from the same or similar material as the pipe to which it is to be welded by the customer. Transition piece ID and OD at the field welding end, and its material, shall be specified on the purchase order.
2. The welding end preparation of the welding neck flange and both ends of the transition piece shall conform to the requirements of IVE-8.
3. The length of the transition piece shall be great enough that the heat from field welding will not affect the metallurgical properties of the shop weld.
4. The API monogram shall be applied to the welding neck flange (solid outline). The API monogram does not apply to the shop weld or the transition piece (dashed outline).

Basic Flange Dimensions							Bolting Dimensions				Facing and Groove Dimensions				
Nominal Size and Bore B	Outside Dia OD	Total Thickness T	Large Dia of Hub J_1	Small Dia of Hub J_2	Length of Hub J_3	Radius at Hub R	Dia of Bolt Circle BC	Number of Bolts	Dia of Bolts	Length of Stud Bolts L_{SSB}***	Raised Face Dia K	Groove OD G	Width of Groove N	Depth of Groove E	Ring Number
10,000 PSI															
1 11/16**	7 3/16	1 21/32	3 5/16	2 13/32	1 27/32	3/8	5 9/16	8	3/4	5	4	2.893	0.450	7/32	BX-150
1 13/16	7 3/8	1 21/32	3 1/2	2 9/16	1 29/32	3/8	5 3/4	8	3/4	5	4 1/8	3.062	0.466	7/32	BX-151
2 1/16	7 7/8	1 47/64	3 15/16	2 15/16	2 1/32	3/8	6 1/4	8	3/4	5 1/4	4 3/8	3.395	0.498	15/64	BX-152
2 9/16	9 1/8	2 1/64	4 3/4	3 5/8	2 1/4	3/8	7 1/4	8	7/8	6	5 3/16	4.046	0.554	17/64	BX-153
3 1/16	10 5/8	2 19/64	5 19/32	4 11/32	2 1/2	3/8	8 1/2	8	1	6 3/4	6	4.685	0.606	19/64	BX-154
4 1/16	12 7/16	2 49/64	7 3/16	5 3/4	2 7/8	3/8	10 3/16	8	1 1/8	8	7 9/32	5.930	0.698	21/64	BX-155
7 1/16	18 7/8	4 1/16	11 7/8	10	3 3/4	5/8	15 7/8	12	1 1/2	11 1/4	11 7/8	9.521	0.921	7/16	BX-156
9	21 3/4	4 7/8	14 3/4	12 7/8	3 11/16	5/8	18 3/4	16	1 1/2	13	14 1/8	11.774	1.039	1/2	BX-157
11	25 3/4	5 9/16	17 3/4	15 3/4	4 1/16	5/8	22 1/4	16	1 3/4	15	16 7/8	14.064	1.149	9/16	BX-158
13 5/8	30 1/4	6 5/8	21 3/4	19 1/2	4 1/2	5/8	26 1/2	20	1 7/8	17 1/4	20 3/8	17.033	1.279	5/8	BX-159
16 3/4	34 5/16	6 5/8	25 13/16	23 11/16	3	3/4	30 9/16	24	1 7/8	17 1/2	22 11/16	18.832	0.705	21/64	BX-162
15,000 PSI															
1 11/16**	7 5/8	1 3/4	3 11/16	2 11/16	1 7/8	3/8	6	8	3/4	5 1/4	3 13/16	2.893	0.450	7/32	BX-150
1 13/16	8 3/16	1 25/32	3 27/32	2 13/16	1 7/8	3/8	6 5/16	8	7/8	5 1/2	4 3/16	3.062	0.466	7/32	BX-151
2 1/16	8 3/4	2	4 3/8	3 1/4	2 1/8	3/8	6 7/8	8	7/8	6	4 1/2	3.395	0.498	15/64	BX-152
2 9/16	10	2 1/4	5 1/16	3 15/16	2 1/4	3/8	7 7/8	8	1	6 3/4	5 1/4	4.046	0.554	17/64	BX-153
3 1/16	11 5/16	2 17/32	6 1/16	4 13/16	2 1/2	3/8	9 1/16	8	1 1/8	7 1/2	6 1/16	4.685	0.606	19/64	BX-154
7 1/16	19 7/8	4 11/16	12 13/16	10 7/8	3 5/8	5/8	16 7/8	16	1 1/2	12 3/4	12	9.521	0.921	7/16	BX-156

* Dimension Q may be omitted on studded connections.
** This flange is inactive; available on special order only.
***L_{SSB} includes thickness of two nuts but not of either point height.

NOTE: *Bolt holes shall be drilled 1/8 in. larger in diameter than the bolt size.*

IVE-8 WELDING ENDS FOR WELDING-NECK FLANGES

NOTE 1: *See Tables IVE-5 and IVE-7 for dimensions and tolerances.*

NOTE 2: *When thickness of the flange welding end is greater than that of the pipe and the additional thickness increases the outside diameter, a taper weld having a slope not exceeding 3 to 1 may be employed or, alternatively, the greater outside diameter may be tapered, at the same maximum slope or less, from a point on the welding bevel equal to the OD of the mating pipe.*

NOTE 3: *When thickness at the flange welding end is greater than that of the pipe and the additional thickness decreases the inside diameter, and exceeds the pipe thickness by 3/32 in., or more, the flange shall be taper-bored from the welding end at a slope not exceeding 3 to 1.*

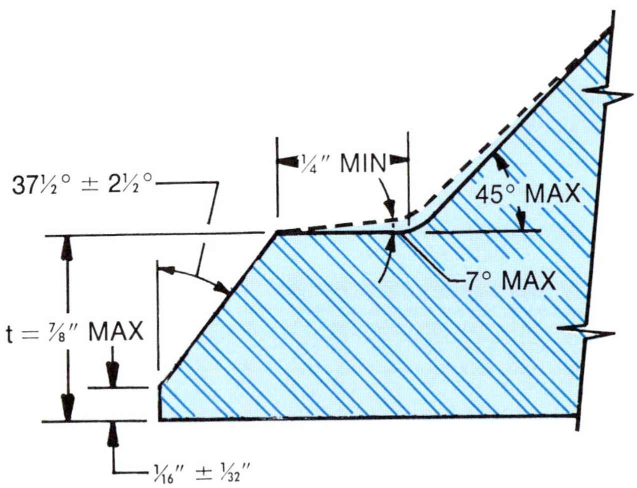

NOTE: WELDING ENDS OF 1½" & 2" 2000 PSI FLANGES MAY BE SQUARE OR CHAMFERED AT MFR'S OPTION.

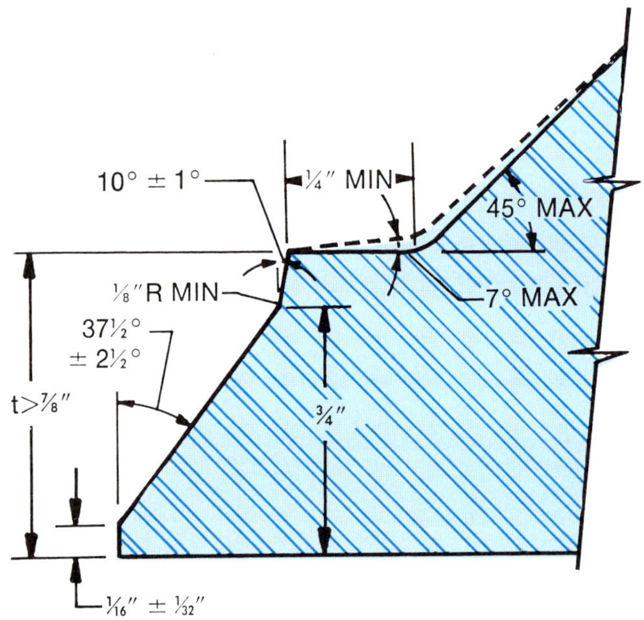

IVE-9 API TYPE 6BX BLIND AND TEST FLANGES FOR 10,000 AND 15,000 PSI MAXIMUM WORKING PRESSURE

All dimensions in inches unless otherwise noted.

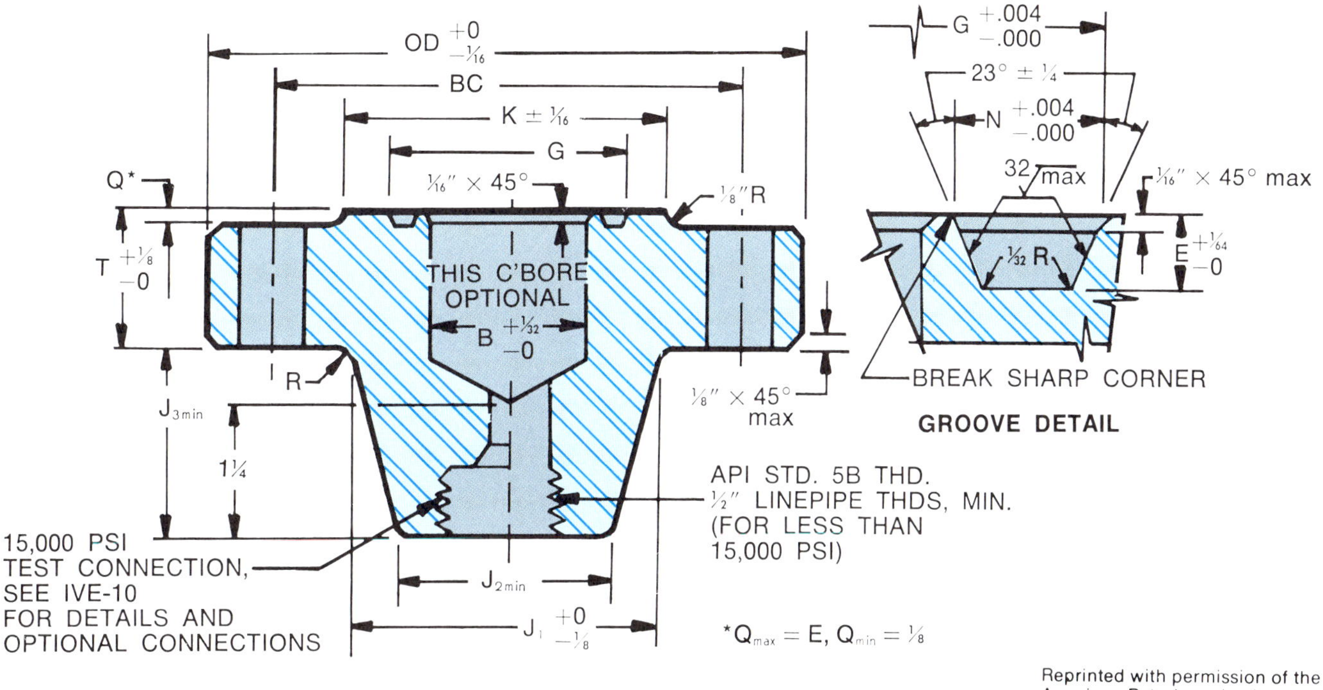

FLANGE SECTION

GROOVE DETAIL

Reprinted with permission of the American Petroleum Institute.

Basic Flange Dimensions							Bolting Dimensions				Facing and Groove Dimensions				
Nominal Size and Bore	Outside Dia	Total Thickness	Large Dia of Hub	Small Dia of Hub	Length of Hub	Radius at Hub	Dia of Bolt Circle	Number of Bolts	Dia of Bolts	Length of Stud Bolts	Raised Face Dia	Groove OD	Width of Groove	Depth of Groove	Ring Number
B	*OD*	*T*	J_1	J_2	J_3	*R*	*BC*			L_{SSB}***	*K*	*G*	*N*	*E*	
10,000 PSI															
1 11/16**	7 3/16	1 21/32	3 5/16	2 13/32	1 27/32	3/8	5 9/16	8	3/4	5	4	2.893	0.450	7/32	BX-150
1 13/16	7 3/8	1 21/32	3 1/2	2 9/16	1 29/32	3/8	5 3/4	8	3/4	5	4 1/8	3.062	0.466	7/32	BX-151
2 1/16	7 7/8	1 47/64	3 15/16	2 15/16	2 1/32	3/8	6 1/4	8	3/4	5 1/4	4 3/8	3.395	0.498	15/64	BX-152
2 9/16	9 1/8	2 1/64	4 3/4	3 5/8	2 1/4	3/8	7 1/4	8	7/8	6	5 3/16	4.046	0.554	17/64	BX-153
3 1/16	10 5/8	2 19/64	5 19/32	4 11/32	2 1/2	3/8	8 1/2	8	1	6 3/4	6	4.685	0.606	19/64	BX-154
4 1/16	12 7/16	2 49/64	7 3/16	5 3/4	2 7/8	3/8	10 3/16	8	1 1/8	8	7 9/32	5.930	0.698	21/64	BX-155
15,000 PSI															
1 11/16**	7 5/8	1 3/4	3 11/16	2 11/16	1 7/8	3/8	6	8	3/4	5 1/4	3 13/16	2.893	0.450	7/32	BX-150
1 13/16	8 3/16	1 25/32	3 27/32	2 13/16	1 7/8	3/8	6 5/16	8	7/8	5 1/2	4 3/16	3.062	0.466	7/32	BX-151
2 1/16	8 3/4	2	4 3/8	3 1/4	2 1/8	3/8	6 7/8	8	7/8	6	4 1/2	3.395	0.498	15/64	BX-152
2 9/16	10	2 1/4	5 1/16	3 15/16	2 1/4	3/8	7 7/8	8	1	6 3/4	5 1/4	4.046	0.554	17/64	BX-153

*Dimension Q may be omitted on studded connections.

NOTE: *Bolt holes shall be drilled 1/8 in. larger in diameter than the bolt size.*

**This flange is inactive; available on special order only.

*** L_{SSB} includes thickness of two nuts but not of either point height.

IVE-10 TEST AND GAGE CONNECTIONS FOR 15,000-LB CHRISTMAS TREES

All dimensions in inches unless otherwise noted.

ITEM NO.	DESCRIPTION
1	COLLAR
2	GLAND
3	PLUG
4	SUB ADAPTER
5	CONNECTOR, ½ NPT

Reprinted with permission of the American Petroleum Institute.

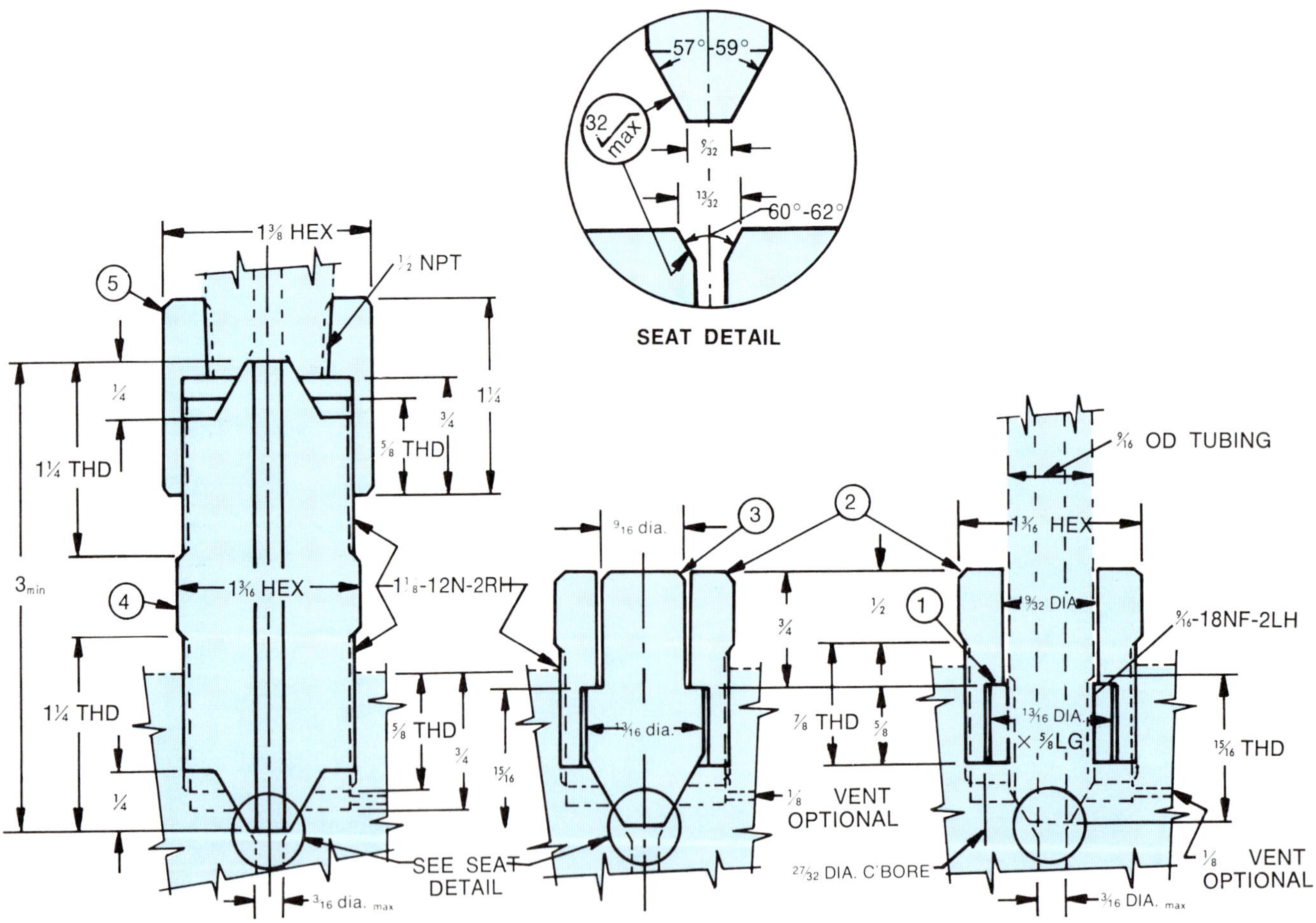

TYPE I CONNECTION **TYPE II CONNECTION** **TYPE III CONNECTION**

IVE-11 API TYPE 6B THREADED FLANGE, COUNTERBORE DIMENSIONS

TABLE 1
LINE-PIPE THREADS IN API 6B FLANGES

All dimensions in inches.

Thread Size: Nominal	Q_4 Diameter	q_4 Depth
1/8	0.47	0.125
1/4	0.60	0.125
3/8	0.74	0.125
1/2	0.93	0.250
3/4	1.14	0.250
1	1.41	0.250
1 1/4	1.75	0.250
1 1/2	1.99	0.250
2	2.50	0.250
2 1/2	3.00	0.375
3	3.63	0.375
3 1/2	4.13	0.375
4	4.63	0.375
5	5.69	0.375
6	6.75	0.375
8	8.75	0.375
10	10.88	0.375
12	12.94	0.375
14 D	14.19	0.375
16 D	16.19	0.375
18 D	18.19	0.375
20 D	20.19	0.375

TABLE 2
LONG CASING THREADS IN API 6B FLANGES

All dimensions in inches.

Thread Size: Outside Diameter	Q_4 Diameter	q_4 Depth
4 1/2	4.63	0.375
5	5.13	0.375
5 1/2	5.63	0.375
6 5/8	6.75	0.375
7	7.13	0.375
7 5/8	7.75	0.375
8 5/8	8.75	0.375
9 5/8	9.75	0.375

TABLE 3
SHORT CASING THREADS IN API 6B FLANGES

All dimensions in inches.

Thread Size: Outside Diameter	Q_4 Diameter	q_4 Depth
4 1/2	4.63	0.375
5	5.13	0.375
5 1/2	5.63	0.375
6 5/8	6.75	0.375
7	7.13	0.375
7 5/8	7.75	0.375
8 5/8	8.75	0.375
9 5/8	9.75	0.375
10 3/4	10.88	0.375
11 3/4	11.88	0.375
13 3/8	13.56	0.375
16	16.19	0.375
20	20.19	0.375

IVE-11 API TYPE 6B THREADED FLANGE, COUNTERBORE DIMENSIONS (CONTINUED)

TABLE 4
NON-UPSET TUBING THREADS IN API 3000-PSI AND 5000-PSI FLANGES

All dimensions in inches.

Thread Size: Outside Diameter	Q_4 Diameter	q_4 Depth
1.050	1.14	0.250
1.315	1.41	0.250
1.660	1.75	0.250
1.900	1.99	0.250
$2\frac{3}{8}$	2.50	0.250
$2\frac{7}{8}$	3.00	0.375
$3\frac{1}{2}$	3.63	0.375
4	4.13	0.375
$4\frac{1}{2}$	4.63	0.375

TABLE 5
EXTERNAL-UPSET REGULAR AND LONG TUBING THREADS IN API 3000-PSI AND 5000-PSI FLANGES

All dimensions in inches.

Thread Size: Outside Diameter	Q_4 Diameter	q_4 Depth
1.050	1.41	0.250
1.315	1.57	0.250
1.660	1.91	0.250
1.900	2.19	0.250
$2\frac{3}{8}$	2.72	0.250
$2\frac{7}{8}$	3.22	0.375
$3\frac{1}{2}$	3.88	0.375
4	4.38	0.375
$4\frac{1}{2}$	4.88	0.375

IVE-12 CAST IRON FLANGE DIMENSIONS AND DRILLING TEMPLATES ANSI B16.1-1975

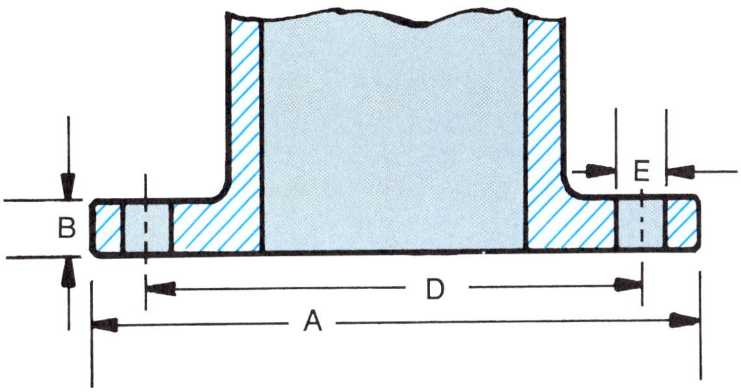

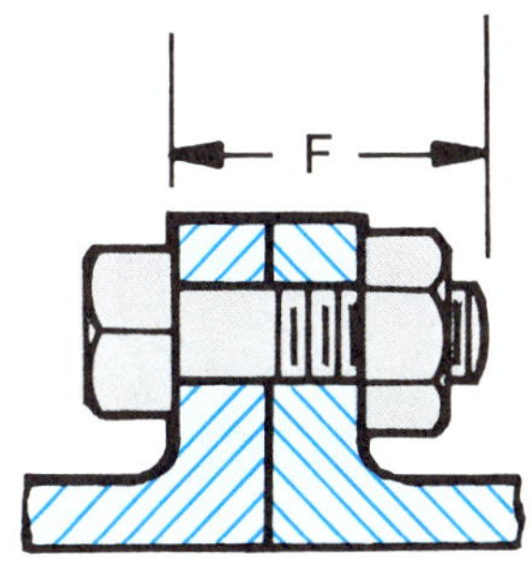

Reprinted with permission from the Walworth Aloyco, Condensed Catalog 68, Walworth Aloyco Div., Walworth Co.

25 LB CAST IRON FLANGES

Dimensions in Inches

Nominal Pipe Size	Flange Diameter A	Flange Thickness B (Min)	Diameter of Bolt Circle D	Diameter of Bolt Holes E	Number of Bolts	Diameter of Bolts	Length of Machine Bolts *25 Lb Flg. to 25 Lb Flg. F[(1)]	Length of Machine Bolts *25 Lb Flg. to 125 Lb Flg. F[(1)]
4	9	¾	7½	¾	8	⅝	2½	2½
5	10	¾	8½	¾	8	⅝	2½	2½
6	11	¾	9½	¾	8	⅝	2½	2½
8	13½	¾	11¾	¾	8	⅝	2½	2¾
10	16	⅞	14¼	¾	12	⅝	2¾	3
12	19	1	17	¾	12	⅝	3	3
14	21	1⅛	18¾	⅞	12	¾	3½	3½
16	23½	1⅛	21¼	⅞	16	¾	3½	3½
18	25	1¼	22¾	⅞	16	¾	3¾	3¾
20	27½	1¼	25	⅞	20	¾	3¾	4
24	32	1⅜	29½	⅞	20	¾	4	4¼
30	38¾	1½	36	1	28	⅞	4½	—
36	46	1⅝	42¾	1	32	⅞	4¾	—
42	53	1¾	49½	1⅛	36	1	5¼	—
48	59½	2	56	1⅛	44	1	5¾	—
54	66¼	2¼	62¾	1⅛	44	1	6¼	—
60	73	2¼	69¼	1¼	52	1⅛	6¼	—
72	86½	2½	82½	1¼	60	1⅛	6¾	—
84	99¾	2¾	95½	1⅜	64	1¼	7½	—
96	113¼	3	108½	1⅜	68	1¼	8	—

* Screwed companion flanges should not be thinner than the 125 lb American Standard thickness on sizes 24-inch and smaller. Other types of flanges may have thicknesses as given in the table.

When flanges are integral with pumps or fittings, the bolt holes, which are in multiples of four are drilled to straddle the center lines unless otherwise ordered.

The 25 lb and 125 lb cast iron flanges have plain faces.

[(1)]These lengths DO NOT appear in ANSI B16.1 — they are only suggested by the Hydraulic Institute.

IVE-12 CAST IRON FLANGE DIMENSIONS AND DRILLING TEMPLATES ANSI B16.1-1975

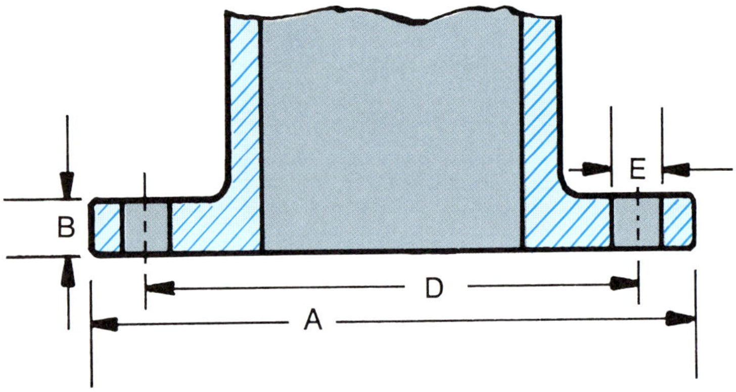

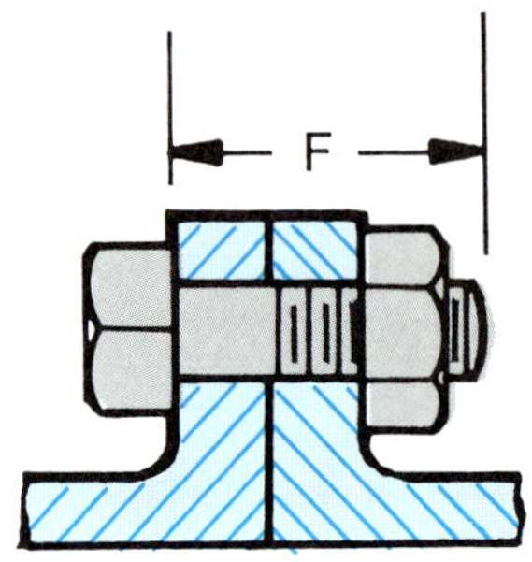

Reprinted with permission from the Walworth Aloyco, Condensed Catalog 68, Walworth Aloyco Div., Walworth Co.

125 LB CAST IRON FLANGES

Dimensions in Inches

Nominal Pipe Size	Flange Diameter A	Flange Thickness B (Min)	Diameter of Bolt Circle D	Diameter of Bolt Holes E	Number of Bolts	Diameter of Bolts	Length of Machine Bolts F
1	$4\frac{1}{4}$	$\frac{7}{16}$	$3\frac{1}{8}$	$\frac{5}{8}$	4	$\frac{1}{2}$	$1\frac{3}{4}$
$1\frac{1}{4}$	$4\frac{5}{8}$	$\frac{1}{2}$	$3\frac{1}{2}$	$\frac{5}{8}$	4	$\frac{1}{2}$	2
$1\frac{1}{2}$	5	$\frac{9}{16}$	$3\frac{7}{8}$	$\frac{5}{8}$	4	$\frac{1}{2}$	2
2	6	$\frac{5}{8}$	$4\frac{3}{4}$	$\frac{3}{4}$	4	$\frac{5}{8}$	$2\frac{1}{4}$
$2\frac{1}{2}$	7	$\frac{11}{16}$	$5\frac{1}{2}$	$\frac{3}{4}$	4	$\frac{5}{8}$	$2\frac{1}{2}$
3	$7\frac{1}{2}$	$\frac{3}{4}$	6	$\frac{3}{4}$	4	$\frac{5}{8}$	$2\frac{1}{2}$
$3\frac{1}{2}$	$8\frac{1}{2}$	$\frac{13}{16}$	7	$\frac{3}{4}$	8	$\frac{5}{8}$	$2\frac{3}{4}$
4	9	$\frac{15}{16}$	$7\frac{1}{2}$	$\frac{3}{4}$	8	$\frac{5}{8}$	3
5	10	$\frac{15}{16}$	$8\frac{1}{2}$	$\frac{7}{8}$	8	$\frac{3}{4}$	3
6	11	1	$9\frac{1}{2}$	$\frac{7}{8}$	8	$\frac{3}{4}$	$3\frac{1}{4}$
8	$13\frac{1}{2}$	$1\frac{1}{8}$	$11\frac{3}{4}$	$\frac{7}{8}$	8	$\frac{3}{4}$	$3\frac{1}{2}$
10	16	$1\frac{3}{16}$	$14\frac{1}{4}$	1	12	$\frac{7}{8}$	$3\frac{3}{4}$
12	19	$1\frac{1}{4}$	17	1	12	$\frac{7}{8}$	$3\frac{3}{4}$
14	21	$1\frac{3}{8}$	$18\frac{3}{4}$	$1\frac{1}{8}$	12	1	$4\frac{1}{4}$
16	$23\frac{1}{2}$	$1\frac{7}{16}$	$21\frac{1}{4}$	$1\frac{1}{8}$	16	1	$4\frac{1}{2}$
18	25	$1\frac{9}{16}$	$22\frac{3}{4}$	$1\frac{1}{4}$	16	$1\frac{1}{8}$	$4\frac{3}{4}$
20	$27\frac{1}{2}$	$1\frac{11}{16}$	25	$1\frac{1}{4}$	20	$1\frac{1}{8}$	5
24	32	$1\frac{7}{8}$	$29\frac{1}{2}$	$1\frac{3}{8}$	20	$1\frac{1}{4}$	$5\frac{1}{2}$
30	$38\frac{3}{4}$	$2\frac{1}{8}$	36	$1\frac{3}{8}$	28	$1\frac{1}{4}$	$6\frac{1}{4}$
36	46	$2\frac{3}{8}$	$42\frac{3}{4}$	$1\frac{5}{8}$	32	$1\frac{1}{2}$	7
42	53	$2\frac{5}{8}$	$49\frac{1}{2}$	$1\frac{5}{8}$	36	$1\frac{1}{2}$	$7\frac{1}{2}$
48	$59\frac{1}{2}$	$2\frac{3}{4}$	56	$1\frac{5}{8}$	44	$1\frac{1}{2}$	$7\frac{3}{4}$

IVE-12 CAST IRON FLANGE DIMENSIONS AND DRILLING TEMPLATES ANSI B16.1-1975

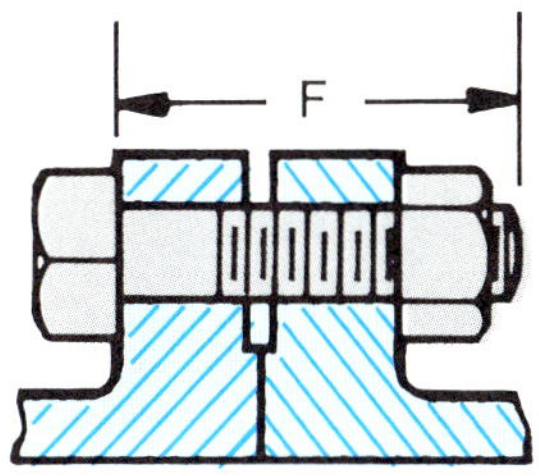

Length of Machine Bolt

Raised Face

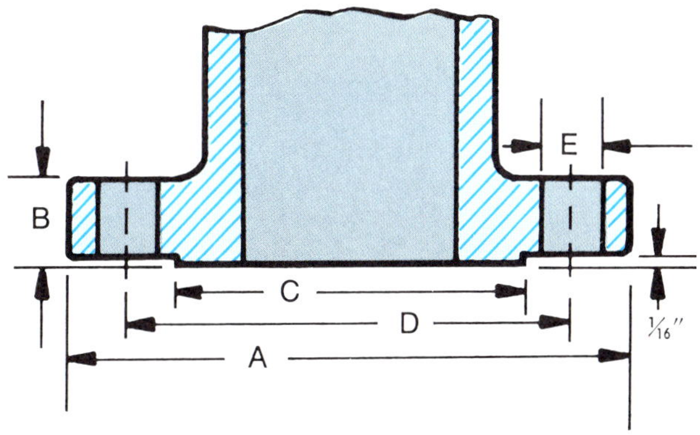

Alternate Flat Face Pump Flange

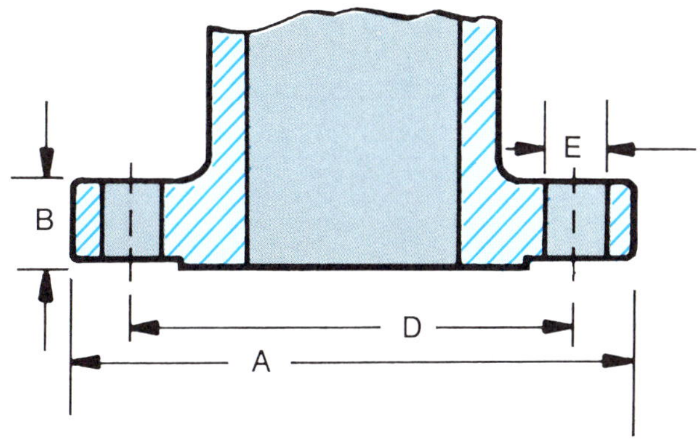

250 LB CAST IRON FLANGES

DIMENSIONS IN INCHES

Nominal Pipe Size	Flange Diameter A	Flange Thickness B (Min)	Diameter of Raised Face C	Diameter of Bolt Circle D	Diameter of Bolt Holes E	Number of Bolts	Diameter of Bolts	Length of Machine Bolts F
1	4 7/8	11/16	2 11/16	3 1/2	3/4	4	5/8	2 1/2
1 1/4	5 1/4	3/4	3 1/16	3 7/8	3/4	4	5/8	2 1/2
1 1/2	6 1/8	13/16	3 9/16	4 1/2	7/8	4	3/4	2 3/4
2	6 1/2	7/8	4 3/16	5	3/4	8	5/8	2 3/4
2 1/2	7 1/2	1	4 15/16	5 7/8	7/8	8	3/4	3 1/4
3	8 1/4	1 1/8	5 11/16	6 5/8	7/8	8	3/4	3 1/2
3 1/2	9	1 3/16	6 5/16	7 1/4	7/8	8	3/4	3 1/2
4	10	1 1/4	6 15/16	7 7/8	7/8	8	3/4	3 3/4
5	11	1 3/8	8 5/16	9 1/4	7/8	8	3/4	4
6	12 1/2	1 7/16	9 11/16	10 5/8	7/8	12	3/4	4
8	15	1 5/8	11 15/16	13	1	12	7/8	4 1/2
10	17 1/2	1 7/8	14 1/16	15 1/4	1 1/8	16	1	5 1/4
12	20 1/2	2	16 7/16	17 3/4	1 1/4	16	1 1/8	5 1/2
14	23	2 1/8	18 15/16	20 1/4	1 1/4	20	1 1/8	6
16	25 1/2	2 1/4	21 1/16	22 1/2	1 3/8	20	1 1/4	6 1/4
18	28	2 3/8	23 5/16	24 3/4	1 3/8	24	1 1/4	6 1/2
20	30 1/2	2 1/2	25 9/16	27	1 3/8	24	1 1/4	6 3/4
24	36	2 3/4	30 5/16	32	1 5/8	24	1 1/2	7 1/2
30	43	3	37 3/16	39 1/4	2	28	1 3/4	8 1/2

When flanges are integral with pumps or fittings, the bolt holes, which are in multiples of four, are drilled to straddle the center lines unless otherwise ordered.

When 250 lb cast iron flanges are bolted to bronze flanges, the 1/16-inch raised face should be removed and a full-face gasket used.

The 1/16-inch raised face on the 250 lb cast iron flanges is included in the dimension B for thickness of flange.

Spotfacing optionally available. (Minimum thickness shall include reduction for spotfacing when applicable.)

Characteristics of Piping Materials
Flanges

IVE-12 CAST IRON FLANGE DIMENSIONS AND DRILLING TEMPLATES ANSI B16.1-1975

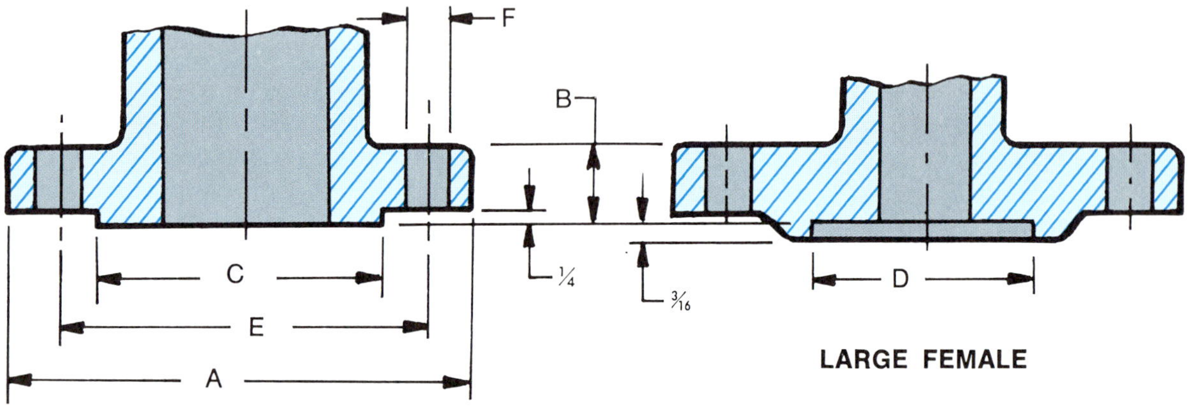

LARGE FEMALE

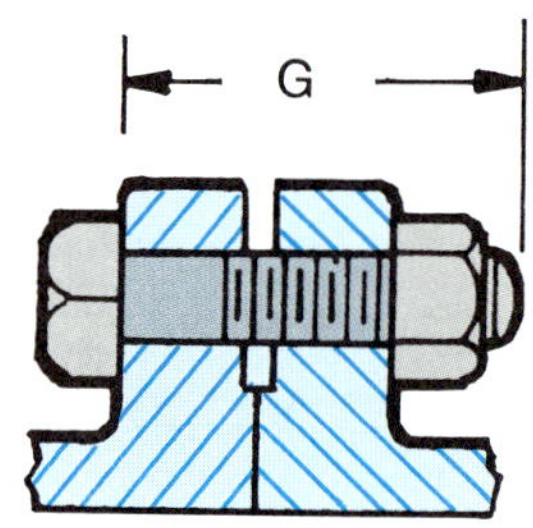

LENGTH OF MACHINE BOLT

800 LB CAST IRON FLANGES

Dimensions in Inches

Nominal Pipe Size	Outside Diameter A	Flange Thick-ness B (Min)	Raised Face O.D. Male C	Raised Face I.D. of Large Female D	Dia. of Bolt Circle E	Dia. of Bolt Holes F	Number of Bolts	Dia. of Bolts	Length of Machine Bolts G
2	6 1/2	1 1/2	3 5/8	3 11/16	5	3/4	8	5/8	4
2 1/2	7 1/2	1 5/8	4 1/8	4 3/16	5 7/8	7/8	8	3/4	4 1/2
3	8 1/4	1 3/4	5	5 1/16	6 5/8	7/8	8	3/4	4 3/4
3 1/2	9	1 7/8	5 1/2	5 9/16	7 1/4	1	8	7/8	5
4	10 3/4	2 1/8	6 3/16	6 1/4	8 1/2	1	8	7/8	5 1/2
5	13	2 3/8	7 5/16	7 3/8	10 1/2	1 1/8	8	1	6
6	14	2 1/2	8 1/2	8 9/16	11 1/2	1 1/8	12	1	6 1/2
8	16 1/2	2 3/4	10 5/8	10 11/16	13 3/4	1 1/4	12	1 1/8	7
10	20	3 1/8	12 3/4	12 13/16	17	1 3/8	16	1 1/4	7 1/2
12	22	3 1/4	15	15 1/16	19 1/4	1 3/8	20	1 1/4	8

When flanges are integral with pumps or fittings, the bolt holes, which are in multiples of four, are drilled to straddle the center lines unless otherwise specified.

The 1/4-inch raised face on the 800 lb cast iron flange is included in the dimension for flange thickness.

Male-female facings are optional with manufacturers.

IVE-13 ASSEMBLY OF STEEL RAISED FACE FLANGES TO CAST IRON, BRASS, BRONZE OR STAINLESS STEEL

The drilling templates of flanges are the same for Class 125 Lb Cast Iron (ANSI B16.1), 150 Lb Steel (ANSI B16.5), 150 Lb Stainless Steel (MSS SP-42) and 150 Lb Brass or Bronze (ANSI B16.24); likewise the templates for 250 Lb Cast Iron (ANSI B16.2), 300 Lb Brass or Bronze (ANSI B16.24) and 300 Lb Steel (ANSI B16.5) are also the same.

In the course of installing piping it is frequently necessary to make joints by bolting together flanges of two different materials.

Bolting a cast iron flange to a steel flange occurs most frequently in the field. Unless unusual care is taken by experienced workmen, there is the probability of breaking the relatively brittle cast iron flange when making such joints, especially if high strength bolts are used. This hazard is primarily due to the cantilever beam stresses resulting from the bolt load acting at a moment arm created by the gap between the O.D. of the raised face on the steel flange and the center line of the bolts.

Bolting brass, bronze or stainless steel flanges to steel flanges having raised faces may cause distortion of the weaker flange and leakage of the joint.

In order to prevent flange distortion or possible breakage, the following practices are recommended for bolting up steel raised face flanges to cast iron, brass, bronze or stainless steel flanges. Data is extracted from Manufacturers Standardization Society of the Valve and Fittings Industry Standard Practice SP-46.

1. Assembly of 125 Lb Cast Iron to 150 Lb Steel Flanges. See Fig. 1 and 2.

1.1 Where 125 Lb cast iron flanges are bolted to 150 Lb steel flanges, the $\frac{1}{16}$-inch raised face on the steel flange shall be removed. See Par. 1.2. When bolting such flanges together using a flat ring gasket extending to the inner edge of the bolts, the bolting shall be of carbon steel without heat treatment other than stress relief equivalent to ASTM A307, Grade B. When bolting such flanges together using a full face gasket, the bolting may be heat treated carbon steel (ASTM A261) or alloy steel (ASTM A193).

1.2 PRACTICE NOT RECOMMENDED. Joints made by using a steel flange with raised face are not recommended. See Par. 1.1.

2. Assembly of 250 Lb Cast Iron to 300 Lb Steel Flanges. See Fig. 3 and 4.

2.1 Where 250 Lb cast iron flanges are bolted to 300 Lb steel flanges, acceptable practice is to make no change in the raised face of either flange. When bolting such flanges together using a flat ring gasket extending to the inner edge of the bolts, the bolting shall be carbon steel without heat treatment other than stress relief equivalent to ASTM A307, Grade B. Preferred practice indicates that the raised face on the steel flange should be removed, but also in this case, bolting shall be of carbon steel without heat treatment other than stress relief equivalent to ASTM A307, Grade B.

3. Assembly of Brass or Bronze, or Stainless Steel Flanges (MSS SP-42) to 150 Lb and 300 Lb Steel Flanges.

3.1 Should occasion arise to make a joint with one of the above combinations, the $\frac{1}{16}$-inch raised face shall be removed from the steel flange and a full face gasket extending to the O.D. of the flange shall be used.

IVE-13 ASSEMBLY OF STEEL RAISED FACE FLANGES TO CAST IRON, BRASS, BRONZE OR STAINLESS STEEL (CONTINUED)

ASSEMBLY OF 125 LB. CAST IRON FLANGES TO 150 LB. STEEL FLANGES

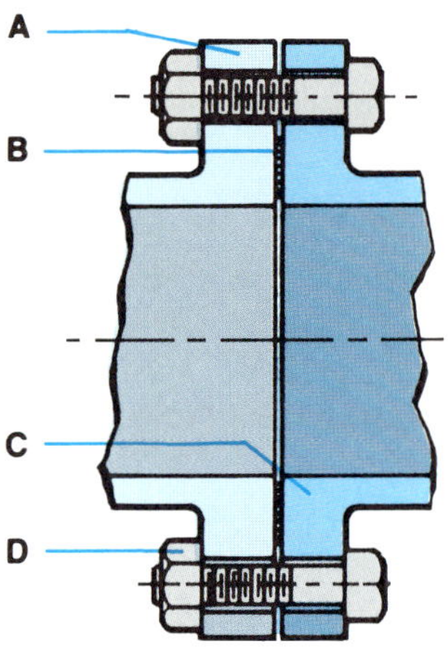

A 125 LB. CAST IRON FLANGE

B FLAT RING GASKET EXTENDING TO THE INNER EDGE OF THE BOLTS

C 150 LB. STEEL FLANGE WITH $\frac{1}{16}$ IN. RAISED FACE REMOVED

D CARBON STEEL BOLTING ONLY ASTM A307 GRADE B OR EQUIVALENT

FIG. 1

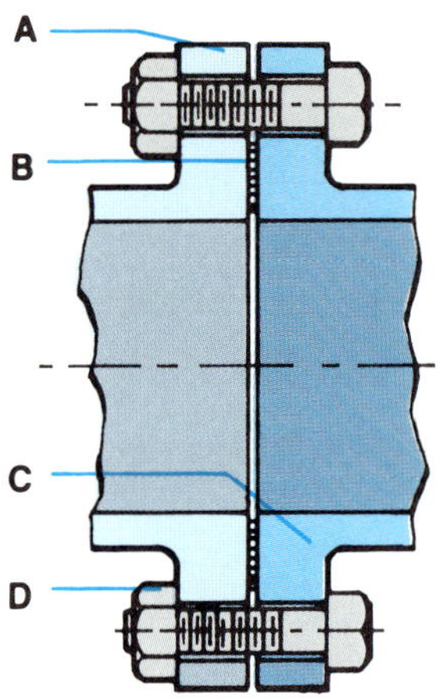

A 125 LB. CAST IRON FLANGE

B FULL FACE GASKET EXTENDING TO OD OF FLANGE

C 150 LB. STEEL FLANGE WITH $\frac{1}{16}$ IN. RAISED FACE REMOVED

D CARBON OR ALLOY STEEL BOLTING MAY BE USED

FIG. 2

ASSEMBLY OF 250 LB. CAST IRON FLANGES TO 300 LB. STEEL FLANGES

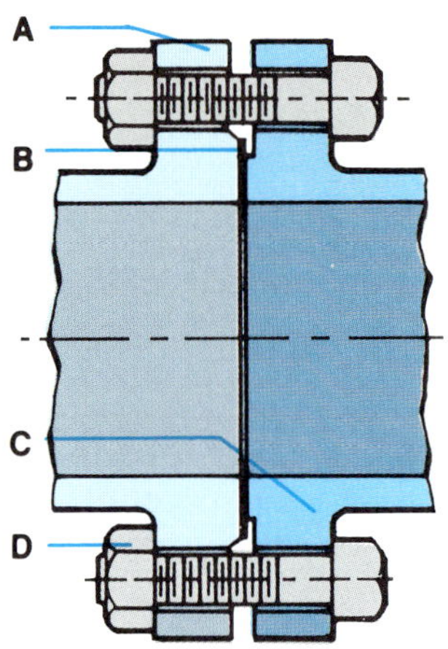

A 250 LB. CAST IRON FLANGE WITH $\frac{1}{16}$ IN RAISED FACE

B FLAT RING GASKET EXTENDING TO THE INNER EDGE OF THE BOLTS

C 300 LB. STEEL FLANGE WITH $\frac{1}{16}$ IN. RAISED FACE NOT REMOVED

D CARBON STEEL BOLTING ONLY ASTM A307 GRADE B OR EQUIVALENT

FIG. 3

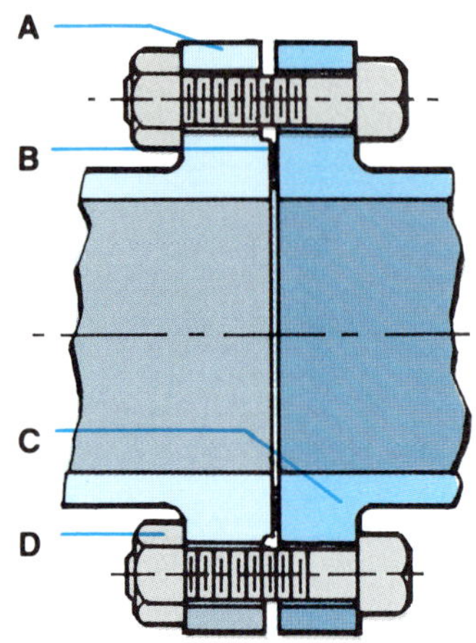

A 250 LB. CAST IRON FLANGE WITH $\frac{1}{16}$ IN. RAISED FACE

B FLAT RING GASKET EXTENDING TO THE INNER EDGE OF THE BOLTS

C 300 LB. STEEL FLANGE WITH $\frac{1}{16}$ IN. RAISED FACE REMOVED

D CARBON STEEL BOLTING ONLY ASTM A307 GRADE B OR EQUIVALENT

FIG. 4

BOLTING PRACTICE FOR FLANGES OF DISSIMILAR METAL

Flange Combination	Facing	Gasket	Bolting Used
125 Lb C.I. to 150 Lb Steel	$\frac{1}{16}$-in. raised face on the steel flange shall be removed.	Flat Ring Type	Carbon Steel without heat treatment other than stress relief, equivalent to ASTM A307, Grade B.
See Fig. 1 & 2		Full Face Type	May be heat treated carbon steel, equivalent to ASTM A261 or Alloy Steel equivalent to A193.
250 Lb C.I. to 300 Lb Steel See Fig. 3 & 4	No change necessary in raised face of either flange. However, preferred practice indicates that raised face on steel flange should be moved.	Flat Ring Type	Carbon Steel, without heat treatment other than stress relief, equivalent to ASTM A307, Grade B.
Brass, Bronze, or Stainless Steel (MSS SP-42) to Steel	$\frac{1}{16}$-in. raised face on the steel flange shall be removed.	Full Face Type *Only*.	Carbon Steel, without heat treatment other than stress relief, equivalent to ASTM A307, Grade B, or heat treated carbon steel, equivalent to ASTM A261 or Alloy Steel equivalent to A193.

Reprinted with permission from the Walworth Aloyco, Condensed Catalog 68, Walworth Aloyco Div., Walworth Co.

Section V

Useful Information

VA-1 CONVERSION TABLE—TEMPERATURE

$$\text{Degrees Celsius} = \frac{(\text{Degrees Fahrenheit} - 32) \times 5}{9}$$

$$\text{Degrees Fahrenheit} = \frac{(\text{Degrees Celsius} \times 9)}{5} + 32$$

Degrees Kelvin (K) = Degrees Celsius + 273.15

Degrees Rankine (R) = Degrees Fahrenheit + 459.69

(0 degrees K or R = absolute zero)

Common equivalents: 0°C = 32°F
20°C = 68°F
25°C = 77°F
100°C = 212°F

VA-2 ACCELERATION OF GRAVITY (AT SEA LEVEL) (g)

Latitude, deg.	0	10	20	30	40	45	50	60	70	80	90
g, ft/sec^2	32.088	32.093	32.108	32.130	32.158	32.174	32.187	32.215	32.238	32.253	32.258
g, m/s^2	9.780	9.782	9.786	9.793	9.802	9.807	9.811	9.819	9.826	9.831	9.832

Note: Correction for altitude above sea level:

−0.003 m/s^2 per 1000 m

−0.003 ft/sec^2 per 1000 ft

VA-3 CONVERSION TABLE—PHYSICAL DIMENSIONS

TABLE A—PHYSICAL DIMENSIONS—LENGTH

U.S. CUSTOMARY UNITS						SI UNITS	OTHER METRIC UNITS	
inch (in.)	foot (ft)	yard (yd)	rod	mile (mi)	naut. mile	meter (m)	centimeter (cm)	kilometer (km)
1	0.0833	0.02778	—	—	—	0.0254	2.54	—
12	1	0.3333	—	—	—	0.3048	30.48	—
36	3	1	0.1818	.000568	.000494	0.9144	91.44	—
198	16.5	5.500	1	.003125	.002716	5.0292	502.9	.00503
63 360	5280	1760	320	1	0.8690	1609	160 935	1.609
72 913	6076.1	2025.4	368.25	1.1508	1	1852	185 200	1.852
39.37	3.281	1.0936	—	—	—	1	100	0.001
0.3937	0.03281	0.01094	—	—	—	0.01	1	0.00001
39 370	3281	1093.6	198.84	0.6214	.5400	1000	100 000	1

TABLE B—PHYSICAL DIMENSIONS—AREA

U.S. CUSTOMARY UNITS						S.I. UNIT	OTHER METRIC UNITS			
square inch ($in.^2$)	square foot (ft^2)	square yard (yd^2)	acre	square mile (mi^2)	circular mil (cir mil)	square meter (m^2)	square millimeter (mm^2)	square centimeter (cm^2)	square hectometer (hectare) (hm^2)	square kilometer (km^2)
1	0.00694	0.0007711	0.1594×10^{-6}	2.4910×10^{-10}	1.2732×10^{6}	6.4517×10^{-4}	645.17	6.4517	6.4517×10^{-8}	6.4517×10^{-10}
144	1	0.11111	2.296×10^{-5}	3.5871×10^{-8}	—	0.0929	92 900	929	0.0929×10^{-4}	0.0929×10^{-6}
1296	9	1	0.000207	0.3228×10^{-6}	—	0.8361	0.8361×10^{6}	0.8361×10^{4}	0.8361×10^{-4}	0.8361×10^{-6}
—	43 560	4840	1	0.001563	—	4.047×10^{3}	—	—	0.4047	4.047×10^{-3}
40.1449×10^{8}	27.878×10^{6}	3.0976×10^{6}	640	1	—	2.5889×10^{6}	—	—	258.89	2.5889
0.7854×10^{-6}	—	—	—	5.1112×10^{-15}	1	0.5067×10^{-9}	0.5067×10^{-3}	0.5067×10^{-5}	—	—
1550	10.76	1.196	—	—	—	1	1×10^{6}	1×10^{4}	1×10^{-4}	—
0.001550	—	—	—	—	1973.6	1×10^{-6}	1	1×10^{-2}	—	—
0.1550	0.001076	—	—	—	197 360	1×10^{-4}	100	1	—	—
—	107 600	11 960	2.471	0.3861×10^{-2}	—	10 000	—	—	1	0.010
—	10.76×10^{6}	1.196×10^{6}	247.1	C.38614	—	1×10^{6}	—	—	1×10^{2}	1

VA-3 CONVERSION TABLE—PHYSICAL DIMENSIONS

TABLE C—PHYSICAL DIMENSIONS—VOLUME

U.S. CUSTOMARY UNITS					S.I. UNIT	OTHER METRIC UNIT
cubic inch (in.^{3})	US gallon (US gal)	Imp. gallon (Imp gal)	cubic foot (ft^3)	barrel* (bbl)	cubic meter (m^3)	liter (l)
1	0.00433	0.00360	0.000579	0.000103	0.00001639	0.01639
231	1	0.8327	0.1337	0.0238	0.003785	3.785
277.42	1.2009	1	0.1606	0.0286	0.004546	4.546
1728	7.481	6.229	1	0.1780	0.02832	28.32
9702	42	34.97	5.615	1	0.15897	158.97
61 024	264.17	220	35.31	6.2898	1	1000
61.024	0.2642	0.2198	0.0353	0.00630	0.001	1

* By trade custom, one barrel petroleum oil is equal to 42 US gal.

VA-4 CONVERSION TABLE—VELOCITY

U.S. CUSTOMARY UNITS			S.I. UNIT	OTHER METRIC UNITS		
feet per min (fpm)	feet per sec (fps)	miles per hr (mph)	meters per sec (m/s)	centimeters per sec (cm/sec)	meters per min (m/min)	kilometers per hr (km/hr)
1	0.01667	0.01136	0.00508	0.5080	0.3048	0.01829
60	1	0.6818	0.3048	30.48	18.29	1.097
88	1.467	1	0.447	44.70	26.82	1.609
196.85	3.281	2.237	1	100	60.0	3.60
1.9685	0.03281	0.02237	0.010	1	0.6	0.036
3.281	0.05468	0.03728	0.01667	1.667	1	0.06
54.68	0.9113	0.6214	0.2778	27.78	16.67	1

NOTE: 1 knot = 1.15155 mph or 1.8532 km/hr.

Angular Velocity

1 revolution per minute (rpm) = 0.1047 radians per second
1 radian per second = 9.55 revolutions per minute

Useful Information
Conversion Tables

VA-5 CONVERSION TABLE—CAPACITY

U.S. CUSTOMARY UNITS					S.I. UNIT	OTHER METRIC UNITS	
Millions of US gallons per day (MGD)	US gallons per min (gpm)	Imp. gallons per min (Igpm)	cubic feet per sec (cfs)	*barrels per hr (bph)	cubic meters per sec (m^3/s)	liters per min (l/min)	cubic meters per hr (m^3/hr)
1	694.4	578	1.547	992	0.0438	2628	157.72
0.00144	1	0.8327	0.002228	1.4286	63.08×10^{-6}	3.785	0.2271
0.00173	1.2009	1	0.00268	1.7156	75.77×10^{-6}	4.5454	0.2728
0.64636	448.86	373.8	1	641.23	2.832×10^{-2}	1699	101.952
0.00100	0.7000	0.5829	0.00156	1	44.166×10^{-6}	2.6495	0.1590
22.824	15 852	13 188	35.316	22 643	1	60 000	3 600
0.000380	0.2642	0.2198	0.000586	0.3774	16.67×10^{-6}	1	0.0600
0.00634	4.4028	3.666	0.00981	6.2898	2.777×10^{-4}	16.667	1

* By trade custom, one barrel petroleum oil is equal to 42 US gal.

$$\text{gpm} = \frac{0.1247}{w} \times \text{lb/hr when } w = \text{density, lb/cu ft}$$

VA-6 CONVERSION TABLE—MASS, WEIGHT AND FORCE

Definitions:

1. Mass is *absolute*, the *pound* (lb) and *kilogram* (kg) are most commonly used units, the kilogram is the SI unit.

U.S. CUSTOMARY UNITS (avoirdupois)					S.I. UNIT	OTHER METRIC UNITS	
grain (gr)	ounce (oz)	pound (lb)	short ton	long ton	kilogram (kg)	gram (g)	metric ton
1	0.002286	—	—	—	—	0.0648	—
437.5	1	0.0625	—	—	0.02835	28.35	—
7 000	16	1	—	—	0.4536	453.6	—
—	—	2 000	1	.8929	907.2	—	0.9072
—	—	2 240	1.12	1	1016	—	1.016
15 432	35.27	2.205	—	—	1	1 000	0.001
15.432	0.03527	0.002205	—	—	0.001	1	—
—	—	2205	1.102	0.9842	1000	1 000 000	1

VA-6 CONVERSION TABLE—MASS, WEIGHT AND FORCE (CONTINUED)

Definitions:

2. *Weight* is a force and is dependent on the acceleration of gravity, and therefore varies with latitude and altitude. The *pound* (lb) and *kilogram* (kg) have been commonly used units, but properly should be termed *pound force* and *kilogram force*. The weight of a body having a mass of 1 pound will be 1 pound force at sea level and 45° latitude, but will be different at other locations where the acceleration of gravity is different than 32.174 ft/sec^2. For instance, at the equator (0° latitude) and sea level, where the acceleration of gravity is 32.088, the weight of a body having a mass of 1 pound will be

$$\frac{1 \times 32.088}{32.174} = 0.9974 \text{ lb force}$$

Or, the weight of a body having a mass of 1 kilogram will be, at sea level,

$$\frac{1 \times 9.780}{9.807} = 0.997 \text{ kg force}$$

3. The *poundal* is the force which, applied to a mass of 1 pound, gives it an acceleration of 1 ft/sec^2; 1 poundal $= \frac{1}{32.174}$ lb

The *slug* is the mass which, if acted on by a force of 1 pound, will accelerate at a rate of 1 ft/sec^2; 1 slug = 32.174 lb

The *dyne* is the force which, applied to a mass of 1 gram (g) gives it an acceleration of 1 cm/sec^2; 1 dyne $= \frac{1}{980.7}$ gram force $= 1.02 \times 10^{-6}$ kg force

The *newton* is the force which, if applied to a mass of 1 kilogram (kg) gives it an acceleration of 1 m/s;

$$1 \text{ newton (N)} = \frac{1 \text{ kg force}}{9.807} = 1.02 \times 10^{-1} \text{ kg force} = 1 \times 10^5 \text{ dynes}$$

Note: Use of the unit "kilogram force" should be avoided.

VA-7 CONVERSION TABLE—DENSITY

U.S. CUSTOMARY UNITS			S.I. UNIT	OTHER METRIC UNIT
pounds per US gallon (lb/US gal)	**pounds per cubic foot (lb/ft^3)**	**pounds per cubic inch (lb/in.^{3})**	**kilograms per cubic meter (kg/m^3)**	**grams per milliliter (g/ml)**
1	7.481	0.004329	119.8	0.1198
0.1337	1	0.0005787	16.02	0.01602
231	1728	1	27 680	27.68
0.008345	0.06243	3.613×10^{-5}	1	0.001
8.345	62.43	0.03613	1 000	1

Note: Density expressed in grams per milliliter = specific gravity = grams per cubic centimeter.

VA-8 CONVERSION TABLE—PRESSURE OR HEAD

U.S. CUSTOMARY UNITS				S.I. UNIT	OTHER METRIC UNITS	
atmospheres	pounds per square inch (psi)	feet of water (60° F) in	inch of mercury (60° F)	newtons per square meter	kilopascal	bar
1	14.096	33.93	30.005	101326	101.31	1.0132
0.06885	1	2.389	2.042	6895	6.895	0.06895
0.02947	0.4331	1	0.8844	2987	2.987	0.02987
0.03333	0.4898	131	1	3376	3.376	0.0376
0.000009869	0.00145	0.000335	0.000296	1	0.001	0.00001
0.009869	0.145	0.335	0.295	1000	1	0.01
0.9869	14.50	33.5	29.5	100000	100	1

VA-9 CONVERSION TABLE—ENERGY, WORK, HEAT

U.S. CUSTOMARY UNITS			S.I. UNIT	OTHER METRIC UNITS	
foot pounds (ft-lb)	British thermal units (Btu)	horsepower-hrs (hp-hr)	joules (J)	calories (cal)	kilowatt-hrs (kWh)
1	0.001285	—	1.356	0.3238	—
778.2	1	.0003930	1055	252.0	0.0002931
—	2544	1	—	641 200	0.7457
0.7376	0.0009478	—	1	0.2388	—
3.088	0.003968	—	4.187	1	—
—	3412	1.341	—	859 900	1

VA-10 CONVERSION TABLE—POWER

U.S. CUSTOMARY UNITS			S.I. UNIT	OTHER METRIC UNITS	
foot-pounds per sec. (ft-lb/sec)	horsepower (hp)	British thermal units per sec (Btu/sec)	watts (W)	kilowatts (kW)	metric horsepower
1	0.00182	0.001285	1.356	0.001356	0.00184
550	1	0.7068	7.457×10^2	0.7457	1.014
778.2	1.415	1	1.055×10^3	1.055	1.434
0.7376	1.341×10^{-3}	9.478×10^{-4}	1	0.001	0.00136
737.6	1.341	0.9478	1000	1	1.360
542.5	0.9863	0.6971	7.355×10^2	0.7355	1

NOTE: For Viscosity Conversions see Tables IIC-3 & IIC-4.

VA-11 CUSTOMARY UNITS TO SI UNITS

VARIABLE	U.S. CUSTOMARY	MULTIPY BY	TO OBTAIN SI UNITS
Length	ft	0.3048	m
Area	ft^2	0.0929	m^2
Volume	ft^3	0.0283	m^3
Time	S	1	S
Angular Velocity Frequency	S^{-1}	1	S^{-1}
Velocity	$\frac{ft}{S}$	0.3048	$\frac{m}{S}$
Kinematic Viscosity	$\frac{ft^2}{S}$	0.0929	$\frac{m^2}{S}$
Volume Flow Rate	$\frac{ft^3}{S}$	0.0283	$\frac{m^3}{S}$
Acceleration	$\frac{ft}{S^2}$	0.3048	$\frac{m}{S^2}$
Density	$\frac{slug}{ft^2}$	515.54	$\frac{kg}{m^3}$
Moment of Inertia	slug-ft^2	1.335	kg - m^2
Dynamic Viscosity	$\frac{slug}{ft\text{-}S}$	47.86	$\frac{kg}{m - S}$
Mass Flow Rate	$\frac{slug}{S}$	14.59	$\frac{kg}{S}$
Moment Impulse	1bf - S	4.449	N - S
Specific Weight	$\frac{1bf}{ft^3}$	157.236	$\frac{N}{m^3}$
Pressure, Stress Modules of Elasticity	$\frac{1bf}{ft^2}$	47.89	$\frac{N}{m^2}$
Force	1bf	4.449	N
Energy, Work, Torque	1bf - ft	1.356	J
Power	$\frac{1bf - ft}{S}$	1.356	WATT

VA-12 CRITICAL TEMPERATURES AND PRESSURES

Compiled by Rudolf Loebel

ORGANIC COMPOUNDS

Formula	Name	Critical temp. T_c °C	Critical press. P_c atm.
$CHClF_2$	Methane, monochlorodifluoro-	96	48.5
$CHCl_2F$	Methane, dichloromonofluoro-	178.5	51
$CHCl_3$	Methane, trichloro- (Chloroform)	263	54
CHF_3	Methane, trifluoro- (Fluoroform)	25.9	46.9
CH_2Cl_2	Methylene chloride	237	60
CH_3NO_2	Methane, nitro-	314.8	62.3
CH_3Br	Methane, monobromide-	194	83.4
CH_3Cl	Methane, monochloro-	143.8	65.9
CH_3F	Methane, monofluoro-	44.6	58
CH_3I	Methane, monoiodo-	254.8	72.7
CH_4	Methane	−82.1	45.8
CH_4O	Methanol (Methyl alcohol)	240	78.5
CH_4S	Methylmercaptan	196.8	71.4
CH_5N	Methylamine	156.9	40.2
$CBrF_3$	Methane, monobromotrifluoro-	67	50.3
$CClF_3$	Methane, monochlorotrifluoro-	28.85	38.2
CCl_2F_2	Methane, dichlorodifluoro-	111.5	39.6
CCl_3F	Methane, trichloromonofluoro-	198	43.2
CCl_4	Methane, tetrachloro- (Carbon tetrachloride)	283.1	45
CF_4	Methane, tetrafluoro-	−45.7	41.4
C_2H_2	Acetylene	35.5	61.6
C_2H_2	Ethyne (see Acetylene)	35.5	61.6
$C_2H_2F_2$	Ethylene, 1,1-difluoro-	30.1	—
C_2H_3N	Acetonitrile	274.7	47.7
$C_2H_3F_3$	Ethane, 1,1,1-trifluoro-	73.1	—
C_2H_4	Ethene	9.9	50.5
C_2H_4	Ethylene (see Ethene)	9.9	50.5
C_2H_4O	Acetaldehyde	187.8	54.7
C_2H_4O	Ethylene oxide	195.8	71
$C_2H_4O_2$	Acetic acid	321.6	57.1
$C_2H_4O_2$	Formic acid, methyl- (Methyl formate)	214	59.2
$C_2H_4Cl_2$	Ethane, 1,1-dichloro-	249.8	50
$C_2H_4F_2$	Ethane, 1,1-difluoro-	386.7	—
C_2H_5Br	Ethane, monobromo-	230.8	61.5
C_2H_5Cl	Ethane, monochloro-	187.2	52
C_2H_5F	Ethane, monofluoro-	102.16	49.6
C_2H_6	Ethane	32.2	48.2
C_2H_6O	Ether, dimethyl-	127	52.6
C_2H_6O	Ethanol (Ethyl alcohol)	243	63
C_2H_6O	Glycol, ethylene-	(374)*	—
C_2H_6S	Dimethylsulfide	229.9	54.5
C_2H_6S	Ethylmercaptan	225.5	54.2
C_2H_7N	Dimethylamine	164.6	52.4
C_2H_7N	Ethylamine	183.2	55.5
$C_2Br_2F_4$	Ethane, dibromotetrafluoro-	214.5	—
C_2ClF_5	Ethane, chloropentafluoro-	80	—
$C_2Cl_2F_4$	Ethane, 1,2-dichlorotetrafluoro-	145.7	—
$C_2Cl_3F_3$	Ethane, trichlorotrifluoro-	214.2	33.7
$C_2Cl_4F_2$	Ethane, tetrachlorodifluoro-	278	32.9
C_2F_6	Ethane, hexafluoro-	24.3	—
C_3H_4	Propadiene	120	43.6
C_3H_4	Allene (see Propadiene)	120	43.6
C_3H_4	Acetylene, methyl-	127.8	52.8
C_3H_4	Propyne (see Acetylene, methyl-)	127.8	52.8
C_3H_5N	Ethyl cyanide	290.8	41.3
C_3H_5N	Propionitrile (see Ethyl cyanide)	290.8	41.3
C_3H_5OCl	Epichlorohydrin	(323)	—
C_3H_5Cl	Propene, 3-chloro- (allyl-chloride)	240.3	46.5
C_3H_6	Propylene	91.9	45.4
C_3H_6	Cyclopropane	124.7	—
C_3H_6O	Acetone	235.5	47
C_3H_6O	Allylalcohol	272	55.5
C_3H_6O	Propylene oxide	209	48.6
$C_3H_6O_2$	Formic acid, ethyl- (Ethyl formate)	235.3	46.3
$C_3H_6O_2$	Acetic acid, methyl- (Methyl acetate)	233.7	46.3
$C_3H_6O_2$	Propanoic acid	337.6	53
C_3H_7Cl	n-Propane, monochloro-	230	45.2
C_3H_8	Propane	96.8	42
C_3H_8O	Ether, ethyl methyl- (Methoxyethane)	164.7	43.4
C_3H_8O	Glycol, 1,2-propylene-	351	—
C_3H_8O	Isopropyl alcohol	235	47
C_3H_8O	n-Propyl alcohol	263.6	51
$C_3H_8O_3$	Glycerol	(452)	—
C_3H_9N	Isopropylamine	209.7	—
C_3H_9N	n-Propylamine	223.8	46.8
	*( ) uncertain		
C_4H_4O	Furan	213.8	52.5
C_4H_4S	Thiophene	307	56.2
C_4H_6	1,2-Butadiene	171	44.4
C_4H_6	1,3-Butadiene	152	42.7
C_4H_8	n-Butene	146	39.7
C_4H_8	2-Butene, cis-	160	40.5
C_4H_8	2-Butene, trans-	155	41.5
C_4H_8O	1,2-Butylene oxide	243	—
C_4H_8O	Ketone, ethyl methyl- (2-Butanone)	262	41
$C_4H_8O_2$	Butanoic acid	355	52
$C_4H_8O_2$	p-Dioxane	314.8	51.4
$C_4H_8O_2$	Acetic acid, ethyl- (Ethyl acetate)	250.4	37.8
$C_4H_8O_2$	Propanoic acid, methyl- (Methyl propionate)	257.4	39.3
$C_4H_8O_2$	Formic acid, propyl-(Propyl formate)	264.9	40.1
C_4H_9Cl	n-Butane, monochloride-	269	—
C_4H_{10}	n-Butane	152	37.5
C_4H_{10}	Isobutane	134.7	35.9
$C_4H_{10}O$	Butanol (n-Butyl alcohol)	289.8	43.6
$C_4H_{10}O$	sec-Butyl alcohol	263	41.4
$C_4H_{10}O$	tert-Butyl alcohol	235	39.2
$C_4H_{10}C$	Ether, diethyl-	192.6	35.6
$C_4H_{10}O$	Isobutyl alcohol	277	42.4
$C_4H_{10}O_3$	Glycol, diethylene-	407	—
$C_4H_{10}S$	Diethyl sulfide	283.8	39.1
$C_4H_{11}N$	Butyl amine	287.9	—
$C_4H_{11}N$	Diethyl amine	223.3	36.6
$C_4H_{11}N$	Isobutyl amine	266.7	—
C_4F_{10}	Butane, perfluoro-	113.2	23
C_5H_5N	Pyridine	346.8	—
C_5H_8	Cyclopentene	232.94	47.2
C_5H_{10}	Cyclopentane	238.6	44.6
C_5H_{10}	1-Pentene	191	39.9
$C_5H_{10}O$	Ketone, diethyl-	287.8	36.9
$C_5H_{10}O$	Propanoic acid, ethyl- (Ethyl propionate)	272.9	33
$C_5H_{10}O_2$	n-Butanoic acid, methyl- (n-Methyl butyrate)	281.3	34.3
$C_5H_{10}O_2$	Acetic acid, n-propyl (n-Propyl acetate)	276	32.9
$C_5H_{10}O_2$	n-Valeric acid	378	37.6
$C_5H_{11}N$	Piperidine	320.8	44.1
C_5H_{12}	Butane, 2-methyl- (See Isopentane)	187.8	32.9
C_5H_{12}	Isopentane	187.8	32.9
C_5H_{12}	Neopentane	160.6	31.6
C_5H_{12}	n-Pentane	196.6	33.3

Reprinted from the CRC Handbook of Chemistry and Physics, 1987.

VA-12 CRITICAL TEMPERATURES AND PRESSURES (Continued)

ORGANIC COMPOUNDS

Formula	Name	Critical temp. T_c °C	Critical press. P_c atm.
C_5H_{12}	Propane, 2,2-dimethyl (see Neopentane)	160.6	31.6
$C_5H_{12}O$	Isoamyl alcohol	309.77	—
C_6H_5Br	Benzene, bromo-	397	44.6
C_6H_5Cl	Benzene, chloro-	359.2	44.6
C_6H_5F	Benzene, fluoro-	286.95	44.6
C_6H_6	Benzene	288.9	48.6
C_6H_6O	Phenol	421.1	60.5
C_6H_7N	Aniline	425.6	52.3
C_6H_7N	α-Picoline	348	—
C_6H_7N	β-Picoline	371.7	—
C_6H_7N	γ-Picoline	372.5	—
C_6H_{10}	Cyclohexene	287.3	—
C_6H_{10}	1,5-Hexadiene	234.4	32.6
C_6H_{12}	Cyclohexane	280.4	40
C_6H_{12}	Cyclopentane, methyl-	259.5	37.4
C_6H_{12}	1-Hexene	231	31.1
$C_6H_{12}O_2$	Formic acid, n-amyl (n-Amyl formate)	302.6	34.1
$C_6H_{12}O_2$	Acetic acid, n-butyl- (n-Butyl acetate)	305.9	30.7
$C_6H_{12}O_2$	Butanoic acid, ethyl- (Ethyl butyrate)	293	30.2
$C_6H_{12}O_2$	Propanoic acid, propyl- (Propyl propionate)	304.8	30.7
$C_6H_{12}O_3$	Paraldehyde	290	—
C_6H_{14}	Butane, 2,3-dimethyl-	226.8	30.9
C_6H_{14}	n-Hexane	234.2	29.9
C_6H_{14}	Pentane, 2-methyl-	224.3	30
$C_6H_{14}O$	Ether, isopropyl-	226.9	28.4
$C_6H_{14}O$	1-Hexyl alcohol	313.5	—
$C_6H_{14}O_4$	Glycol, triethylene-	(437)	—
$C_6H_{15}N$	Dipropyl amine	277	31
$C_6H_{15}N$	Hexyl amine	318.8	—
$C_6H_{15}N$	Triethyl amine	258.9	30
C_7H_5N	Benzonitrile	426.2	41.6
C_7H_8	Benzene, methyl- (see Toluene)	320.8	41.6
C_7H_8	Toluene	320.8	41.6
C_7H_8O	Anisole, (see Benzene, methoxy-)	368.5	41.3
C_7H_8O	Benzene, methoxy-	368.5	41.3
C_7H_8O	o-Cresol	424.4	49.4
C_7H_8O	m-Cresol	432	45
C_7H_8O	p-Cresol	431.4	50.8
C_7H_9N	Aniline, methyl-	428.4	51.3
C_7H_9N	2,3-Lutidine	382.3	—
C_7H_9N	2,4-Lutidine	374	—
C_7H_9N	2,6-Lutidine	350.6	—
C_7H_9N	3,4-Lutidine	410.6	—
C_7H_9N	3,5-Lutidine	394.1	—
C_7H_{14}	Cyclohexane, methyl-	299.1	34.3
C_7H_{14}	Cyclopentane, ethyl-	296.3	33.5
C_7H_{14}	1-Heptene	264.1	—
$C_7H_{14}O_2$	Acetic acid, isoamyl- (Isoamyl acetate)	326.1	28
C_7H_{16}	Butane, 2,2,3-trimethyl-	258.3	29.8
C_7H_{16}	n-Heptane	267.1	27
C_7H_{16}	Hexane, 2-methyl-	257.9	27.2
C_7H_{16}	Pentane, 3-ethyl-	267.6	28.6
C_7H_{16}	Pentane, 2,4-dimethyl-	247.1	27.4
$C_7H_{16}O$	1-Heptyl alcohol	365.3	29.4
C_7F_{16}	n-Heptane, perfluoro-	201.7	16
C_8H_8	Styrene	374.4	39.4
C_8H_{10}	Benzene, ethyl-	343.9	36.9
C_8H_{10}	o-Xylene	359	35.7
C_8H_{10}	m-Xylene	346	34.7
C_8H_{10}	p-Xylene	345	33.9
$C_8H_{10}O$	Xylenol	449.7	56.4
$C_8H_{10}O$	Phenetol (see Benzene, ethoxy-)	374	33.8
$C_8H_{10}O$	Benzene, ethoxy-	374	33.8
$C_8H_{11}N$	Aniline, N,N-dimethyl-	414.4	35.8
$C_8H_{11}N$	Aniline, N-ethyl-	425.4	—
C_8H_{16}	n-Octene	305	25.5
C_8H_{18}	Butane, 2,2,3,3-tetramethyl-	270.8	24.5
C_8H_{18}	Heptane, 4-methyl-	290	25.6
C_8H_{18}	Hexane, 2,4-dimethyl-	282	25.8
C_8H_{18}	n-Octane	296	24.8
C_8H_{18}	Pentane, 2,2,3-trimethyl-	294	28.2
$C_8H_{18}O$	1-Octyl alcohol	385.5	26.5
$C_8H_{19}N$	Dibutyl amine	322.6	—
C_9H_7N	Quinoline	508.8	—
C_9H_{12}	Benzene, n-propyl-	365	31.2
C_9H_{12}	Benzene, 1,2,3-trimethyl-	391.3	31
C_9H_{12}	Cumene (see Benzene, isopropyl-)	362.7	31.2
C_9H_{12}	Benzene, isopropyl-	362.7	31.2
C_9H_{20}	n-Nonane	321	22.5
C_9H_{21}	Tripropyl amine	304.3	—
$C_{10}H_8$	Naphthalene	474.8	40.6
$C_{10}H_{12}O$	Ether, diphenyl-	494	30.9
$C_{10}H_{14}$	Benzene, 1-isopropyl-4-methyl-	385.5	27.7
$C_{10}H_{14}$	Benzene, 1,2,3,5-tetramethyl-	402.8	28.6
$C_{10}H_{14}$	p-Cymene, (see Benzene, 1-isopropyl-4-methyl-)	—	—
$C_{10}H_{14}$	Isodurene, (see Benzene, 1,2,3,5-tetramethyl-)	—	—
$C_{10}H_{14}O$	3-p-Cymenol	425.1	33
$C_{10}H_{14}O$	Thymol, (see 3-p-Cymenol)	425.1	33
$C_{10}H_{18}$	Decalin, cis-	418	28.7
$C_{10}H_{18}$	Decalin, trans-	408	28.7
$C_{10}H_{22}$	n-Decane	344.4	20.8
$C_{11}H_{10}$	Naphthalene, 1-methyl-	498.8	32.1
$C_{12}H_{10}$	Biphenyl	495	31.8
$C_{12}H_{11}N$	Diphenylamine	615.5	—
$C_{12}H_{18}$	Benzene, hexamethyl-	494	23.5
$C_{12}H_{18}$	Mellitine, (see Benzene, hexamethyl-)	—	—
$C_{12}H_{26}$	n-Dodecane	386	17.9
$C_{12}H_{27}N$	Tributyl amine	365.2	—

Reprinted from the CRC Handbook of Chemistry and Physics, 1987.

VA-12 CRITICAL TEMPERATURES AND PRESSURES (Continued)
INORGANIC COMPOUNDS

Name	Formula	Critical temp. T_c °C	Critical press. P_c atm.
Ammonia	NH_3	132.5	112.5
Argon	Ar	−122.3	48
Boron tribromide	BBr_3	300	—
Boron trichloride	BCl_3	178.8	38.2
Boron trifluoride	BF_3	−12.26	49.2
Carbon dioxide	CO_2	31	72.9
Carbon disulfide	CS_2	279	78
Carbon monoxide	CO	−140	34.5
Carbonyl sulfide	COS	104.8	65
Chlorine	Cl_2	144	76.1
Cyanogen	C_2N_2	126.6	—
Deuterium	D_2	−234.8	16.4
Fluorine	F_2	−129	55
Germanium tetrachloride	$GeCl_4$	276.9	38
Helium	He	−267.9	2.26
Hydrogen	H_2	−239.9	12.8
Hydrogen bromide	HBr	90	84.5
Hydrogen chloride	HCl	51.4	82.1
Hydrogen deuteride	HD	237.3	14.6
Hydrogen cyanide	HCN	183.5	48.9
Hydrogen fluoride	HF	188	64

Name	Formula	Critical temp. T_c °C	Critical press. P_c atm.
Hydrogen iodide	HI	150	81.9
Hydrogen sulfide	H_2S	100.4	88.9
Hydrazine	N_2H_2	380	—
Iodine	I_2	512	116
Krypton	Kr	−63.8	54.3
Neon	Ne	−228.7	26.9
Nitric oxide	NO	−93	64
Nitrogen dioxide	NO_2	157.8	100
Nitrogen	N_2	−147	33.5
Nitrous oxide	N_2O	36.5	71.7
Oxygen	O_2	−118.4	50.1
Ozone	O_3	−5.16	67
Phosphine	PH_3	51.3	64.5
Radon	Rn	104.04	62
Silane, chlorotrifluoro-	$SiClF_3$	34.5	34.2
Silane	SiH_4	−3.46	47.8
Silicon tetrachloride	$SiCl_4$	232.8	—
Silicon tetrafluoride	SiF_4	−14.06	36.7
Sulfur dioxide	SO_2	157.8	77.7
Stannic chloride	$SnCl_4$	318.7	37
Water	H_2O	374.1	218.3
Xenon	Xe	16.6	58

Volume of Round and Rectangular Tanks

VB-1 CONVERSION TABLES—VOLUME OF ROUND AND RECTANGULAR TANKS

NUMBER OF U.S. GALLONS IN ROUND TANKS FOR ONE FOOT IN DEPTH

Dia. of Tanks	No. U. S. gals.	Cubic ft. and Area in sq. ft.	Dia. of Tanks	No. U. S. gals.	Cubic ft. and Area in sq. ft.	Dia. of Tanks	No. U. S. gals.	Cubic ft. and Area in sq. ft.	Dia. of Tanks	No. U. S. gals.	Cubic ft. and Area in sq. ft.	Dia. of Tanks	No. U. S. gals.	Cubic ft. and Area in sq. ft.	Dia. of Tanks	No. U. S. gals.	Cubic ft. and Area in sq. ft.
1′	5.87	.785	3′- 4″	65.28	8.727	5′- 8″	188.66	25.22	12′	846.03	113.10	19′	2120.90	283.53	26′	3971.60	530.93
1′- 1″	6.89	.922	3′- 5″	68.58	9.168	5′- 9″	194.25	25.97	12′-3″	881.65	117.86	19′-3″	2177.10	291.04	26′-3″	4048.40	541.19
1′- 2″	8.	1.069	3′- 6″	71.97	9.621	5′-10″	199.92	26.73	12′-6″	918.	122.72	19′-6″	2234.	298.65	26′-6″	4125.90	551.55
1′- 3″	9.18	1.227	3′- 7″	75.44	10.085	5′-11″	205.67	27.49	12′-9″	955.09	127.68	19′-9″	2291.70	306.35	26′-9″	4204.10	562.
1′- 4″	10.44	1.396	3′- 8″	78.99	10.559	6′	211.51	28.27	13′	992.91	132.73	20′	2350.10	314.16	27′	4283.	572.56
1′- 5″	11.79	1.576	3′- 9″	82.62	11.045	6′ -3″	229.50	30.68	13′-3″	1031.50	137.89	20′-3″	2409.20	322.06	27′-3″	4362.70	583.21
1′- 6″	13.22	1.767	3′-10″	86.33	11.541	6′- 6″	248.23	33.18	13′-6″	1070.80	143.14	20′-6″	2469.10	330.06	27′-6″	4443.10	593.96
1′- 7″	14.73	1.969	3′-11″	90.13	12.048	6′- 9″	267.69	35.78	13′-9″	1110.80	148.49	20′-9″	2529.60	338.16	27′-9″	4524.30	604.81
1′- 8″	16.32	2.182	4′	94.	12.566	7′	287.88	38.48	14′	1151.50	153.94	21′	2591.	346.36	28′	4606.20	615.75
1′- 9″	17.99	2.405	4′- 1″	97.96	13.095	7′- 3″	308.81	41.28	14′-3″	1193.	159.48	21′-3″	2653.	354.66	28′-3″	4688.80	626.80
1′-10″	19.75	2.640	4′- 2″	102.	13.635	7′- 6″	330.48	44.18	14′-6″	1235.30	165.13	21′-6″	2715.80	363.05	28′-6″	4772.10	637.94
1′-11″	21.58	2.885	4′- 3″	106.12	14.186	7′- 9″	352.88	47.17	14′-9″	1278.20	170.87	21′-9″	2779.30	371.54	28′-9″	4856.20	649.18
2′	23.50	3.142	4′- 4″	110.32	14.748	8′	376.01	50.27	15′	1321.90	176.71	22′	2843.60	380.13	29′	4941.	660.52
2′- 1″	25.50	3.409	4′- 5″	114.61	15.321	8′- 3″	399.88	53.46	15′-3″	1366.40	182.65	22′-3″	2908.60	388.82	29′-3″	5026.60	671.96
2′- 2″	27.58	3.687	4′- 6″	118.97	15.90	8′- 6″	424.48	56.75	15′-6″	1411.50	188.69	22′-6″	2974.30	397.61	29′-6″	5112.90	683.49
2′- 3″	29.74	3.976	4′- 7″	123.42	16.50	8′- 9″	449.82	60.13	15′-9″	1457.40	194.83	22′-9″	3040.80	406.49	29′-9″	5199.90	695.13
2′- 4″	31.99	4.276	4′- 8″	127.95	17.10	9′	475.89	63.62	16′	1504.10	201.06	23′	3108.	415.48	30′	5287.70	706.86
2′- 5″	34.31	4.587	4′- 9″	132.56	17.72	9′- 3″	502.70	67.20	16′-3″	1551.40	207.39	23′-3″	3175.90	424.56	30′-3″	5376.20	718.69
2′- 6″	36.72	4.909	4′-10″	137.25	18.35	9′- 6″	530.24	70.88	16′-6″	1599.50	213.82	23′-6″	3244.60	433.74	30′-6″	5465.40	730.62
2′- 7″	39.21	5.241	4′-11″	142.02	18.99	9′- 9″	558.51	74.66	16′-9″	1648.40	220.35	23′-9″	3314.	443.01	30′-9″	5555.40	742.64
2′- 8″	41.78	5.585	5′	146.88	19.63	10′	587.52	78.54	17′	1697.90	226.98	24′	3384.10	452.39	31′	5646.10	754.77
2′- 9″	44.43	5.940	5′- 1″	151.82	20.29	10′- 3″	617.26	82.52	17′-3″	1748.20	233.71	24′-3″	3455.	461.86	31′-3″	5737.50	766.99
2′-10″	47.16	6.305	5′- 2″	156.83	20.97	10′- 6″	642.74	86.59	17′-6″	1799.30	240.53	24′-6″	3526.60	471.44	31′-6″	5829.70	779.31
2′-11″	49.98	6.681	5′- 3″	161.93	21.65	10′- 9″	678.95	90.76	17′-9″	1851.10	247.45	24′-9″	3598.90	481.11	31′-9″	5922.60	791.73
3′	52.88	7.069	5′- 4″	167.12	22.34	11′	710.90	95.03	18′	1903.60	254.47	25′	3672.	490.87	32′	6016.20	804.25
3′- 1″	55.86	7.467	5′- 5″	172.38	23.04	11′- 3″	743.58	99.40	18′-3″	1956.80	261.59	25′-3″	3745.80	500.74	32′-3″	6110.60	816.86
3′- 2″	58.92	7.876	5′- 6″	177.72	23.76	11′- 6″	776.99	103.87	18′-6″	2010.80	268.80	25′-6″	3820.30	510.71	32′-6″	6205.70	829.58
3′- 3″	62.06	8.296	5′- 7″	183.15	24.48	11′- 9″	811.14	108.43	18′-9″	2065.50	276.12	25′-9″	3895.60	520.77	32′-9″	6301.50	842.39

To find the capacity of tanks greater than the largest given in the table, look in the table for a Tank of one-half the given size and multiply its capacity by 4, or one of one-third its size and multiply its capacity by 9, etc.

NUMBER OF U.S. GALLONS IN RECTANGULAR TANKS FOR ONE FOOT IN DEPTH

Width of Tank	Length of Tank																		
	3′	3′-6″	4′	4′-6″	5′	5′-6″	6′	6′-6″	7′	7′-6″	8′	8′-6″	9′	9′-6″	10′	10′-6″	11′	11′-6″	12′
2′	44.88	52.36	59.84	67.32	74.81	82.29	89.77	97.25	104.73	112.21	119.69	127.17	134.65	142.13	149.61	157.09	164.57	172.05	179.53
2′-6″	56.10	65.45	74.80	84.16	93.51	102.86	112.21	121.56	130.91	140.26	149.61	158.96	168.31	177.66	187.01	196.36	205.71	215.06	224.41
3′	67.32	78.54	89.77	100.99	112.21	123.43	134.65	145.87	157.09	168.31	179.53	190.75	201.97	213.19	224.41	235.63	246.86	258.07	269.30
3′-6″	—	91.64	104.73	117.82	130.91	144.00	157.09	170.18	183.27	196.36	209.45	222.54	235.63	248.73	261.82	274.90	288.00	301.09	314.18
4′	—	—	119.69	134.65	149.61	164.57	179.53	194.49	209.45	224.41	239.37	254.34	269.30	284.26	299.22	314.18	329.14	344.10	359.06
4′-6″	—	—	—	151.48	168.31	185.14	201.97	218.80	235.63	252.47	269.30	286.13	302.96	319.79	336.62	353.45	370.28	387.11	403.94
5′	—	—	—	—	187.01	205.71	224.41	243.11	261.82	280.52	299.22	317.92	336.62	355.32	374.03	392.72	411.43	430.13	448.83
5′-6″	—	—	—	—	—	226.28	246.86	267.43	288.00	308.57	329.14	349.71	370.28	390.85	411.43	432.00	452.57	473.14	493.71
6′	—	—	—	—	—	—	269.30	291.74	314.18	336.62	359.06	381.50	403.94	426.39	448.83	471.27	493.71	516.15	538.59
6′-6″	—	—	—	—	—	—	—	316.05	340.36	364.67	388.98	413.30	437.60	461.92	486.23	510.54	534.85	559.16	583.47
7′	—	—	—	—	—	—	—	—	366.54	392.72	418.91	445.09	471.27	497.45	523.64	549.81	575.99	602.18	628.36
7′-6″	—	—	—	—	—	—	—	—	—	420.78	448.83	476.88	504.93	532.98	561.04	589.08	617.14	645.19	673.24
8′	—	—	—	—	—	—	—	—	—	—	478.75	508.67	538.59	568.51	598.44	628.36	658.29	688.20	718.12
8′-6″	—	—	—	—	—	—	—	—	—	—	—	540.46	572.25	604.05	635.84	667.63	699.42	731.21	763.00
9′	—	—	—	—	—	—	—	—	—	—	—	—	605.92	639.58	673.25	706.90	740.56	774.23	807.89
9′-6″	—	—	—	—	—	—	—	—	—	—	—	—	—	675.11	710.65	746.17	781.71	817.24	852.77
10′	—	—	—	—	—	—	—	—	—	—	—	—	—	—	748.05	785.45	822.86	860.26	897.66
10′-6″	—	—	—	—	—	—	—	—	—	—	—	—	—	—	—	824.73	864.00	903.26	942.56
11′	—	—	—	—	—	—	—	—	—	—	—	—	—	—	—	—	905.14	946.29	987.43
11′-6″	—	—	—	—	—	—	—	—	—	—	—	—	—	—	—	—	—	989.29	1032.3
12′	—	—	—	—	—	—	—	—	—	—	—	—	—	—	—	—	—	—	1077.2

Useful Information
Volume of Tanks

VB-2 VOLUME OF HORIZONTAL ROUND TANKS WITH HEMISPHERICAL HEADS

1. All dimensions, in feet, do not include wall thickness.
2. This table lists commonly used tank dimensions. Use formula on page 00 for calculating any special tank volumes.

A	2.5	5	7.5	10	12.5	15	17.5	20	22.5	25
R	1	2.5	5	5	5	10	10	10	10	15
X LEVEL / VOLUME	ft^3 US GAL	ft^3 US GAL	ft^3 US GAL	ft^3 US GAL	ft^3 US GAL	ft^3 US GAL	ft^3 US GAL	ft^3 US GAL	ft^3 US GAL	ft^3 US GAL
1.25	8 60	29 217	65 486	79 591	93 696	170 1272	190 1421	211 1578	231 1728	324 2424
2.5	— —	82 613	197 1474	235 1758	274 2050	520 3890	577 4316	633 4735	690 5161	982 7345
5	— —	164 1227	556 4159	654 4892	753 5632	1576 11788	1729 12833	1883 14085	2036 15229	2983 22313
7.5	— —	— —	916 6852	1074 8034	1232 9215	2939 21984	3208 23996	3477 26008	3746 28020	5664 42367
10	— —	— —	1113 8325	1309 9791	1505 11257	4451 33293	4843 36226	5236 39165	5629 42105	8822 65989
12.5	— —	— —	— —	— —	— —	5962 44596	6478 48455	6995 52323	7511 56182	12287 91907
15	— —	— —	— —	— —	— —	7325 54791	7957 59518	8589 64246	9221 68973	15904 118962
17.5	— —	— —	— —	— —	— —	8381 62690	9110 68143	9839 73596	10567 79041	19521 146017
20	— —	— —	— —	— —	— —	8901 66579	9687 72459	10472 78331	11257 84202	22987 171943
22.5	— —	— —	— —	— —	— —	— —	— —	— —	— —	26145 195565
25	— —	— —	— —	— —	— —	— —	— —	— —	— —	28826 215618

VB-2 VOLUME OF HORIZONTAL ROUND TANKS WITH HEMISPHERICAL HEADS (Continued)

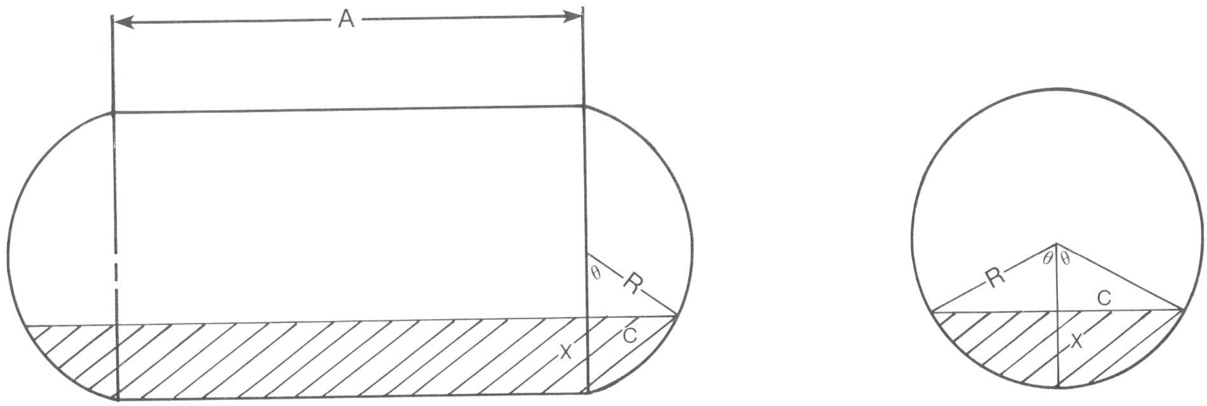

TANK WITH HEMISPHERICAL HEADS
ALL DIMENSIONS DO NOT INCLUDE WALL THICKNESS

VOLUME OF SHADED AREA = VOLUME OF STRAIGHT TANK + VOLUME OF HEMISPHERICAL HEADS

R: Radius of tank, hemispherical head

X: Liquid level for horizontal tank

A: Length of horizontal tank, not including hemispherical heads.

θ: $\sin^{-1} \frac{C}{R}$

C: $R^2 - CR - X)^2$

1. If $O < x \leq R$

$$\text{Shaded Volume} = \pi R^2 A \frac{\theta}{180} - C (R - X) A + \frac{1}{6} \pi x (3C^2 + X^2)$$

2. If $R < x \leq 2R$

$$\text{Shaded Volume} = (\pi R^2 A) \quad 1 - \frac{\theta}{180} + \frac{4}{3} \pi R^3 + C (X - R) A - \frac{\pi}{6} (2R - X) \bullet$$
$$\left[3 R^2 - 3 (X - R)^2 + (2R - X)^2\right]$$

Useful Information

Barometric Pressure — Effect of Altitude

VC BAROMETRIC PRESSURE—EFFECT OF ALTITUDE

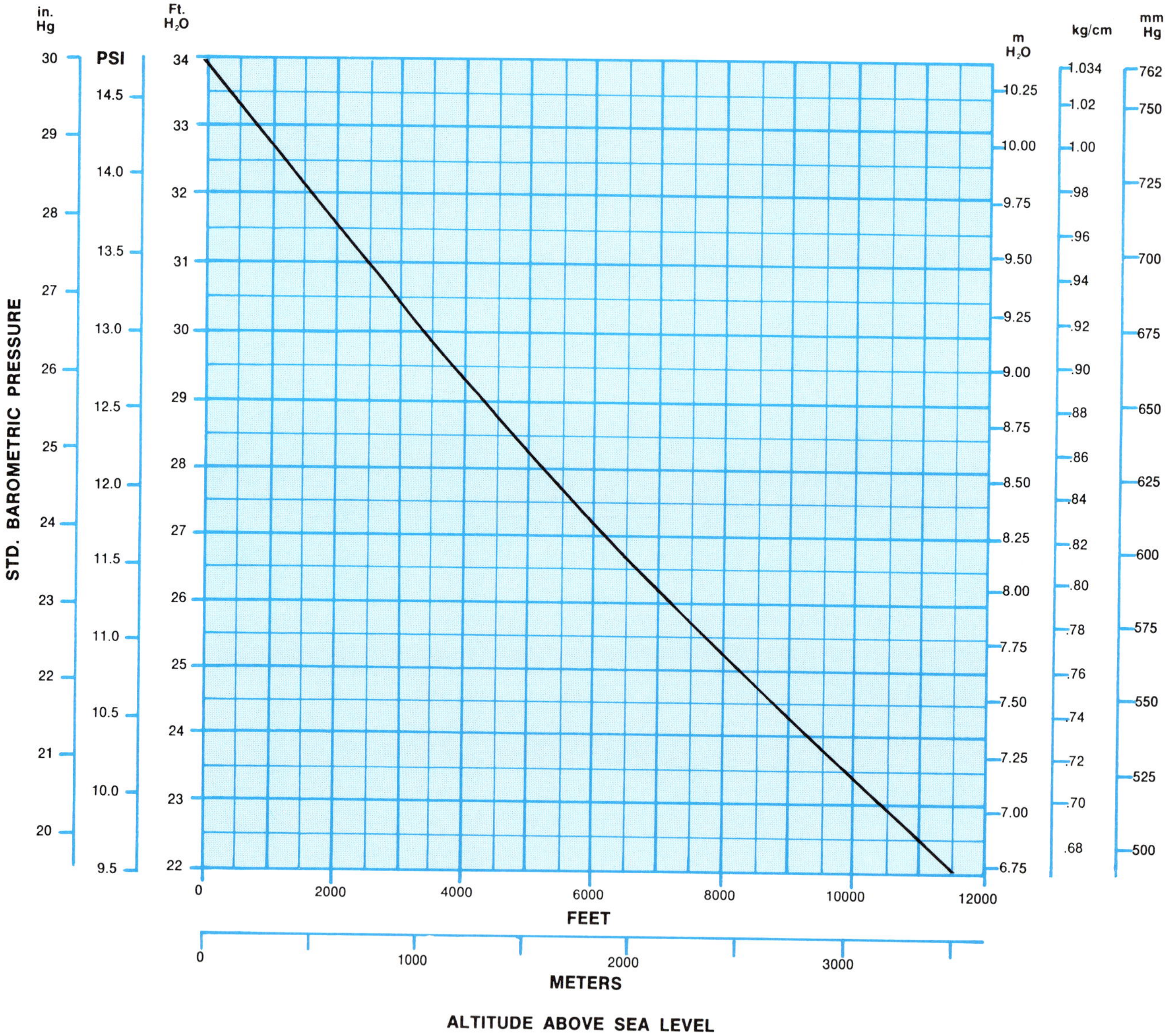

Circumference and Areas of Circles

VD-1 CIRCUMFERENCES AND AREAS OF CIRCLES

Diameter	Circum.	Area	Diameter	Circum.	Area	Diameter	Circum.	Area	Diameter	Circum.	Area
⅛	.3927	.0123	8⅛	25.525	51.849	16⅛	50.653	204.22	24⅛	75.791	457.11
¼	.7854	.0491	8¼	25.918	53.456	16¼	51.051	207.39	24¼	76.184	461.86
⅜	1.1781	.1104	8⅜	26.311	55.088	16⅜	51.444	210.60	24⅜	76.576	466.64
½	1.5708	.1963	8½	26.704	56.745	16½	51.836	213.82	24½	76.969	471.44
⅝	1.9635	.3068	8⅝	27.096	58.426	16⅝	52.229	217.08	24⅝	77.362	476.26
¾	2.3562	.4417	8¾	27.489	60.132	16¾	52.622	220.35	24¾	77.754	481.11
⅞	2.7489	.6013	8⅞	27.882	61.862	16⅞	53.014	223.65	24⅞	78.147	485.98
1	3.1416	.7854	9	28.274	63.617	17	53.407	226.98	25	78.540	490.87
1⅛	3.5343	.9940	9⅛	28.667	65.397	17⅛	53.800	230.33	25⅛	78.933	495.79
1¼	3.9270	1.2272	9¼	29.060	67.201	17¼	54.192	233.71	25¼	79.325	500.74
1⅜	4.3197	1.4849	9⅜	29.452	69.029	17⅜	54.585	237.10	25⅜	79.718	505.71
1½	4.7124	1.7671	9½	29.845	70.882	17½	54.978	240.53	25½	80.111	510.71
1⅝	5.1051	2.0739	9⅝	30.238	72.760	17⅝	55.371	243.98	25⅝	80.503	515.72
1¾	5.4978	2.4053	9¾	30.631	74.662	17¾	55.763	247.45	25¾	80.896	520.77
1⅞	5.8905	2.7612	9⅞	31.023	76.589	17⅞	56.156	250.95	25⅞	81.289	525.84
2	6.2832	3.1416	10	31.416	78.540	18	56.549	254.47	26	81.681	530.93
2⅛	6.6759	3.5466	10⅛	31.809	80.516	18⅛	56.941	258.02	26⅛	82.074	536.05
2¼	7.0686	3.9761	10¼	32.201	82.516	18¼	57.334	261.59	26¼	82.467	541.19
2⅜	7.4613	4.4301	10⅜	32.594	84.541	18⅜	57.727	265.18	26⅜	82.860	546.35
2½	7.8540	4.9087	10½	32.987	86.590	18½	58.119	268.80	26½	83.252	551.55
2⅝	8.2467	5.4119	10⅝	33.379	88.664	18⅝	58.512	272.45	26⅝	83.645	556.76
2¾	8.6394	5.9396	10¾	33.772	90.763	18¾	58.905	276.12	26¾	84.038	562.00
2⅞	9.0321	6.4918	10⅞	34.165	92.886	18⅞	59.298	279.81	26⅞	84.430	567.27
3	9.4248	7.0686	11	34.558	95.033	19	59.690	283.53	27	84.823	572.56
3⅛	9.8175	7.6699	11⅛	34.950	97.205	19⅛	60.083	287.27	27⅛	85.216	577.87
3¼	10.210	8.2958	11¼	35.343	99.402	19¼	60.476	291.04	27¼	85.608	583.21
3⅜	10.603	8.9462	11⅜	35.736	101.62	19⅜	60.868	294.83	27⅜	86.001	588.57
3½	10.996	9.6211	11½	36.128	103.87	19½	61.261	298.65	27½	86.394	593.96
3⅝	11.388	10.321	11⅝	36.521	106.14	19⅝	61.654	302.49	27⅝	86.786	599.37
3¾	11.781	11.045	11¾	36.914	108.43	19¾	62.046	306.35	27¾	87.179	604.81
3⅞	12.174	11.793	11⅞	37.304	110.75	19⅞	62.439	310.24	27⅞	87.572	610.27
4	12.566	12.566	12	37.699	113.10	20	62.832	314.16	28	87.965	615.75
4⅛	12.959	13.364	12⅛	38.092	115.47	20⅛	63.225	318.10	28⅛	88.357	621.26
4¼	13.352	14.186	12¼	38.485	117.86	20¼	63.617	322.06	28¼	88.750	626.80
4⅜	13.744	15.033	12⅜	38.877	120.28	20⅜	64.010	326.05	28⅜	89.143	632.36
4½	14.137	15.904	12½	39.270	122.72	20½	64.403	330.06	28½	89.535	637.94
4⅝	14.530	16.800	12⅝	39.663	125.19	20⅝	64.795	334.10	28⅝	89.928	643.55
4¾	14.923	17.721	12¾	40.055	127.68	20¾	65.188	338.16	28¾	90.321	649.18
4⅞	15.315	18.665	12⅞	40.448	130.19	20⅞	65.581	342.25	29⅞	90.713	654.84
5	15.708	19.635	13	40.841	132.73	21	65.973	346.36	29	91.106	660.52
5⅛	16.101	20.629	13⅛	41.233	135.30	21⅛	66.366	350.50	29⅛	91.499	666.23
5¼	16.493	21.648	13¼	41.626	137.89	21¼	66.759	354.66	29¼	91.892	671.96
5⅜	16.886	22.691	13⅜	42.019	140.50	21⅜	67.152	358.84	29⅜	92.284	677.71
5½	17.279	23.758	13½	42.412	143.14	21½	67.544	363.05	29½	92.677	683.49
5⅝	17.671	24.850	13⅝	42.804	145.80	21⅝	67.937	367.28	29⅝	93.070	689.30
5¾	18.064	25.967	13¾	43.197	148.49	21¾	68.330	371.54	29¾	93.462	695.13
5⅞	18.457	27.109	13⅞	43.590	151.20	21⅞	68.722	375.83	29⅞	93.855	700.98
6	18.850	28.274	14	43.982	153.94	22	69.115	380.13	30	94.248	706.86
6⅛	19.242	29.465	14⅛	44.375	156.70	22⅛	69.508	384.46	30¼	95.033	718.69
6¼	19.635	30.680	14¼	44.768	159.48	22¼	69.900	388.82	30½	95.819	730.62
6⅜	20.028	31.920	14⅜	45.160	162.30	22⅜	70.293	393.20	30¾	96.604	742.64
6½	20.420	33.183	14½	45.553	165.13	22½	70.686	397.61	31	97.389	754.77
6⅝	20.813	34.472	14⅝	45.946	167.99	22⅝	71.079	402.04	31¼	98.175	766.99
6¾	21.206	35.785	14¾	46.338	170.87	22¾	71.471	406.49	31½	98.960	779.31
6⅞	21.598	37.122	14⅞	46.731	173.78	22⅞	71.864	410.97	31¾	99.746	791.73
7	21.991	38.485	15	47.124	176.71	23	72.257	415.48	32	100.531	804.25
7⅛	22.384	39.871	15⅛	47.517	179.67	23⅛	72.649	420.00	32¼	101.316	816.86
7¼	22.776	41.282	15¼	47.909	182.65	23¼	73.042	424.56	32½	102.102	829.58
7⅜	23.169	42.720	15⅜	48.302	185.66	23⅜	73.435	429.13	32¾	102.887	842.39
7½	23.562	44.180	15½	48.695	188.69	23½	73.827	433.74	33	103.673	855.30
7⅝	23.955	45.664	15⅝	49.087	191.75	23⅝	74.220	438.36	33¼	104.458	868.31
7¾	24.347	47.173	15¾	49.480	194.83	23¾	74.613	443.01	33½	105.243	881.41
7⅞	24.740	48.707	75⅞	49.873	197.93	23⅞	75.006	447.69	33¾	106.029	894.62
8	25.133	50.265	16	50.265	201.06	24	75.398	452.39	34	106.814	907.92

Useful Information

Circumference and Areas of Circles

VD-1 CIRCUMFERENCES AND AREAS OF CIRCLES (Continued)

Diameter	Circum.	Area	Diameter	Circum.	Area	Diameter	Circum.	Area	Diameter	Circum.	Area
34¼	107.600	921.32	50¼	157.865	1983.2	66¼	208.131	3447.2	93½	293.739	6866.1
34½	108.385	934.82	50½	158.650	2003.0	66½	208.916	3473.2	94	295.310	6939.8
34¾	109.170	948.42	50¾	159.436	2022.8	66¾	209.701	3499.4	94½	296.881	7013.8
35	109.956	962.11	51	160.221	2042.8	67	210.487	3525.7	95	298.451	7088.2
35¼	110.741	975.91	51¼	161.007	2062.9	67¼	211.272	3552.0	95½	300.022	7163.0
35½	111.527	989.80	51½	161.792	2083.1	67½	212.058	3578.5	96	301.593	7238.2
35¾	112.312	1003.8	51¾	162.577	2103.3	67¾	212.843	3605.0	96½	303.164	7313.8
36	113.097	1017.9	52	163.363	2123.7	68	213.628	3631.7	97	304.734	7389.8
36¼	113.883	1032.1	52¼	164.148	2144.2	68¼	214.414	3658.4	97½	306.305	7466.2
36½	114.668	1046.3	52½	164.934	2164.8	68½	215.199	3685.3	98	307.876	7543.0
36¾	115.454	1060.7	52¾	165.719	2185.4	68¾	215.984	3712.2	98½	309.447	7620.1
37	116.239	1075.2	53	166.504	2206.2	69	216.770	3739.3	99	311.018	7697.7
37¼	117.024	1089.8	53¼	167.290	2227.0	69¼	217.555	3766.4	99½	312.588	7775.6
37½	117.810	1104.5	53½	168.075	2248.0	69½	218.341	3793.7	100	314.159	7854.0
37¾	118.596	1119.2	53¾	168.861	2269.1	69¾	219.126	3821.0	101	317.30	8011.85
38	119.381	1134.1	54	169.646	2290.2	70	219.911	3848.5	102	320.44	8171.28
38¼	120.166	1149.1	54¼	170.431	2311.5	70¼	220.697	3876.0	103	323.58	8332.29
38½	120.951	1164.2	54½	171.217	2332.8	70½	221.482	3903.6	104	326.73	8494.87
38¾	121.737	1179.3	54¾	172.002	2354.3	70¾	222.268	3931.4	105	329.87	8659.01
39	122.522	1194.6	55	172.788	2375.8	71	223.053	3959.2	106	333.01	8824.73
39¼	123.308	1210.0	55¼	173.573	2397.5	71½	224.624	4015.2	107	336.15	8992.02
39½	124.093	1225.4	55½	174.358	2419.2	72	226.195	4071.5	108	339.29	9160.88
39¾	124.878	1241.0	55¾	175.144	2441.1	72½	227.765	4128.2	109	342.43	9331.32
40	125.664	1256.6	56	175.929	2463.0	73	229.336	4185.4	110	345.58	9503.32
40¼	126.449	1272.4	56¼	176.715	2485.0	73½	230.907	4242.9	111	348.72	9676.89
40½	127.235	1288.2	56½	177.500	2507.2	74	232.478	4300.8	112	351.86	9852.03
40¾	128.020	1304.2	56¾	178.285	2529.4	74½	234.049	4359.2	113	355.00	10028.75
41	128.805	1320.3	57	179.071	2551.8	75	235.619	4417.9	114	358.14	10207.03
41¼	129.591	1336.4	57¼	179.856	2574.2	75½	237.190	4477.0	115	361.28	10386.89
41½	130.376	1352.7	57½	180.642	2596.7	76	238.761	4536.5	116	364.42	10568.32
41¾	131.161	1369.0	57¾	181.427	2619.4	76½	240.332	4596.3	117	367.57	10751.32
42	131.947	1385.4	58	182.212	2642.1	77	241.903	4656.6	118	370.71	10935.88
42¼	132.732	1402.0	58¼	182.998	2664.9	77½	243.473	4717.3	119	373.85	11122.02
42½	133.518	1418.6	58½	183.783	2687.8	78	245.044	4778.4	120	376.99	11309.73
43¾	134.303	1435.4	58¾	184.569	2710.9	78½	246.615	4839.8	121	380.13	11499.01
43	135.088	1452.2	59	185.354	2734.0	79	248.186	4901.7	122	383.27	11689.87
43¼	135.874	1469.1	59¼	186.139	2757.2	79½	249.757	4963.9	123	386.42	11882.29
43½	136.659	1486.2	59½	186.925	2780.5	80	251.327	5026.5	124	389.56	12076.28
43¾	137.445	1503.3	59¾	187.710	2803.9	80½	252.898	5039.6	125	392.70	12271.85
44	138.230	1520.5	60	188.496	2827.4	81	254.469	5153.0	126	395.84	12468.98
44¼	139.015	1537.9	60¼	189.281	2851.0	81½	256.04	5216.8	127	398.98	12667.68
44½	139.801	1555.3	60½	190.066	2874.8	82	257.611	5281.0	128	402.12	12867.96
44¾	140.586	1572.8	60¾	190.852	2898.6	82½	259.181	5345.6	129	405.27	13069.81
45	141.372	1590.4	61	191.637	2922.5	83	260.752	5410.6	130	408.41	13273.23
45¼	142.157	1608.2	61¼	192.423	2946.5	83½	262.323	5476.0	131	411.55	13478.22
45½	142.942	1626.0	61½	193.208	2970.6	84	263.894	5541.8	132	414.69	13684.78
45¾	143.728	1643.9	61¾	193.993	2994.8	84½	265.465	5607.9	133	417.83	13892.91
46	144.513	1661.9	62	194.779	3019.1	85	267.035	5674.5	134	420.97	14102.61
46¼	145.299	1680.0	62¼	195.564	3043.5	85½	268.606	5741.5	135	424.12	14313.88
46½	146.084	1698.2	62½	196.350	3068.0	86	270.177	5808.8	136	427.26	14526.72
46¾	146.869	1716.5	62¾	197.135	3092.6	86½	271.748	5876.5	137	430.40	14741.14
47	147.655	1734.9	63	197.920	3117.2	87	273.319	5944.7	138	433.54	14957.12
47¼	148.440	1753.5	63¼	198.706	3142.0	87½	274.889	6013.2	139	436.68	15174.68
47½	149.226	1772.1	63½	199.491	3166.9	88	276.460	6082.1	140	439.82	15393.80
47¾	150.011	1790.8	63¾	200.277	3191.9	88½	278.031	6151.4	141	442.96	15614.50
48	150.796	1809.6	64	201.062	3217.0	89	279.602	6221.1	142	446.11	15836.77
48¼	151.582	1828.5	64¼	201.847	3242.2	89½	281.173	6291.2	143	449.25	16060.61
48½	152.367	1847.5	64½	202.633	3267.5	90	282.743	6361.7	144	452.39	16286.02
48¾	153.153	1866.5	64¾	203.418	3292.8	90½	284.314	6432.6	145	455.53	16513.00
49	153.938	1885.7	65	204.204	3318.3	91	285.885	6503.9	146	458.67	16741.55
49¼	154.723	1905.0	65¼	204.989	3343.9	91½	287.456	6575.5	147	461.81	16971.67
49½	155.509	1924.4	65½	205.774	3369.6	92	289.027	6647.6	148	464.96	17203.36
49¾	156.294	1943.9	65¾	206.560	3395.3	92½	290.597	6720.1	149	468.10	17436.62
50	157.080	1963.5	66	207.345	3421.2	93	292.168	6792.9	150	471.24	17671.46

Reprinted with permission from the Allis-Chalmers Corporation.

Cylinder/Piston Displacements

VD-2 CYLINDER/PISTON DISPLACEMENTS

DIA. OF CYL. IN.	AREA, SQUARE INCHES	LENGTH OF STROKE IN INCHES													
		2	3	4	5	6	7	8	9	10	12	14	15	16	18
		DISPLACEMENT PER STROKE IN GALLONS OR DECIMAL PARTS THEREOF													
½	.196	.002	.003	.003	.004	.005	.006	.007	.008	.009	.010	.012	.013	.014	.015
⅜	.307	.003	.004	.005	.007	.008	.009	.011	.012	.013	.016	.019	.020	.021	.024
¾	.442	.004	.006	.008	.010	.011	.013	.015	.017	.019	.023	.027	.029	.031	.034
⅞	.601	.005	.008	.010	.013	.016	.018	.021	.023	.026	.031	.036	.039	.042	.047
1	.785	.007	.010	.014	.017	.020	.024	.027	.031	.034	.041	.048	.051	.054	.061
1⅛	.994	.009	.013	.017	.022	.026	.030	.034	.039	.043	.052	.060	.065	.069	.077
1¼	1.227	.011	.016	.021	.027	.032	.037	.043	.048	.053	.064	.074	.079	.085	.096
1⅜	1.485	.013	.019	.026	.032	.039	.044	.051	.058	.064	.077	.089	.096	.103	.116
1½	1.767	.015	.023	.031	.038	.046	.054	.061	.069	.077	.092	.107	.115	.122	.138
1¾	2.405	.021	.031	.042	.052	.063	.073	.083	.094	.104	.125	.146	.156	.170	.187
2	3.142	.027	.041	.054	.068	.082	.095	.109	.122	.136	.163	.190	.204	.218	.245
2¼	3.976	.034	.052	.069	.086	.103	.121	.138	.155	.172	.206	.241	.258	.275	.309
2½	4.909	.043	.004	.085	.106	.128	.149	.170	.191	.213	.255	.298	.319	.340	.383
2¾	5.940	.051	.077	.103	.129	.154	.180	.206	.231	.257	.309	.360	.386	.411	.463
3	7.069	.061	.092	.122	.153	.184	.214	.245	.275	.306	.367	.428	.459	.489	.551
3¼	8.296	.072	.108	.144	.179	.215	.251	.287	.323	.359	.431	.503	.539	.575	.647
3½	9.621	.083	.125	.167	.208	.249	.292	.333	.375	.417	.499	.583	.625	.666	.750
3¾	11.045	.095	.143	.191	.239	.287	.335	.382	.430	.478	.574	.669	.717	.765	.861
4	12.566	.109	.163	.218	.272	.326	.381	.435	.490	.544	.653	.762	.816	.870	.979
4¼	14.186	.123	.184	.246	.307	.368	.429	.491	.553	.614	.737	.860	.921	.982	1.105
4½	15.904	.138	.207	.275	.344	.413	.482	.551	.619	.689	.826	.964	1.033	1.102	1.239
4¾	17.721	.153	.230	.307	.384	.460	.537	.614	.690	.767	.920	1.073	1.150	1.227	1.380
5	19.635	.170	.255	.340	.425	.510	.595	.680	.765	.850	1.020	1.190	1.275	1.360	1.530
5½	21.648	.187	.281	.375	.469	.562	.656	.750	.843	.937	1.124	1.311	1.405	1.499	1.686
5½	23.758	.206	.309	.411	.514	.617	.720	.823	.926	1.029	1.234	1.440	1.543	1.646	1.851
5¾	25.967	.225	.337	.449	.562	.674	.787	.899	1.011	1.124	1.348	1.573	1.686	1.798	2.022
6	28.274	.245	.367	.489	.612	.734	.857	.979	1.102	1.224	1.469	1.714	1.836	1.958	2.203
6¼	30.680	.266	.398	.531	.664	.797	.929	1.062	1.195	1.328	1.593	1.859	1.992	2.124	2.390
6½	33.183	.287	.431	.574	.718	.861	1.005	1.149	1.293	1.436	1.723	2.011	2.155	2.298	2.589
6¾	35.785	.309	.465	.619	.775	.929	1.084	1.239	1.394	1.549	1.858	2.168	2.323	2.479	2.788
7	38.485	.333	.499	.666	.833	.999	1.166	1.333	1.499	1.666	1.999	2.332	2.499	2.666	2.999
7¼	41.282	.358	.536	.715	.894	1.073	1.253	1.432	1.610	1.789	2.147	2.505	2.684	2.863	3.221
7½	44.179	.383	.574	.765	.956	1.148	1.339	1.530	1.721	1.913	2.295	2.678	2.869	3.060	3.443
7¾	47.173	.408	.613	.817	1.021	1.225	1.429	1.633	1.837	2.042	2.450	2.858	3.063	3.267	3.675
8	50.266	.435	.652	.870	1.088	1.306	1.523	1.741	1.958	2.176	2.611	3.046	3.264	3.482	3.917
8¼	53.456	.462	.694	.924	1.156	1.389	1.621	1.852	2.084	2.316	2.779	3.242	3.474	3.703	4.174
8½	56.745	.491	.736	.982	1.225	1.473	1.719	1.964	2.210	2.456	2.947	3.438	3.684	3.929	4.420
8¾	60.132	.520	.780	1.041	1.301	1.561	1.822	2.082	2.342	2.603	3.123	3.644	3.904	4.164	4.685
9	63.617	.551	.826	1.101	1.377	1.652	1.928	2.203	2.479	2.754	3.305	3.856	4.131	4.406	4.957
9¼	67.201	.582	.873	1.164	1.456	1.746	2.067	2.329	2.620	2.909	3.493	4.069	4.375	4.657	5.240
9½	70.882	.613	.920	1.227	1.534	1.840	2.147	2.454	2.761	3.068	3.681	4.295	4.620	4.908	5.523
9¾	74.662	.646	.969	1.293	1.616	1.939	2.264	2.585	2.908	3.232	3.878	4.525	4.848	5.171	5.817
10	78.540	.680	1.020	1.360	1.700	2.040	2.380	2.720	3.060	3.400	4.080	4.760	5.100	5.440	6.120
10¼	82.516	.716	1.074	1.432	1.790	2.148	2.506	2.864	3.222	3.580	4.296	5.012	5.370	5.728	6.444
10½	86.590	.750	1.125	1.500	1.875	2.250	2.625	3.000	3.375	3.750	4.500	5.250	5.625	6.000	6.750
10¾	90.762	.786	1.179	1.572	1.965	2.358	2.751	3.144	3.537	3.930	4.716	5.502	5.895	6.288	7.074
11	95.033	.823	1.234	1.645	2.057	2.464	2.879	3.291	3.726	4.114	4.937	5.760	6.171	6.582	7.405
11½	99.402	.860	1.290	1.720	2.152	2.580	3.012	3.440	3.873	4.300	5.160	6.024	6.454	6.880	7.746
11½	103.869	.899	1.348	1.798	2.248	2.697	3.147	3.597	4.046	4.496	5.395	6.295	6.744	7.194	8.093
11¾	108.434	939	1.409	1.878	2.349	2.818	3.288	3.756	4.227	4.695	5.636	6.576	7.046	7.512	8.454
12	113.098	.979	1.468	1.958	2.448	2.938	3.422	3.917	4.406	4.896	5.875	6.854	7.314	7.833	8.813
12¼	117.859	1.018	1.527	2.036	2.545	3.054	3.563	4.072	4.580	5.090	6.108	7.126	7.635	8.144	9.160
12½	122.718	1.062	1.593	2.124	2.655	3.186	3.717	4.244	4.779	5.310	6.372	7.434	7.965	8.488	9.558
12¾	127.676	1.104	1.656	2.208	2.760	3.312	3.864	4.416	4.967	5.520	6.624	7.728	8.280	8.832	9.934
13	132.733	1.149	1.723	2.297	2.872	3.445	4.022	4.596	5.170	5.745	6.894	8.042	8.619	9.193	10.342

Reprinted from the GASO Reciprocating Pump Manual.

VD-3 CAPACITY OF PUMPS OF GIVEN DIAMETER AND LENGTH STROKE

Diameter of Cylinder, Inches	LENGTH OF STROKE IN INCHES, AND CAPACITY PER STROKE IN GALLONS, OF PUMP CYLINDER, WITH GIVEN DIAMETER																
	2	3	4	5	6	7	8	9	10	12	14	15	16	18	20	22	24
1¼	.0106	.0159	.0212	.0266	.0319	.0372	.0425	.0478	.0531	.0637	.0743	.0797	.0848	.0955	.1062	.1168	.1274
1⅜	.0128	.0192	.0256	.0321	.0385	.0419	.0513	.0578	.0642	.077	.089	.0963	.1027	.1156	.1280	.1408	.1541
1½	.0153	.0229	.0306	.0382	.0459	.0535	.0612	.0688	.0765	.0918	.1071	.1147	.1224	.1377	.1530	.1683	.1836
1¾	.0208	.0312	.0416	.0521	.0625	.0729	.0833	.0937	.1041	.1249	.1457	.1562	.1666	.1874	.2082	.2290	.2499
2	.0272	.0408	.0544	.068	.0816	.0952	.1088	.1224	.136	.1632	.1904	.204	.2176	.2448	.2720	.2992	.3264
2¼	.0344	.0516	.0688	.086	.1033	.1205	.1377	.1548	.1721	.2063	.241	.258	.21754	.3096	.34	.3784	.4128
2½	.0425	.0637	.0850	.1062	.1275	.1487	.17	.1912	.2125	.255	.2975	.3187	.344	.3825	.4252	.4677	.51
2¾	.0514	.0771	.1028	.1285	.1543	.1799	.2057	.2313	.2571	.3085	.3598	.3855	.4114	.4626	.5142	.5656	.617
3	.0612	.0918	.1224	.1530	.1836	.2142	.2448	.2754	.306	.3672	.4284	.459	.4896	.5508	.612	.6732	.7344
3¼	.0718	.1077	.1436	.1795	.2154	.2513	.2872	.3231	.3594	.4312	.503	.5385	.5748	.6466	.7182	.79	.8624
3½	.0833	.1249	.1666	.2082	.2499	.2915	.3332	.3748	.4165	.4998	.5831	.6247	.6664	.7497	.833	.9163	.9996
3¾	.0956	.1434	.1912	.239	.2868	.3346	.3824	.4302	.478	.5736	.6692	.687	.7648	.8605	.9561	1.0517	1.147
4	.1088	.1632	.2176	.272	.3264	.3808	.4352	.4896	.544	.6528	.7616	.816	.8904	.9792	1.088	1.1968	1.3056
4¼	.1228	.1842	.2456	.307	.3684	.4298	.4912	.5526	.6141	.7368	.8596	.921	.9824	1.105	1.228	1.3508	1.473
4½	.1377	.2065	.2754	.3442	.4131	.4819	.5508	.6196	.6885	.8262	.9639	1.0327	1.1016	1.2393	1.377	1.5147	1.6524
4¾	.1534	.2301	.3068	.3835	.4602	.5369	.6136	.6903	.7671	.9204	1.073	1.15	1.227	1.380	1.534	1.6874	1.84
5	.17	.2550	.34	.425	.51	.595	.68	.765	.85	1.02	1.19	1.275	1.36	1.53	1.7	1.87	2.04
5¼	.1874	.2811	.3748	.4685	.5622	.6559	.7496	.8433	.9371	1.124	1.311	1.405	1.499	1.686	1.874	2.0614	2.248
5½	.2057	.3085	.4114	.5142	.6171	.7199	.8228	.9256	1.0285	1.2342	1.4399	1.5427	1.6456	1.8513	2.057	2.2627	2.4684
5¾	.2248	.3372	.4496	.562	.6744	.7868	.8992	1.011	1.124	1.348	1.573	1.686	1.789	2.022	2.248	2.4728	2.696
6	.2448	.3672	.4896	.612	.7344	.8568	.9792	1.106	1.2240	1.4688	1.7136	1.8362	1.9584	2.2032	2.448	2.6928	2.9376
6¼	.2656	.3984	.5312	.6640	.7968	.9296	1.062	1.195	1.328	1.593	1.859	1.992	2.124	2.39	2.656	2.9216	3.186
6½	.2872	.4308	.5744	.7182	.8610	1.0052	1.1488	1.2926	1.4364	1.7955	2.0109	2.1546	2.2982	2.5885	2.8728	3.16	3.4473
6¾	.3098	.4647	.6196	.7745	.9294	1.084	1.239	1.394	1.549	1.858	2.168	2.323	2.479	2.788	3.098	3.4078	3.716
7	.3332	.4998	.6664	.833	.9996	1.1662	1.3328	1.4994	1.666	1.9992	2.3324	2.499	2.6656	2.9988	3.332	3.6652	3.9984
7¾	.4084	.6126	.8168	1.021	1.225	1.429	1.633	1.837	2.042	2.45	2.858	3.063	3.266	3.674	4.084	4.4924	4.9
8	.4352	.6528	.8704	1.088	1.3056	1.5232	1.7408	1.9584	2.176	2.6112	3.0464	3.264	3.4816	3.9168	4.352	4.7872	5.2224
9	.5508	.8262	1.1010	1.377	1.6524	1.9278	2.2032	2.4786	2.754	3.3048	3.8556	4.131	4.4064	5.0572	5.508	6.0588	6.6096
10	.68	1.02	1.36	1.7	2.04	2.38	2.72	3.06	3.4	4.08	4.76	5.1	5.44	6.12	6.8	7.48	8.16
11	.8227	1.2341	1.6451	2.057	2.464	2.879	3.2911	3.7258	4.1139	4.9367	5.7595	6.1709	6.5823	7.4051	8.2279	9.0506	9.8735
12	.9792	1.468	1.9584	2.448	2.9376	3.4222	3.9168	4.4064	4.896	5.8752	6.8544	7.344	7.833	8.8128	9.792	10.7712	11.7504
13	1.149	1.723	2.297	2.872	3.445	4.022	4.596	5.170	5.745	6.894	8.042	8.616	9.192	10.34	11.49	12.639	13.78
14	1.332	1.998	2.665	3.331	3.997	4.664	5.33	5.996	6.663	7.994	9.328	9.993	10.66	11.99	13.32	14.652	15.98
15	1.529	2.294	3.059	3.824	4.589	5.354	6.119	6.884	7.649	9.178	10.70	11.47	12.23	13.76	15.29	16.819	18.35
16	1.74	2.61	3.48	4.35	5.22	6.09	6.96	7.83	8.703	10.44	12.18	13.05	13.92	15.66	17.40	19.14	20.88
18	2.202	3.303	4.404	5.505	6.606	7.707	8.808	9.909	11.01	13.21	15.41	16.51	17.61	19.81	22.02	24.22	26.42
20	2.720	4.08	5.440	6.8	8.16	9.52	10.88	12.24	13.6	16.32	19.04	20.4	21.76	24.48	27.2	29.92	32.6

Reprinted with permission from MetPro Corporation, Dean Pump Division.

VE MECHANICAL FRICTION IN LINE SHAFTS

Mechanical Friction

The mechanical-friction chart should be used to determine the added horsepower due to mechanical friction in rotating the line shaft. The chart was compiled from test data submitted by representative turbine pump manufacturers. Variations in designs used by individual manufacturers may affect the figures slightly.

AMERICAN NATIONAL STANDARD—DEEP WELL VERTICAL TURBINE PUMPS

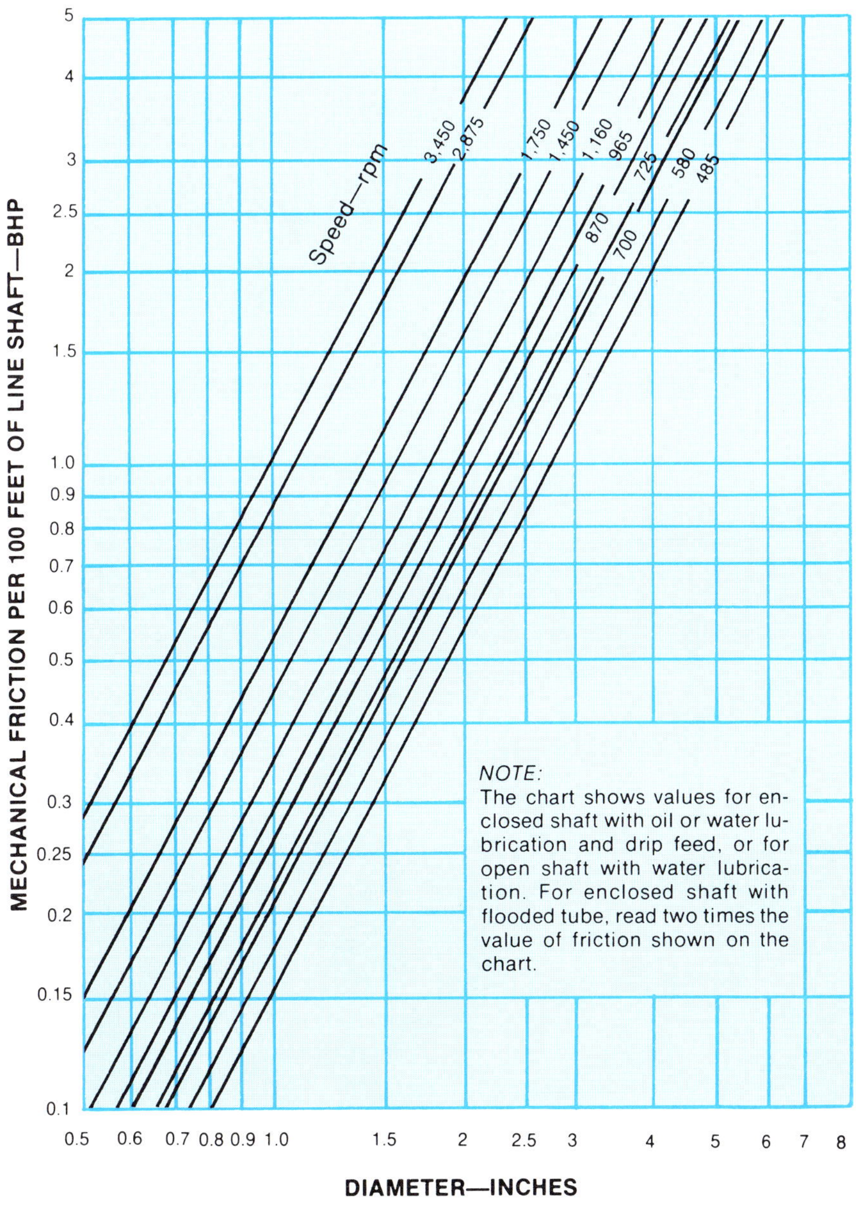

Reprinted with permission from the American Water Works Association, 6666 West Quincy Ave., Denver, Colo. 80255 —ANSI B58.1-1971.

Section VI

Bibliography

1. Experimental Investigation of Friction Losses in Wrought Iron Pipe When Installed with Couplings," by Kessler, University of Wisconsin, Bulletin 82-1935.
2. "Friction Factors for Pipe Flow," by Moody, Transactions of the ASME, 1944.
3. "Experiments Upon the Flow of Water in Pipes and Pipe Fittings," by John R. Freeman, published in 1941 by the ASME.
4. "Flow Test on Asbestos—Cement Pipe," by Edgar K. Wilson; Public Works, April, 1935.
5. "Turbulent Flow in Pipes, With Particular Reference to the Transition Region below the Smooth and Rough Pipe Laws," by C. F. Colebrook, Journal of the Institute of Civil Engineers, London, England, Paper No. 5204, Vol. 11, 1938–39, pp. 133–156.
6. "Comparative Tests of Friction Losses in Cement-Lined and Tar-Coated Cast Iron Pipe," by M. L. Enger, Journal of the American Water Works Association, Vol. 18, No. 4, Oct. 1927, pp 409–416.
7. "Der Stromungswiderstand von geraden Gasrohren," Von Ivo Vusković, Mitteilungen des Hydraulischen Institutes der Technischen Hochschule, München, Heft 9, R. Oldenbourg, 1939. ("The Resistance to Flow in Straight Gas Pipes," by Ivo Vusković, Transactions of the Munich Hydraulic Institute, Bulletin 9, R. Oldenbourg, 1939).
8. "Age Effect in Friction Factors for Coated Cast-Iron Pipes," by M. S. Carter, unpublished thesis of the Civil Engineering Department of the Massachusetts Institute of Technology, May 1936.
9. "The Reduction of Carrying Capacity of Pipes with Age," by C. F. Colebrook and C. M. White, Journal of the Institution of Civil Engineers, London, England, Paper No. 5137, Vol. 10, 1937–1938, pp 99-118.
10. "Report of Committee on Pipe Line Friction Coefficients," Journal of the New England Water Works Association, Vol. 49, No. 3, Sept. 1935.
11. "Hydraulics," by George E. Russell, 5th edition, Henry Holt and Company, p 187, also p 149.
12. Transactions of the Munich Hydraulic Institute, Bulletin 3, Authorized Translation, ASME 1945.
    a) "Loss in 90-degree Pipe Bends of Constant Circular Cross-Section," by A. Hoffman, pp 29-41.
    b) "Energy Loss in Smooth- and Rough-Surfaced Bends and Curves in Pipe Lines," by Werner Schubert, pp 81-99.
13. "Verluste in Glatten Rohrkrümmern mit Kreisrundem Querschnitt bei Weniger als 90° Ablenkung," von Rudolf Wasielewski, Mitteilungen des Hydraulischen Institutes der Technischen Hochschule, München, Heft 5, R. Oldenbourg, 1932. ("Losses in Smooth Pipe Bends of Circular Cross-Section with Less Than 90° Change in Direction," by Rudolf Wasielewski, Transactions of the Munich Hydraulic Institute, Bulletin 5, R. Oldenbourg, 1932.)
14. "Hydraulics and its Applications" by A.H. Gibson, 3rd Edition, Corstable and Company, London, England. 1976; p.p. 92-95.